RADIOANALYTICAL
CHEMISTRY I

ELLIS HORWOOD SERIES IN ANALYTICAL CHEMISTRY
Series Editors: Dr. R. A. CHALMERS and Dr. MARY MASSON, University of Aberdeen
Consultant Editor: Prof. J. N. MILLER, Loughborough University of Technology

S. Allenmark	**Chromatographic Enantioseparation – Methods and Applications**
G. E. Baiulescu & V.V. Coşofreţ	**Application of Ion Selective Membrane Electrodes in Organic Analysis**
G. E. Baiulescu, P. Dumitrescu & P. Gh. Zugravescu	**Sampling**
G. E. Baiulescu, C. Patroescu & R. A. Chalmers	**Education and Teaching in Analytical Chemistry**
G. I. Bekov & V. S. Letokhov	**Laser Resonant Photoionization Spectroscopy for Trace Analysis**
S. Bance	**Handbook of Practical Organic Microanalysis**
H. Barańska, A. Łabudzińska & J. Terpiński	**Laser Raman Spectrometry**
K. Beyermann	**Organic Trace Analysis**
O. Budevsky	**Foundations of Chemical Analysis**
J. Buffle	**Complexation Reactions in Aquatic Systems: An Analytical Approach**
D. T. Burns, A. Townshend & A. G. Catchpole	**Inorganic Reaction Chemistry, Volume 1: Systematic Chemical Separation**
D. T. Burns, A. Townshend & A. H. Carter	**Inorganic Reaction Chemistry, Volume 2: Reactions of the Elements and their Compounds: Part A: Alkali Metals to Nitrogen, Part B: Osmium to Zirconium**
E. Casassas & S. Aligret	**Solvent Extraction**
S. Caroli	**Improved Hollow Cathode Lamps for Atomic Spectroscopy**
J. Churáček	**New Trends in the Theory & Instrumentation of Selected Analytical Methods**
E. Constantin, A. Schnell & A. Pape	**Mass Spectrometry**
R. Czoch & A. Francik	**Instrumental Effects in Homodyne Electron Paramagnetic Resonance Spectrometers**
T. E. Edmonds	**Interfacing Analytical Instrumentation with Microcomputers**
J. K. Foreman & P. B. Stockwell	**Automatic Chemical Analysis**
Z. Galus	**Fundamentals of Electrochemical Analysis, Second Edition**
J. Gasparič & J. Churáček	**Laboratory Handbook of Paper and Thin Layer Chromatography**
S. Görög	**Steroid Analysis in the Pharmaceutical Industry**
T. S. Harrison	**Handbook of Analytical Control of Iron and Steel Production**
J. P. Hart	**Electroanalysis of Biologically Important Compounds**
T. F. Hartley	**Computerized Quality Control: Programs for the Analytical Laboratory**
Saad S. M. Hassan	**Organic Analysis using Atomic Absorption Spectrometry**
M. H. Ho	**Analytical Methods in Forensic Chemistry**
Z. Holzbecher, L. Diviš, M. Král & F. Vláčil	**Handbook of Organic Reagents in Inorganic Chemistry**
A. Hulanicki	**Reactions of Acids and Bases in Analytical Chemistry**
David Huskins	**Electrical and Magnetic Methods in On-line Process Analysis**
David Huskins	**General Handbook of On-line Process Analysers**
David Huskins	**Optical Methods in On-line Process Analysis**
David Huskins	**Quality Measuring Instruments in On-line Process Analysis**
J. Inczédy	**Analytical Applications of Complex Equilibria**
Z. K. Jelínek	**Particle Size Analysis**
M. Kaljurand & E. Küllik	**Computerized Multiple Input Chromatography**
R. Kalvoda	**Operational Amplifiers in Chemical Instrumentation**
I. Kerese	**Methods of Protein Analysis**
S. Kotrlý & L. Šůcha	**Handbook of Chemical Equilibria in Analytical Chemistry**
J. Kragten	**Atlas of Metal-ligand Equilibria in Aqueous Solution**
A. M. Krstulović	**Quantitative Analysis of Catecholamines and Related Compounds**
F. J. Krug & E. A. G. Zagatto	**Flow Injection Analysis in Agriculture & Environmental Science**
V. Linek, V. Vacek, J. Sinkule & P. Beneš	**Measurement of Oxygen by Membrane-Covered Probes**
C. Liteanu & S. Gocan	**Gradient Liquid Chromatography**
C. Liteanu, E. Hopîrtean & R. A. Chalmers	**Titrimetric Analytical Chemistry**
C. Liteanu & I. Rîcă	**Statistical Theory and Methodology of Trace Analysis**
Z. Marczenko	**Separation and Spectrophotometric Determination of Elements**
M. Meloun, J. Havel & E. Högfeldt	**Computation of Solution Equilibria**
M. Meloun, J. Militky & M. Forina	**Chemometrics in Instrumental Analysis: Solved Problems for IBM PC**
O. Mikeš	**Laboratory Handbook of Chromatographic and Allied Methods**
J. C. Miller & J. N. Miller	**Statistics for Analytical Chemistry, Second Edition**
J. N. Miller	**Fluorescence Spectroscopy**
J. N. Miller	**Modern Analytical Chemistry**

RADIOANALYTICAL CHEMISTRY I

J. TÖLGYESSY
Professor of Nuclear and Environmental Chemistry
Slovak Technical University
Bratislava, Czechoslovakia

M. KYRŠ
Institute of Nuclear Research
Řež near Prague, Czechoslovakia

ELLIS HORWOOD LIMITED
Publishers · Chichester

Halsted Press: a division of
JOHN WILEY & SONS
New York · Chichester · Brisbane · Toronto

VEDA
Bratislava

First published in 1989 by
ELLIS HORWOOD LIMITED in co-operation with VEDA
Market Cross House, Cooper Street,
Chichester, West Sussex, PO19 1EB, England
The publisher's colophon is reproduced from James Gillison's drawing of the ancient Market Cross, Chichester.

Distributors:
Australia and New Zealand:
JACARANDA WILEY LIMITED
GPO Box 859, Brisbane, Queensland 4001, Australia

Canada:
JOHN WILEY & SONS CANADA LIMITED
22 Worcester Road, Rexdale, Ontario Canada

Europe and Africa:
JOHN WILEY & SONS LIMITED
Baffins Lane, Chichester, West Sussex, England

North and South America and the rest of the world:
Halsted Press: a division of
JOHN WILEY & SONS
605 Third Avenue, New York, NY 10158, USA

Soust-East Asia:
JOHN WILEY & SONS (SEA) PTE LIMITED
37 Jalan Pemimpin # 05–04
Block B, Union Industrial Building, Singapore 2057

Indian Subcontinent:
WILEY EASTERN LIMITED
4835/24 Ansari Road
Daryaganj, New Delhi 110002, India

East European countries, Democratic People's Republic of Korea, People's Republic of China, People's Republic of Mongolia, Republic of Cuba, Socialist Republic of Vietnam:
VEDA, Publishing House of the Slovak Academy of Sciences
Klemensova 19
814 30 Bratislava, Czechoslovakia

© **1989 J. Tölgyessy, M. Kyrš/Ellis Horwood Limited**

British Library Cataloguing in Publication Data
Tölgyessy, J (Juraj)
Radioanalytical chemistry.
(Ellis Horwood series in analytical chemistry)
1. Chemical analysis. Samples. Preparation
I. Title II. Kyrš, Miroslav
543′.0

Library of Congress Card No. 88-32857 (vol. 1)

ISBN 0-85312-745-X v. 1 (Ellis Horwood Limited)
ISBN 0-470-21301-9 (Halsted Press)
ISBN 80-224-0185-4 (Set) (Veda)
ISBN 80-224-0009-2 v. 1

Printed in Czechoslovakia

Table of contents

1

Introduction

The first workers to use radionuclides for analytical purposes were Hevesy and Paneth, who in 1913 determined the solubility of lead sulphide in water by using a natural lead isotope (RaD) as an indicator [1]. In 1923 Hahn used ^{231}Pa to find the yield of isolation of ^{234}Pa, and this is considered the first example of isotope dilution analysis [2]. In a whole series of papers in the middle twenties of this century Ehrenberg used ^{212}Pb to determine, by direct or indirect precipitation methods, several anions forming lead salts of low solubility, and also some cations. These papers [3] belong to the branch of radioanalytical chemistry nowadays called 'radio-reagent methods'.

The discovery of artificial radioactivity resulted in a substantial increase in the number of radionuclides suitable for analytical purposes and made activation analysis possible.

The past two decades have witnessed the appearance and/or development of new analytical procedures based on the principles above, and on other examples of the interaction of ionizing (nuclear) radiation with matter. Therefore, today we may speak of a new scientific branch, namely that of radioanalytical chemistry.

1.1 DEFINITION OF RADIOANALYTICAL CHEMISTRY

Radioanalytical chemistry can be defined as a special field of chemistry, which belongs simultaneously both to the area of *analytical chemistry* and to the realm of *applied radiochemistry*. In other words, radioanalytical chemistry is concerned primarily with the use of radioactive nuclides and nuclear radiation for analytical purposes. It is a question of convention whether we consider in this connection analytical purposes exclusively as the qualitative or quantitative determination of substances, or also include other experimental work important for the analyst, such as the determination of solubility and complexation constants, distribution of elements in co-precipitation, solvent extraction, etc. In our opinion the first approach, leading to a narrower definition of nuclear analytical chemistry, is slightly preferable.

Some authors propose the expression 'analytics' or 'analytical sciences' as a general category including both chemical and physical methods of analysis. In that sense this book could have been entitled 'Nuclear Analytics' or 'Analytical Nucleonics'. The main reason why it is called 'Radioanalytical Chemistry' is not only the fact that the term 'analytics' is not yet very common but also the emphasis on chemical methods and the chemical aspects of the procedures described, together with the somewhat shorter treatment of nuclear physical methods.

The main advantages of radioanalytical chemistry stem from the extraordinary energy of ionizing particles compared to that of common chemical reactions. As a result, extremely small amounts of substances can be detected and monitored in the course of the analytical procedure. This usually brings about high sensitivity, speed and simplicity, easy automation and, in many cases, elimination of the chemical handling of the sample (non-destructive analysis).

Radioanalytical chemistry has reached maturity, having survived the usual previous successive stages of exaggerated enthusiasm and unfounded rejection. Radioanalytical procedures are and should be used nowadays for the solution of those problems for which they are superior to other analytical approaches, such as atomic absorption analysis, polarography, emission spectroscopy and many others. To enable the reader to discern such problems or areas of superiority of radioanalytical chemistry is one of the aims of the present work.

1.2 CLASSIFICATION OF RADIOANALYTICAL METHODS

The definition of radioanalytical chemistry can be complemented by a classification of this branch of applied radiochemistry.

In our opinion the early classification given in 1955 by Broda and Schönfeld [4] still remains valid.

It splits radioanalytical chemistry into five main areas:
(a) indicator analysis;
(b) radio-reagent methods;
(c) isotope dilution analysis;
(d) activation analysis;
(e) analysis based on the interaction of nuclear radiation with matter.

Indicator analysis is based on the proportionality between concentration of a radioactive substance and the associated count-rate under constant conditions. Some examples are given in Section 2.3.

Radio-reagent methods are based, as follows from the wording, on chemical *reactions,* the extent of which is measured by means of the *radioactivity* of one component of the initial substances or of the products. This area is covered by Chapters 4–7.

Isotope dilution analysis is characterized by the fact that a change in *specific activity* can be used for determination of the concentration of a substance. The method can be used when the isolation of the total amount of the substance for final determination purposes is impractical (see Chapter 3).

Activation analysis requires *inducing, by irradiation, radioactivity* in originally inactive components of mixtures, and relies on the proportionality or relationship between the component concentration and the radioactivity induced. The method is described in Chapter 8 (Vol. II).

The last area is comprised of methods based on particle-scattering, back-scattering, X-rays, or ionization generated by nuclear radiation (see Chapter 9, Vol. II).

Recently Shamaev [5] proposed a modified classification of radioanalytical methods. It can be summarized as follows.

Radioanalytical methods:
1 Radioisotope methods
   1.1 Radioindicator methods
     ⋮
   1.1.9 Concentration-dependent distribution
     ⋮
   1.1.12 Methods using radioactive reagents
   1.1.13 Method of isotope exchange

1.1.14 Radiometric titration method
1.2 Isotope dilution
2 Nuclear physical methods
2.1 Activation analysis
2.2 Absorption and scattering of radiation

This new and interesting attempt at classification removes some disadvantages of older conventional classifications, but in our opinion also suffers from certain drawbacks. The term radioindicator methods has been widely used in a different sense, so confusion can occur. Some methods that have been used only once or twice or by one author (e.g. 1.1.3 Method of isolation with different quantities of reagents) are classified on the same level as widely published well-known methods, which have been reviewed extensively (e.g. 1.1.12 or 1.1.14). Furthermore, the classification is given in a graphical form involving the use of three types of arrows, so a clear-cut hierarchy between various methods is not achieved. For example, the radio-release method is not classified as a special case of radio-reagent methods, but only a 'transition' from the latter to the former is envisaged, or the latter is characterized as a 'modernization' of the former. Shamaev's ideas, however, will prove very useful and inspiring in any serious future efforts to improve the detailed classification of radioanalytical methods.

Another attempt at innovation in the classification of some areas of radioanalytical chemistry has been made by Perezhogin (see p. 116). The question of the detailed classification of some branches of radioanalytical chemistry is treated in the corresponding sections (3.1, 4.1, etc.).

REFERENCES

[1] G. Hevesy and F. Paneth, *Z. Anorg. Chem.*, 1913, **82**, 323.
[2] O. Hahn, *Z. Phys. Chem.*, 1923, **103**, 461.
[3] R. Ehrenberg, *Biochem. Z.*, 1925, **164**, 183; 1928, **197**, 467.
[4] E. Broda and T. Schönfeld, *Radiochemische Methoden der Mikrochemie*, Springer, Vienna, 1955.
[5] V. I. Shamaev, *J. Radioanal. Chem.*, 1979, **51**, 205.

SELECTED BIBLIOGRAPHY

J. Tölgyessy and Š. Varga, *Nuclear Analytical Chemistry* II, University Park Press, Baltimore; Veda, Bratislava, 1972.
J. Tölgyessy and Š. Varga, *Nuclear Analytical Chemistry* III, University Park Press, Baltimore; Veda, Bratislava, 1974.
M. Kyrš and G. N. Bilimovich (eds.), *Novye metody radioanaliticheskoi khimii* (*New Methods of Radioanalytical Chemistry*), Energoizdat, Moscow, 1982.

M. Haïssinsky, *La chimie nucléaire et ses applications,* Masson, Paris, 1957 (*Nuclear Chemistry and its Applications,* Addison-Wesley, London, 1964).

G. Friedländer and J. Kennedy, *Nuclear and Radiochemistry,* Wiley, New York, 1956.

W. Schulze, *Radiochemie,* de Gruyter, Berlin, 1971.

J. Tölgyessy and E. H. Klehr, *Nuclear Environmental Chemical Analysis,* Ellis Horwood, Chichester, 1987.

2

Analysis by means of activity

In this chapter the simplest cases of the use of radionuclides in analytical chemistry are treated, i.e. the determination of naturally occurring isotopes, and analysis by using radioactive tracers, the concentration of a substance then being proportional to its radioactivity. Section 2.2 deals briefly with the most common methods for the analysis of mixtures of radioisotopes. Many principles described in this chapter form a component of more elaborate procedures described in subsequent chapters, so that a good comprehension of the material in Chapter 2 will help the reader to fully understand the more complicated procedures described in the later parts of the book, such as isotope and radio-reagent methods.

2.1 ANALYSIS BY MEANS OF NATURALLY OCCURRING RADIOACTIVE ELEMENTS

A number of naturally occurring elements are radioactive, because they contain at least one natural radionuclide. Table 2.1 provides information on some naturally occurring radionuclides. The nuclides listed either decay to stable nuclides, e.g.

Table 2.1 — Naturally occurring radionuclides. (Reproduced from [1] by permission of the copyright holders, Academic Press Inc., New York.)

A_X	Half-life (yr)	Specific activity (Bq/g)
Alpha emitters		
^{231}Pa	32000	1.8×10^9
^{210}Po	0.38	1.7×10^{14}
^{226}Ra	1620	3.7×10^{10}
^{220}Rn	55 sec	variable
^{222}Rn	0.01	5.5×10^{15}
^{147}Sm	1.1×10^{11}	120
^{232}Th	1.4×10^{10}	4100
^{235}U	7.1×10^8	560
^{238}U	4.5×10^9	12000
Beta/gamma emitters		
^{40}K	1.3×10^9	31
^{138}La	1.1×10^{11}	0.78
^{87}Rb	4.7×10^{10}	900
^{187}Re	7×10^{10}	630

or give rise to radioactive decay chains, of which the two most important are those from ^{238}U and ^{232}Th:

$$^{238}\text{U} \xrightarrow{\alpha} {}^{234}\text{Th} \xrightarrow{\beta,\gamma} {}^{234}\text{Pa} \xrightarrow{\beta,\gamma} {}^{234}\text{U} \xrightarrow{\alpha} {}^{230}\text{Th} \xrightarrow{\alpha}$$

$$^{226}\text{Ra} \xrightarrow{\alpha} {}^{222}\text{Rn} \xrightarrow{\alpha} {}^{218}\text{Po} \xrightarrow{\alpha} {}^{214}\text{Pb} \xrightarrow{\beta,\gamma} {}^{214}\text{Bi} \xrightarrow{\beta,\gamma}$$

$$^{214}\text{Po} \xrightarrow{\alpha} {}^{210}\text{Pb} \xrightarrow{\beta,\gamma} {}^{210}\text{Po} \xrightarrow{\beta,\gamma} {}^{210}\text{Bi} \xrightarrow{\alpha} {}^{206}\text{Pb}$$

$$^{232}\text{Th} \xrightarrow{\alpha} {}^{228}\text{Ra} \xrightarrow{\beta} {}^{228}\text{Ac} \xrightarrow{\beta,\gamma} {}^{228}\text{Th} \xrightarrow{\alpha} {}^{224}\text{Ra} \xrightarrow{\alpha}$$

$$^{220}\text{Rn} \xrightarrow{\alpha} {}^{216}\text{Po} \xrightarrow{\alpha} {}^{212}\text{Pb} \xrightarrow{\beta,\gamma} {}^{212}\text{Bi} \xrightarrow{\beta,\gamma} {}^{212}\text{Po} \xrightarrow{\alpha}$$

$$^{208}\text{Tl} \xrightarrow{\beta,\gamma} {}^{208}\text{Pb}$$

Naturally occurring radioactive elements can be determined by measurement of their radioactivity. This measurement makes it immediately possible to determine the quantity of the radioactive element which has constant isotopic composition (potassium, rubidium, samarium, lutetium, rhenium, francium).

When a naturally occurring polyisotopic radioactive element contains more than one radioactive isotope, however, the radiometric determination of the total quantity of the element has several variants. The naturally occurring radioactive elements may contain one radioactive isotope with a relatively long half-life, and one, or more, isotope(s) with short half-life (radium, thorium, polonium, radon, actinium, protactinium). One possibility is to start the radioactivity measurements only after decay of the short-lived radionuclides. The drawback of this procedure is that the decay of radionuclides, even those with a relatively short half-life, may go on for months and even years. In favourable cases it is more convenient to measure the activity of the decay products (e.g. radon, thoron), or to correct the activity of short-lived radioactive isotopes after determination of the isotopic composition of the element.

2.1.1 Emanation analysis

Emanation analysis provides a quantitative determination of radon (^{222}Rn), thoron (^{220}Rn), actinon (^{219}Rn), and their parent substances, and of elements which are in radioactive equilibrium with them. Such indirect determinations require both components (e.g. radium and radon) to be in equilibrium, or in a generally well defined ratio. Hence, the ratio of their amounts must be known. If these conditions are satisfied, the total quantity of the parent element in the sample can be calculated from the measured inert-gas activity.

Emanation analysis requires quantitative removal of the emanation (inert gas) from the sample, before measurement of its radioactivity. To this end, the solid substance (e.g. mineral, rock, soil) to be analysed is brought into the liquid state, either by direct melting or by fusion with a suitable flux, such as potassium or sodium carbonate, borax, sodium phosphate, lithium metaborate. The inert gas escapes from the melt, either completely or partially. In the latter case it is necessary to know the ratio of the total to the released amount of inert gas.

To release the inert gas from the melt, various procedures may be employed. In one, the vessel containing the melt is connected to an evacuated ionization chamber; air is bubbled through the melt and carries along the released inert gas, which is accumulated in the ionization chamber. For short-lived radioactive inert gases the air-stream method is more suitable; in this, a stream of air at constant flow-rate is passed through the melt and then through the activity detector. The agitation of the melt and hence increase of its contact area with air speeds up the release of inert gas. The melt may also be sprayed into the carrier gas by means of a small propellor, or can be shaken with the carrier gas.

Determination of individual radioactive inert gases is governed by their half-lives. Determination of ^{222}Rn presents no problem, despite its relatively long half-life. It is not technically possible to measure thoron (^{220}Rn) by integration, i.e. by filling the ionization chamber, on account of its short half-life. However, thoron may be measured by means of the air-stream method. Measurement of the activity of actinon (^{219}Rn) is even more difficult because of its extremely short half-life (3.96 sec).

Emanation analysis is suitable for samples with a low activity, for which gamma-activity determination would not be feasible. The method is characterized by a high degree of sensitivity and relatively good accuracy, because of the favourable conditions of ionization detection of alpha radiation with suitable detecting geometry.

2.1.2 Determination of individual elements

2.1.2.1 Potassium

Naturally occurring potassium is a polyisotopic element with three components: ^{39}K (93.08%), ^{41}K (6.91%) and ^{40}K (0.0119%). Potassium-40 decays by a complex scheme which includes beta decay with a half-life of 1.39×10^9 years, $E_{max} = 1.33$ MeV, and emission of a 1.46 MeV gamma ray.

Potassium determination by measurement of its natural radioactivity (^{40}K) is based on the stable isotope composition of naturally occurring potassium. Beta or gamma counting of salts and solutions is used, with the aid of Geiger–Müller (GM) or scintillation detectors with 4π-detection geometry.

When the beta activity of a sample is measured, self-absorption of beta radiation in the sample must be taken into account. If the sample layer is thicker than the saturation thickness, the beta activity will be lower than that corresponding to the true potassium content. The determination can be done with a calibration graph prepared by measuring the radioactivity of standard samples with known potassium content.

In the analysis of liquid samples, the measured radioactivity increases non-linearly with increasing potassium content in the solution, owing to self-absorption. Experiments indicate that the following empirical equation is valid:

$$\log A = b - a\varrho + \log C$$

where A is the activity (cpm), a and b are constants (a corresponding to the mass attenuation coefficient, and b expressing the geometric conditions and efficiency of detection), ϱ is the specific gravity of the solution, and C is the molar concentration of the potassium.

Equipment for potassium determination is commercially available. A typical outfit for potassium determination in liquid samples is shown in Fig. 2.1. The measuring cuvette is lead-shielded to reduce the background. The detector is made up of 10 GM counters to enhance its detection efficiency. This equipment enables 0–20 % of potassium to be determined with an error of $\pm 1.5\%$, within 10–20 min.

Potassium is determined by measuring its natural radioactivity in various substances, such as soil, minerals, rocks, artificial fertilizers, foodstuffs, etc.

Fig. 2.1 — A view of the equipment for radiometric analysis of potassium samples in solution. (Reproduced from [2] by permission of the copyright holders, SVTL, Bratislava, SNTL, Prague.)

2.1.2.2 Polonium

The long-lived isotope of polonium, ^{210}Po, is of analytical interest. This radionuclide occurs in the uranium decay series (RaF). ^{210}Po decays with a half-life of 138.4 days by emission of 5.3 MeV alpha particles.

Although many methods of separation of polonium are available (spontaneous deposition, electro-deposition, co-precipitation, adsorption, etc.), the spontaneous deposition or electro-deposition of the element are the simplest techniques for its separation. In spontaneous deposition, some co-deposition of bismuth will be encountered; this co-deposition is avoided in controlled potential electro-deposition. The alpha activity of separated polonium is often measured with a window-less gas ionization proportional counter.

2.1.2.3 *Radium*

Various minerals, rocks, soils and water contain radium. However, radium cannot be directly determined by a radiometric procedure in the mineral or rock where it occurs; the sample first has to be brought into a suitable state. Hence, the first step in radium determination consists of dissolving the sample. The next steps are the measurement of (a) the alpha activity of radium, (b) the gamma activity of radium, and (c) the radioactivity of radon.

Determination of radium by measuring its alpha activity consists of evaporating a radium salt solution on a flat dish or co-precipitating ^{226}Ra with barium sulphate, and measuring the alpha activity of the residue or precipitate with an alpha-scintillation counter, or an alpha-ionization chamber. Measurement of alpha activity presupposes that the sample is extremely thin, or is thicker than the saturation thickness. This method is characterized by simplicity and extraordinary sensitivity. Its drawback, however, is that the measured values of the radiation intensity, which are essentially proportional to the radium content in the sample, are strongly affected by self-absorption of radium alpha radiation within the sample itself.

Determination of radium by measuring its gamma activity is generally of comparative nature. The gamma activity of the sample, and of a standard containing a known amount of radium, is measured under identical conditions with an ionization chamber, GM tube, or scintillation counter. Soft radiation components in these measurements are filtered out by means of some suitable screening material (generally a 5-mm layer of lead). However, comparison of the sample and the standard is influenced by the state of radioactive equilibrium in both cases. Equilibrium must have been established, since the decay series involves many gamma emitters: ^{214}Pb and ^{214}Bi which are the major sources of gamma radiation (98%), and ^{214}Po, ^{210}Pb, ^{210}Bi and ^{210}Po. Radium samples attain a state of equilibrium in about 5 weeks when sealed in an airtight container.

Determination of radium by measurement of radon activity is described in the next section (2.1.2.4).

2.1.2.4 *Radon*

Radon (^{222}Rn) is found in numerous sites in nature, since it usually occurs with radium and uranium. As an indirect decay product of uranium, radon is found in uranium ores. Radon is absorbed from rocks and minerals containing uranium and radium into underground waters which may then appear as radioactive springs.

Radon samples attain radioactive equilibrium in about 3 hr. In measurement of radon activity, its decay during the measurement must be taken into account. When radioactive equilibrium with short-lived products is attained, the number of alpha particles measured will be about three times that of those due to the radon itself.

Determination of radon (resulting from radium decay) by the emanation method can also be used to calculate the amount of radium. Radon is purged from the solution by bubbling an inert gas through the solution after attainment of equilibrium. The released radon is transferred from the solution to the alpha-radiation detector (ionization chamber, scintillation counter). The method is comparative in character and the standard used is a solution containing ^{226}Ra.

2.1.2.5 Actinium
Owing to its similarity to the lanthanides, actinium is normally coprecipitated with lanthanum fluoride, oxalate, carbonate or hydroxide. The weak beta activity (0.046 MeV) from ^{227}Ac is usually counted in a windowless gas ionization proportional counter. The alpha radiation is of very low intensity; for its measurement a sufficiently sensitive semiconductor detector must be used.

Actinon (^{219}Rn) has a very short half-life (3.92 sec), so the emanation method may be used only in the dynamic mode, and the activity of the sample is compared with that of a standard.

2.1.2.6 Thorium
All the isotopes of thorium are radioactive. The principal isotope is ^{232}Th, which is the parent member of the thorium decay chain. The determination of thorium by measuring its radioactivity is not simple, because its isotopic composition is variable.

Thorium determination is based on the measurement of its alpha activity, or the activity of thoron.

Determination of thorium by measuring its alpha activity is actually based on measurement of the activity of thorium decay products. Various procedures can be adopted for this purpose. The greatest energy is possessed by the alpha particles emitted by ^{212}Po. Consequently, all alpha particles having a lower energy can be absorbed, and only the alpha radiation of ^{212}Po measured. Once radioactive equilibrium is attained, the alpha activity of ^{212}Po is proportional to the amount of ^{232}Th in the sample. Another procedure measures the total alpha activity of the powdered sample; this activity includes the alpha radiation emitted by members of the uranium and actinium decay chains, so the total activity has to be corrected for this in order to obtain the amount of ^{232}Th.

Determination of thorium by measurement of thoron activity is done by the dynamic carrier-gas stream method, which is comparative. A solution with a known content of the element (in equilibrium with the decay products) may be used as a standard.

Because of their selectivity, alpha and gamma spectrometry are important in the determination of thorium.

2.1.2.7 Protactinium
The most significant isotope of protactinium is ^{231}Pa, a member of the ^{235}U decay chain. Its radiometric determination is based on the measurement of alpha activity, after a chemical separation of the element. The activity is measured with a scintillation or semiconductor counter.

2.1.2.8 Uranium
Determination of uranium by measuring its activity is of considerable practical significance in searching for uranium deposits, in its technical processing, and also when monitoring the environment.

Methods for radiometric determination of uranium are divided into the following classes:
(a) direct methods, based on measurement of particles or photons emitted by uranium isotopes;
(b) indirect methods, based on measurement of other members of the decay chain.

If a direct radiometric determination of uranium on the basis of its alpha activity is performed, the sample must be free of all radioactive and inactive impurities. The presence of active impurities, particularly those with alpha activity, increase the activity of the sample, while inactive impurities may decrease it by self-absorption. Samples with at least saturation thickness, or extremely thin samples, are used. Saturation samples are used for materials of constant isotopic composition, since the alpha radiation of different isotopes has a different energy, and hence also a different range. Saturation thickness is of the order of 0.1 mm. Extremely thin samples are those in which absorption of alpha radiation is negligible; their thickness corresponds to a surface density less than 0.1 mg/cm^2.

The direct measurement of uranium in macroamounts is possible by use of the 0.184 MeV gamma ray of ^{235}U. This technique also serves to determine the enrichment of uranium with ^{235}U.

Indirect methods are frequently used for uranium determination, even though difficulties can arise from the possible lack of equilibrium with decay products. It is therefore important either to use methods that do not require equilibrium conditions, or to know the degree of equilibrium.

Prospecting for uranium deposits with a GM or scintillation counter represents a field application of the indirect determination of uranium. Equilibrium is usually assumed, and the results are expressed as 'uranium equivalent' after comparison with a U_3O_8 standard.

For the determination of uranium in biological material, the World Health Organization [3] recommends the following procedure. The organic material containing uranium is mineralized to bring the uranium into aqueous solution. An aliquot is then dried in a calibrated dish and its alpha activity is measured. This activity is compared with that of a uranium standard prepared identically. The sample and the standard should have at least approximately the same isotopic and chemical composition.

Under conditions of radioactive equilibrium between uranium and its decay products, the emanation procedure can also be used.

2.2 ANALYSIS OF MIXTURES OF RADIONUCLIDES, GAMMA SPECTROSCOPY, ALPHA SPECTROSCOPY, ANALYSIS WITH PREVIOUS CHEMICAL SEPARATION

The nuclear analysis of mixtures can include (a) identification of the number and nature of the radionuclides present in the mixture; (b) determination of the relative radioactivity of each, e. g. the ratio of the count-rates of one radionuclide in two complex mixtures of different composition; (c) quantitative measurements, i.e. the determination of the quantity of each radionuclide (in becquerels or curies).

The first of these tasks can be approached by considering the method of producing the mixture, the chemical behaviour of the radioactive species compared to that of various inactive carriers, the half-life values, the type and energy of the radiation emitted, and the branching ratios of the radiation.

The second task is the one most frequently met in radioanalytical work. It can be solved in many different ways, but basically there are two typical approaches: (a) to add inactive carrier in known amounts, allow for isotope exchange, separate and purify until further purification does not influence the characteristics of the radiation, and measure the radioactivity and chemical yield; the measurement of radioactivity in this case need not be discriminatory, (b) without chemical pretreatment, to measure the intensity at one energy only (if no other radionuclides emitting at the same energy are present) or to use the difference in half-lives to differentiate between two (or more) radionuclides, as shown in Fig. 2.2.

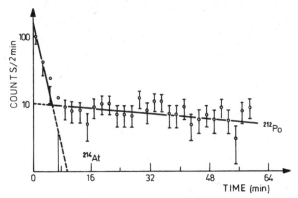

Fig. 2.2 — Decay of ^{214}At and ^{212}Po alpha peak 8.78 MeV, 10 min irradiation of Th. Half-life of ^{212}Po = 60.6 min, of ^{214}At = 1.28 $\pm$ 0.19 min. (Reproduced from [4] by permission of the copyright holder, Akadémiai Kiadó, Budapest.)

The third task, quantitative measurement, is equivalent to the determination of the number of disintegrations of a given radionuclide per second. The essential step is to find the overall efficiency (ε) of the counting procedure. Radiation from the isotope of interest may escape being counted for one of the following reasons:

(a) the trajectory of the particles may prevent contact with the counter; here the solid angle subtended by an isotropically emitting source to the detector is important;

(b) the particles can be absorbed or scattered by the finite mass of the source;

(c) the particle can be absorbed by any covering material over the sample, scattered or absorbed by the air between source and counter, or absorbed in the window of the counter;

(d) it can arrive at the counter at almost the same time as another particle, so that the two of them produce only one response (correction for dead-time is needed for high count-rates);

(e) it can hit the sensitive part of the detector, but by chance not produce enough ionization to result in a count.

On the other hand, some particles that originally depart from the source in a 'wrong' direction can arrive at the detector because of (a) the effect of back-scattering (of electrons) by the material supporting the source or (b) scatter by the counter supports and shield.

Therefore, absolute quantitative calibration includes the determination of these individual efficiency factors, which can be expressed as percentages and multiplied together to give the overall efficiency.

2.2.1 Gamma-ray spectrometry

In gamma-ray spectrometry, a mixture of gamma-ray emitting radio-nuclides can be resolved quantitatively by pulse-height analysis. The analysis is based on the fact that the pulse-heights (in volts) produced by a phototube are proportional to the amounts of gamma-ray energy arriving at the scintillation detector. The amplification of the pulse is proportional to the voltage applied to the phototube, which can be adjusted so that the entire gamma spectrum can be examined.

In a 'single-channel' counting procedure, only pulses of a certain height are counted; pulses of smaller or larger height are not registered ('seen') by the pulse-height analyser. This pulse-height 'window' is set by two pulse-height selectors.

In 'multichannel' analysis the pulses are sorted according to height by a digital computer technique, and after a certain period of measurement, the distribution of the number of pulses according to height is obtained.

The number of counts registered in a certain channel (energy interval), plotted against channel number, is called a gamma (or alpha, or beta) spectrum. The gamma radiation of a simple radionuclide having a unique gamma energy does not produce a sharp spectral line, but a peak, owing to the statistical nature of the interaction of gamma rays with matter.

An NaI(Tl) scintillation detector spectrum for ^{137}Cs is shown in Fig. 2.3. The peak at channel No. 100 is due to back-scatter and corresponds to 0.184 MeV; the edge at channel No. 180 is the Compton edge with energy 0.478 MeV, and the peak at channel No. 280 is the photopeak (0.662 MeV) which is characteristic for ^{137}Cs. The average width of the photopeak is 32 channels. The resolution (in per cent) characterizing the discriminating power of a detector is defined as $100\ \delta E/E$, where δE is the average width [full width of the peak at half of the maximum (FWHM) count level] of a photopeak, and E is its energy. For the purpose of finding the resolution of a detector (relative to a certain energy or source) the energy can be expressed in channel numbers. Therefore, in this case resolution with respect to ^{137}Cs is $100\ \delta E/E = 3200/280 = 11.5\%$.

An energy calibration for a detector consists of finding the linear dependence between particle energy (MeV) and channel number.

A block diagram of the electronics of a spectrometric system is given in Fig. 2.4.

It is evident that analyses of mixtures of radionuclides are not simple, for the multiplicity of types of interaction of gamma rays with the detector material generally leads to complicated gamma-ray spectra. Besides the peak due to total absorption (photopeak at E) peaks are caused by the following: pair production and escape of one or both annihilation photons (E 0.51 or 1.02 MeV), 180° Compton scattering, single Compton scattering, multiple Compton scattering, external Compton scattering, and iodine K X-rays (from the NaI(Tl) crystal). For analysis and measurement of spectra of mixed gamma radiation, it is convenient and precise to measure only the contribution to the spectra resulting from total energy absorption.

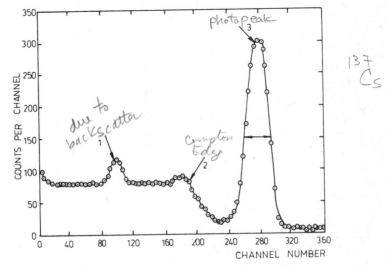

Fig. 2.3 — NaI(Tl) spectrum of ^{137}Cs. Constructed from data given in [5].

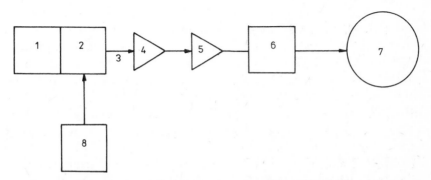

Fig. 2.4 — A block diagram of the electronics of a spectrometric system. Data from [5] were used for construction of the figure. 1 — detector and phototube, 2 — photomultiplier, 3 — dynode, 4 — scintillation preamplifier, 5 — amplifier, 6 — multichannel analyser, 7 — teletype or other readout accessory, 8 — HV power supply.

If a semiconductor (such as lithium-drifted germanium) is used instead of an NaI(Tl) scintillation detector, the resolution is improved by a factor of 30 or more. The tremendous increase in resolution compared to scintillation detectors is shown in Fig. 2.5.

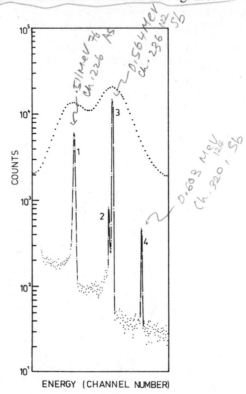

Fig. 2.5 — Gamma spectrum of an irradiated lead bullet. Constructed from data given in [5].

Here a spectrum is shown of lead bullet containing approximately 0.8 % Sb and 350 ppm As, that has been irradiated for 2 min at $\sim 10^{12}$ neutrons/cm^2/sec. Peak 1 corresponds to 0.511 MeV (channel 226) from ^{76}As; peak 3 to 0.564 MeV (channel 236) from ^{122}Sb; peak 4 to 0.603 MeV (channel 320) from ^{124}Sb. The upper curve is an expanded portion of a spectrum taken with a 3 × 3 inch NaI(Tl) crystal, with a resolution of 7 % for 0.662 MeV. The lower curve is the spectrum of the same energy range taken with a Ge(Li) low-energy photon detector, with an integral cooled preamplifier. On the other hand, the efficiency of Ge(Li) detectors is only 10–30 % of that of a 3 × 3 inch NaI(Tl) detector. The efficiency of one of the largest Ge(Li) detectors ever manufactured is 50 % of the value that would be obtained from the standard 3 × 3 inch

NaI(Tl) detector. The price of Ge(Li) detectors, at least for the present, is related to their photopeak efficiency.

The advent of high-resolution semiconductor gamma-ray spectrometers opened a new era in the acquisition of accurate analytical data, which otherwise could not be obtained without great effort spent on chemical separations.

On the other hand, the interpretation of these spectra remains rather complicated, and analysis of the spectra is done by computer techniques. The major problem is the occurrence of interferences. Each peak used for quantitative analysis may have its own characteristic problems. The analysis begins with the integration of the counts in each peak of interest, and subtraction of the background counts, which are obtained from regions of the spectrum adjacent to the peaks.

The search for possible interferences is complicated because they may affect one spectrum, but not another.

Currently, most authors are of the opinion that the most reliable way to analyse complex spectra is to treat each peak, its background and its interferences as an individual problem.

The main steps in interpretation of gamma-ray spectra are: (a) peak search, (b) peak integration, (c) peak-broadening correction, (d) background determination, (e) interference detection and correction.

Except for the peak-broadening correction, the steps are self-explanatory. The term peak-broadening correction refers to the number of counts not included in the integration step because the integration involves a fixed number of channels, which may not encompass the entire peak. The fraction of counts not included will vary if the peak shapes are not constant. To the extent that a peak is Gaussian, peak width is a reliable index of the fraction of counts lost in integration over a fixed number of channels, and this is used in making the correction.

The present state of the art allows a precision of $\sim 2\%$ in the final result for the most precisely measured elements, if the counting error is sufficiently low ($< 1.5\%$). For higher counting errors the overall precision of the determination is set mainly by the counting error.

It is noteworthy that the International Atomic Energy Agency organized (for the period 1984–1985) a gamma-ray spectroscopy service for quality-control of new methods for processing Ge(Li) spectra. All participants were provided with a set of Ge(Li) spectra on magnetic tape, or in other form, and were asked to perform an evaluation of peak parameters (detectability, position and area for singlet peaks, position and area for doublet peaks) by their new methods.

Beta-ray spectroscopy, as a means of identifying and measuring mixtures of beta-emitting nuclides by beta-energy resolution, is very difficult because of the continuous energy spectra of beta radiation. On the other hand, mono-energetic conversion electrons yield spectra which can be relatively easily interpreted, and it is even possible to discriminate the peaks corresponding to the different binding energies of the electrons. With a Ge(Li) p-n junction radiation detector at liquid-nitrogen temperature, the peaks of ^{207}Bi conversion electrons ($E_K = 974$ keV, $E_L = 1.050$ MeV, $E_{M+N+...} = 1.065$ MeV) can be discerned if energy resolution of the order of 0.3–3 % for K electrons is achieved.

2.2.2 Alpha spectrometry

For alpha spectrometry, silicon charged-particle detectors (surface-barrier detectors) have been used extensively for about 15 years. During this time they have revolutionized nuclear-particle detection. They can be used over an extensive range of energies (20 keV–200 MeV). The inherent resolution of these surface-barrier detectors is surpassed only by that of magnetic spectrometers.

Their resolution in a spectrum is calculated as $\delta(\Delta E/\Delta C)$, where $\Delta E/\Delta C$ is the slope of the linear region of the calibration graph for alpha spectroscopy [i.e. the slope of a plot of energy E (MeV) against channel number C] and δ is again the FWHM of the peak (in channel numbers). The resolution is then given in keV. In a typical ^{210}Po alpha spectrum, the energy peak at 5.305 MeV occurs in channels 512–528 with its maximum in channel 520; the resolution is $(5.305/520) \times 8 = 0.082$ MeV.

Another advantage of these detectors is the fact that the detector output pulses rise rapidly, and hence they are well-suited for fast (~ 1 nsec) timing with coincidence circuitry in time-to-pulse-height converters. The efficiency of the active volume of these detectors is essentially 100 %, and $\Delta E/\Delta C$ is linear over a rather broad range. Compared to scintillation counters, gas proportional counters or ionization chambers, they also have good long-term pulse-height stability. Moreover, they are relatively inexpensive.

There are three main parameters that define a silicon surface-barrier detector: (a) resolution, (b) active area, (c) depletion depth, which is synonymous with the sensitive depth of the detector and must be sufficient to completely stop all the charged particles that are to be measured (a 50-μm thickness is adequate to stop all natural alpha-particles, since they are usually less than 8 MeV in energy).

A typical detector may exhibit a resolution of 16 keV for ^{241}Am alpha particles, have an active area of 50 mm^2, and a depletion depth of 100 μm.

Owing to the 100 % efficiency of the detector and the comparatively large mass of alpha particles, it is easy to determine the absolute activity of an alpha source by using the formula

$$\text{activity} = (\Sigma_a/t)(4\pi s^2/\pi r^2) \text{ alpha particles/sec,}$$

where s is the distance from source to detector, r the radius of the detector, t the time (in sec), and Σ_a the area under the alpha peak.

A portion of an alpha spectrum can be expanded with a biased amplifier. In this procedure each pulse that has an amplitude less than the bias level setting will be entirely eliminated from the spectrum. Each pulse that has an amplitude greater than the bias level will have the bias level subtracted from it and the portion above the bias level can be amplified by a factor of up to 20.

It is possible to chose any minimum energy level for the amplification of the range above it, because the bias level adjustment range is sufficiently large (0–10 V).

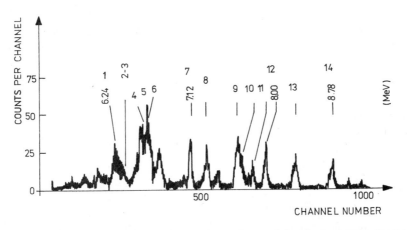

Fig. 2.6 — Alpha spectrum of spallation products after 1 min irradiation of Th with 600 MeV protons: 1 — ^{226}Th; 2 — ^{226}Th/^{221}Fr; 3 — ^{222}Ra/^{223}At; 4 — ^{223}Ac/^{221}Bi; 5 — ^{222m}Ac/^{225}Th/^{221}Ra; 6 — ^{218}Ra; 7 — ^{219}Fr; 8 — ^{214}Po; 9 — ^{217}Ra; 10 — ^{218}Fr, 11 — ^{215}At; 12 — ^{213}Po; 13 — ^{214}At/^{212}Po. Data given in [4] were used for construction of the figure.

As an example of the great possibilities of alpha spectrometry, the study of the decay ratio of ^{230}Ra can be mentioned: 24 % of ^{230}Th nuclei decay to the excited state of radium (0.068 MeV), the remaining 76 % to the ground state (0 MeV). The corresponding alpha energies ($E_{\alpha\ \text{ground}} = 4.682$ MeV, $E_{\alpha\ \text{excit}} = 4.614$ MeV) can easily be discriminated in an alpha spectrum since the FWHM of the peaks is ~ 22 keV.

An example of a complicated alpha spectrum is given in Fig. 2.6.

2.2.3 Analysis with prior chemical separation

Prior chemical separation is used to simplify the subsequent determination based on physical principles. On the other hand, the introduction of this chemical operation renders the overall determination more complicated. Therefore, a decision to include chemical separations should be made carefully.

Prior separation can be either complete (i.e. individual substances are separated) or partial (with enrichment of some components in certain fractions). The latter is more frequently used when sophisticated measuring equipment is available. In many cases, prior chemical separation is done after the addition of isotopic carriers. This type of analysis is described in connection with reverse isotope dilution (p.55).

In analysis by means of activity, no problems are generally caused by contamination of the sample during the chemical separation, because the presence of traces of impurities in the chemical reagents used does not affect the result. Only when the radionuclide isolated can be found in the laboratory atmosphere or the reagents, etc., are special measures needed (work in a dry-box or clean-room, special purification of reagents, etc.). Currently, most analysts endorse the opinion that even the tremendous progress in the resolution of detectors for gamma and alpha spectroscopy cannot render prior chemical separation superfluous in all cases.

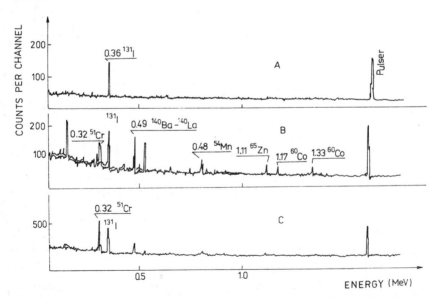

Fig. 2.7 — Gamma-ray spectra of paper disc impregnated with Ag(I). The disc had been used to filter radioactive solutions prior to radioactivity measurement. Constructed from data given in [6].

An example of the influence of prior chemical separation on the shape of gamma spectra is given in Fig. 2.7. Figure 2.7B shows the spectrum of a disc which had been used for filtering neutral reactor water 'cooled' for several weeks and spiked with additional ^{131}I. Under these conditions all nuclides are adsorbed and no chemical separation occurs. Figure 2.7C corresponds to an analogous filtration procedure where the solution had been made $0.1M$ in hydrochloric acid. The filter disc adsorbs only ^{131}I and ^{51}Cr. The simplification of the spectrum is clearly visible. For comparison, the spectrum of a disc after filtration of neutral distilled water containing ^{131}I is shown in Fig. 2.7A.

An example of practically complete separation and of isolation of some fractions with a high-pressure ion-exchange column is given in Fig. 2.8.

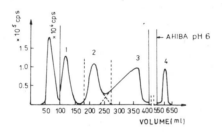

Fig. 2.8 — Separation of an irradiated Am target: 1 — Eu; 2 — Cm; 3 — Am; 4 — Ce + Pr; 8 g of Dowex W X8 with 10–14 µm grains, pH 3.65, pressure 145 atm, flow-rate = 16 cm/min. AHIBA — 0.5 M α-hydroxyisobutyric acid. Data from [7] are plotted.

2.3 ANALYSIS WITH RADIOACTIVE TRACERS

Under this heading, we will deal with processes in which the concentration of a substance c_i can be determined by use of the equation $c_i = kA_i$, where A_i is the radioactivity measured for the substance, and k a known proportionality constant which can be adjusted by the operator before the study is undertaken.

As an example, consider a study of the dependence of the solubility of a rather insoluble substance (such as $BaSO_4$) on the composition of the aqueous phase. One of the most simple ways begins by addition of a negligible concentration of radioactive ^{133}Ba to a standardized barium chloride solution, thorough mixing, and measurement of the radioactivity of an aliquot under strictly standardized conditions, to obtain the value of k. The radioactively labelled solution is then used to prepare

solid $BaSO_4$ by precipitation with sulphuric acid. The precipitate is
collected and thoroughly washed (or purified by any suitable method
without regard to yield). An excess of the solid $BaSO_4$ is then shaken
for a sufficiently long time with solutions of various compositions, to
yield a series of aqueous phases saturated with respect to barium.
Measurement of the radioactivity of aliquots of these saturated solutions (after filtration) under the same conditions as before, allows
determination of the concentration of barium in the saturated solutions
from the equation $c_i = kA_i$.

Let us consider some errors which might occur in this type of
determination.

(1) The solubility of $^{133}BaSO_4$ might in principle be different from that
    of natural barium. This is the problem of identity of chemical
    behaviour of different isotopes of the same element.

(2) The solubility of $BaSO_4$ might be different in a radiation field
    caused by the presence of large amounts of ^{133}Ba. This is the
    problem of dependence of physico-chemical properties of sub-
    stances upon internal radiation (autoradiolysis).

(3) The sulphuric acid used for the precipitation of $BaSO_4$ might con-
    tain enough barium to increase the mass of inactive barium in the
    precipitate sufficiently for k to be higher than for the ideal case. This
    is the problem of undesired isotope exchange. If sulphuric acid
    contained organic-bound barium (OrgBa), no error in the deter-
    mination would occur because the isotope exchange reaction

$$^{133}Ba^{2+} + OrgBa \rightleftharpoons Ba^{2+} + Org^{133}Ba$$

    exhibits a negligible rate.

(4) Let us assume that in the radioactive preparation of ^{133}Ba, a small
    amount of a similar organic complexing substance is present and
    the complex OrgBa is not as effectively precipitated by SO_4^{2-} as
    Ba^{2+}. Then washing of the precipitate would probably remove
    more ^{133}Ba than inactive barium, and the required constant k would
    again be higher. This is the problem of incomplete isotope exchange
    in the process of labelling.

Let us now treat these four problems in more detail. The identity of
chemical behaviour of two isotopes of the same nuclide is nearly
absolute. A certain difference is attributable to the difference in mass of
the two isotopes, unless they are nuclear isomers. Isomeric tracers have
the same thermodynamic and kinetic properties. The difference in the
properties of isotopes or their compounds is sometimes called the
isotope effect. As a rule, the isotope effect increases with an increase in

mass ratio of the corresponding isotopes. Therefore, the heavier the atoms, the smaller the isotope effect. The isotope effect for atoms is usually larger than for molecules because the presence of identical atoms in both type of molecules reduces the effect.

Two examples of the isotope effect are as follows.

(1) The solubility of inorganic salts is usually lower in heavy water, D_2O, than in H_2O. The difference is rather small for KCl at 100 °C (3.5 %) but amounts to 26 % for CdI_2 at 25 °C and 36 % for $PbCl_2$ at the same temperature.

(2) The equilibrium constant $K = [^{14}NH_3][^{15}NH_4^+]/[^{15}NH_3][^{14}NH_4^+]$ at 25 °C is equal to 1.035. In most practical work with isotopes other than the lightest, the isotope effect can be neglected, unless, of course, the chemical operation is repeated many times (e.g. in a chromatographic column). For one stage in separations applied to inorganic systems, the isotope effect is seldom larger than 1 %.

The problem of autoradiolysis is probably the most important part of a more general question, namely that of chemical transformations which take place in every system containing a radioactive indicator. The radiation energy is imparted to the system either through excitation or ionization, and leads to the formation of highly reactive free radicals or excited molecules. Typical effects include redox reactions in aqueous solutions, decomposition, and polymerization or isomerization of organic substances. In most tracer applications the effects described above are negligible. The effect may become significant if the specific activity is high, or in biological and colloidal systems with extremely high radiation sensitivity.

The problem of undesired isotope exchange occurs if the radioactive isotope in the labelled compound undergoes measurable isotope exchange with the inactive isotopes of the same element.

It would, for instance, be impossible to measure by tracer analysis the solubility of lead sulphate by using radioactive lead (Pb*) in aqueous solutions of tetraethyl-lead if the isotope exchange

$$Pb^{*2+} + Pb(C_2H_5)_4 \rightleftharpoons Pb^2 + Pb^*(C_2H_5)_4$$

occurs to a measurable extent. The reason is that, because the reaction goes to the right, the specific activity of the lead ions would be lower than the value corresponding to the precipitate. This decrease would be partly reflected in the decrease in specific activity of the surface layers of the precipitate, and measuring the total activity $Pb^{*2+} + Pb^*(C_2H_5)_4$ of the aqueous phase would not yield the correct value of the solubility.

Generally, the presence of any substances capable of isotope exchange with the tracer alters the result of the determination and has either to be eliminated or taken into account. If this cannot be achieved, the method of radioactive tracers cannot be applied.

Incomplete isotope exchange causes differences between the behaviour of the non-radioactive and the radioactive substances, and therefore can be a serious source of errors. Isotope exchange of molecules, atoms, or simple ions in the gaseous phase usually occurs rapidly enough, and does not cause any difficulties in the determination. Problems may arise with heterogeneous isotope exchange (gas–solid, solid–liquid phases), where colloidal particles are involved, or where the element occurs in different oxidation states or is bound strongly in a complex. Consider some examples. The isotope exchange of bromide ions between $PtBr_6^{2-}$ and simple Br^- exhibits a half-time of 10 min under certain conditions; the exchange of radioactive iodine between IO_3^- and I^- in neutral solutions can take more than 30 years (half-time), that between IO_4^- and IO_3^-, 150 hr. The half-time of the reaction of isotope exchange between $Co(NO_2)_3(NH_3)_3$ and $Co(II)$ is greater than 1 day. In analytical practice the following ways of securing full isotope exchange can be used: (a) transformation of both forms into gaseous compounds or simple ions, (b) transformation, successively, of both forms into all their chemical species occurring in the system (conversion into all oxidation states or complex forms), (c) increasing the temperature, and allowing sufficient time for isotope exchange to be complete.

In all cases where incomplete isotope exchange is suspected, it is necessary to verify the completeness.

The most common ways for verification are to check that (a) the result does not change if more time is allowed for isotope exchange, or if heating and cooling are applied, (b) the result does not depend on the original states of the labelled and inactive forms (e.g. the pH of the original solutions, or the oxidation state of the radioactive substance, etc.), (c) classical determination without the aid for radioactive substances gives the same results as those obtained with radioactive tracers, (d) any previous drastic chemical treatment of the test mixture (treatment with acids, transformation into gaseous compounds, etc.) does not change the final results of the determination.

The most common areas of application of tracer analysis in analytical chemistry are:

(a) determination of solubility;

(b) distribution of substances between contiguous phases as in solvent extraction, ion exchange, adsorption, co-precipitation, etc.;

(c) checking for completeness of precipitation;

(d) finding the yield of a complex purification process, or conversely determination of losses of a substance;

(e) following the distribution of a substance in a chromatographic system, or tracing the elution peaks;

(f) determination of vapour concentrations in air and/or losses of substances during evaporation, and

(g) determination of dissociation constants of analytically important complexes.

The principle of tracer analysis is a vital element in the methods of isotope dilution and radio-reagent methods but these two techniques include additional principles (change of specific activity in IDA, and the use of a chemical reaction in which the concentration of one reaction component is measured by one of the reaction products in radio-reagent procedures). These additional principles are absent in what is referred to as tracer analysis.

The radiochemical study of hydrolysable carbides is an example of the efficient use of tracer analysis [8]. The decomposition of Group IIIb dicarbides by water or aqueous solutions leads to a mixture of hydrogen and hydrocarbons up to C_{12}. The exact yields of individual hydrocarbons are important in the determination of the hydrolysis mechanism. To enhance the sensitivity of the determination of higher hydrocarbons, rare-earth carbides labelled with ^{14}C were prepared, and decomposed with water or tritium oxide. Samples of hydrocarbons labelled with tritium were analysed by use of radio liquid-chromatography (simultaneous detection of mass and activity). It was found that each class of carbide gives characteristic hydrocarbons, and the calculated overall H/C ratio allows the original carbide phase structure or composition to be deduced. The hydrocarbons CH_4, C_2H_2 and C_3H_4 are typical of the carbide phase composition, and can be used for phase analysis. The radiochemical method used is simple and convenient; the total activity in the analysis is only 3.7 MBq, and a gas flow proportional counter is used for activity measurements. The use of T_2O is justified because the usual flame-ionization detector is insensitive to hydrogen.

REFERENCES

[1] H. J. M. Bowen, *Environmental Chemistry of the Elements*, Academic Press, London, 1979.

[2] J. Tölgyessy, *Jadrové žiarenie v chemickej analýze*, SVTL, Bratislava; SNTL, Prague, 1962; Müszaki Könyvkiadó, Budapest, 1965.

[3] *Methods of Radiochemical Analysis*, World Health Organization, Geneva, 1966.

[4] D. Molzahn and R. Brandt, *J. Radioanal. Chem.*, 1974, **21**, 461.

[5] *The Instruction Manual of Using Detectors*, Ortec, New York, 1976.
[6] G. I. Csada, O. Gimesi, E. Bányai and M. Ördögh, *J. Radioanal. Chem.*, 1974, **21**, 427.
[7] S. Specht, B. O. Schütz and H.-J. Born, *J. Radioanal. Chem.*, 1974, **21**, 167.
[8] B. Hájek, V. Brožek, M. Matucha, M. Popl and J. Mostecký, *J. Radioanal. Chem.*,
     1974, **21**, 407.

SELECTED BIBLIOGRAPHY

Š. Varga, *Principles of Nuclear Tracer Methods*, in J. Tölgyessy, Š. Varga and V. Kriváň,
   Nuclear Analytical Chemistry, Vol. I, University Park Press, Baltimore; Veda, Bratis-
   lava, 1971.
D. A. Lambie, *Techniques for the Use of Radioisotopes in Analysis*, Spon, London, 1964.
T. T. Gorsuch, *Radioactive Tracers in Chemical Analysis*, The Radiochemical Centre,
   Amersham, 1966.
G. W. Stoppard, *Basic Principles of the Tracer Method*, Wiley, New York, 1962.
K. B. Zaborenko, B. Z. Iofa, V. B. Lukyanov and I. O. Bogatyrev, *Metod radioaktivnykh
   indikatorov v khimii*, Izd. Vysshaya Shkola, Moscow, 1964.
J. W. McMillan, *The Use of Tracers in Inorganic Analysis*, in *Radiochemical Methods in
   Analysis*, D. I. Coomber (ed.), Plenum Press, New York, 1975.
R. P. Ekins, *Tracer Methods in Organic and Biochemical Analysis*, in *Radiochemical
   Methods in Analysis*, D. I. Coomber (ed.), Plenum Press, New York, 1975.
V. K. Tya, A. T. Ralaban, A. R. Palmer and J. C. Maynard (eds.), *Radioisotope
   Production and Quality Control*, IAEA, Vienna, 1971.
G. Sheppard, *Radiochromatography of Labelled Compounds*, Review Booklet 14, The
   Radiochemical Centre, Amersham, 1972.
A. Evans and M. Muramatsu (eds.), *Radiotracer Techniques and Applications*, Vol. 1,
   Dekker, New York, 1977.
V. Balek and J. Tölgyessy, *Emanation Thermal Analysis and Other Radiometric Emana-
   tion Methods*, Elsevier, Amsterdam, 1984.
E. Broda and T. Schönfeld, *Radiochemische Methoden in der Mikrochemie*, in *Handbuch
   der mikrochemischen Methoden*, Springer-Verlag, Vienna, 1955.
V. Majer, *Radiometrické rozbory*, Čs. Spol. Chem., Prague, 1949.
A. J. Moses, *Nuclear Techniques in Analytical Chemistry*, Pergamon Press, Oxford, 1964.
V. L. Shashkin, *Metody analiza estestvennych radioaktivnykh elementov*, Gosatomizdat,
   Moscow, 1961.
J. Tölgyessy and Š. Varga. *Nukleárna analytická chémia*, Alfa, Bratislava, 1976.
P. Kruger, *Principles of Activation Analysis*, Wiley, New York, 1971.
I. Perlman, *Analysis of Gamma Spectra*, in *Non-destructive Activation Analysis*, S. Amiel
   (ed.), Elsevier, Amsterdam, 1981.
D. F. Covell, *Gamma-Ray Spectrometry*, in *Radiochemical Methods in Analysis*, D. I.
   Coomber (ed.), Plenum Press, New York, 1975.
C. T. Peng, *Liquid Scintillation and Cherenkov Counting in Radiochemical Methods in
   Analysis*, Application Note 34, 2nd Ed., Ortec, USA, 1976.
C. E. Crouthamel, F. Adams and R. Dams, *Applied Gamma-ray Spectrometry*, 2nd Ed.,
   Pergamon Press, Oxford, 1970.
J. Starý, M. Kyrš, M. Marhol, O. Navrátil and J. Hála, *Separační metody v radiochemii*,
   Academia, Prague, 1975.
A. C. Wahl and N. A. Bonner: *Ispoľzovanie radioaktivnosti pri khimicheskikh is-
   sledovaniyakh*, Izd. inostr. lit., Moscow, 1954. (Translation of *Radioactivity Applied
   to Chemistry*, Wiley, New York, 1951.)

J. Tölgyessy, Š. Varga and V. Kriváň, *Nuclear Analytical Chemistry*, Vols 1, 2, University
  Park Press, Baltimore; Veda, Bratislava, 1971, 1972.
T. Braun and E. Bujdosó (eds.), *Radiochemical Separation Methods*, Proceedings of the
  7th Radiochemical Conference, 27 April 1973, Mariánske Lázně (Czechoslovakia),
  Akadémiai Kiadó, Budapest, 1975.
H. P. Weise and C. J. Segebade, *Radioanal. Chem.*, 1977, **37**, 195.
W. S. Lyon and H. H. Ross, *Anal. Chem.*, 1978, **50**, 80R; 1980, **52**, 69R; 1982, **54**, 227R.
T. N. Dragnev, Zh. S. Karamanova and B. P. Damyanov, *Radiokhimiya*, 1979, **21**, 228.
J. F. Lemming and D. D. Rakel, *Guide to Plutonium Isotopic Measurements Using
  Gamma Ray Spectrometry*, USA Report MLM-2981, 1982.
D. L. Horrocks and C. T. Peng (eds.), *Organic Scintillators and Liquid Scintillation
  Counting*, Academic Press, New York, 1971.
P. Stanley and B. Scoggins (eds.), *Liquid Scintillation Counting – Recent Advances*,
  Academic Press, New York, 1974.
G. R. Gilmore and G. W. A. Newton, *Radiochemistry*, 1975, **2**, 175, Special Periodical
  Report, Royal Society of Chemistry, London.
Radiation Detection + Measurement + Analysis, Ortec, Oak Ridge, 1983.

3

Isotope dilution analysis (IDA)

3.1 INTRODUCTION, PRINCIPLES AND CLASSIFICATION

3.1.1 Introduction and principles

The analyst is often faced with the problem that a component in a mixture cannot be determined without a preceding isolation, and a compromise must be sought between adequate purity and a sufficiently quantitative yield. If the number of purification steps is large, appreciable losses of the substance to be determined are likely to occur and the result will be too low. Conversely, if the purification is carried out so as to ensure a very high yield, some impurities which have not been entirely removed can seriously lower the precision of the determination.

In such instances a solution can be sought by using the principle of isotope dilution analysis (IDA). In this technique W_0 g of radioactive nuclide of known radioactivity A_0 (an isotope of the inactive substance to be determined), is added to the original mixture containing W_x g of the substance to be determined. Complete homogenization of the active and inactive atoms (isotope exchange) is attained, and the substance is adequately purified and isolated. In this process, the radioactive and inactive atoms behave identically, so the percentage losses of each isotope are the same. Then the isolated substance is weighed (W_2 g) and its radioactivity (A_2) is measured. W_2 is related to the fraction of $(W_x + W_0)$ recovered, which is found from the radioactivity recovered.

That is,

$$\frac{W_2}{W_x + W_0} = \frac{A_2}{A_0}$$

which gives

$$W_x = \frac{W_2 A_0}{A_2} - W_0 \qquad (3.1)$$

Equation (3.1) allows W_x to be calculated, because all the other quantities are known. One important consequence of Eq. (3.1) is that if W_0 is much larger than W_x, W_x is obtained as a small difference between two large numbers, and the error involved may be prohibitive.

Equation (3.1) contains the ratio W_2/A_2 which is the reciprocal of the specific activity S_2 (radioactivity per unit weight). It has proved useful to replace activities in Eq. (3.1) by specific activities, the following expression being obtained:

$$W_x = W_0 \left(\frac{A_0 W_2}{W_0 A_2} - 1 \right) = W_0 \left(\frac{S_0}{S_2} - 1 \right). \qquad (3.2)$$

Formulae such as Eq. (3.2) are not always used to calculate the result of an analysis. It is also possible to add to each of a series of solutions containing various values of W_x of the pure substance to be determined, W_0 g of radioactive tracer with fixed specific activity S_0. Then a calibration graph $[W_x = f(1/S_2)]$is obtained which is linear for sufficiently high dilutions. A graphical representation of the principle of IDA and examples of the use of Eqs. (3.1) and (3.2) are given in Fig. 3.1.

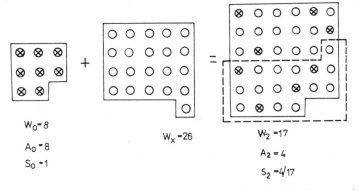

Fig. 3.1 — Graphical representation of the principle of IDA. Activity is proportional to the number of circles with a cross, mass is proportional to the number of all circles (both crossed and open).

Equation (3.2) is the most frequently used relationship in IDA. It shows that the ratio W_x/W_0 can be calculated from the change in specific activity of the radioactive substance arising from the addition of the unknown amount of inactive identical substance. Consequently, quantitative separation is not required and yields of 1% or even lower may be satisfactory in many instances. This allows the analyst to include a much wider variety of methods in his separation scheme. Simple, time- and labour-saving elegant separation schemes can be used, which would be absolutely precluded in the customary methods of analysis because of the concomitant loss of the substance to be analysed.

It is evident from the concept of specific activities that IDA can also be performed with stable isotopes if a sample substance with an isotopic abundance different from that in the unknown sample is available. In this instance, isotopic abundances are used instead of specific activities.

IDA is applied mainly in the following instances.
(1) A substance in a mixture with chemically similar substances is to be determined, and a quantitative isolation is impossible (e.g. mixtures of rubidium and potassium, or of optical isomers).
(2) The substance to be determined occurs at a very low concentration, so that losses by adsorption on the walls of vessels, on precipitates, etc., during the purification procedure are inevitable.
(3) The analysis has to be performed very quickly, e.g. because of radioactive decay or a shift of equilibrium.
(4) The substance to be determined is contained in a reservoir and only a portion of it is available (e.g. the water in a large living animal, the exchangeable potassium in a human body).

It follows from these applications that IDA is a means of improving quantitative analysis by using radioisotopes or isotopically enriched elements, but it is by no means an autonomous analytical method as are gravimetry, colorimetry, neutron activation, polarography, etc. The use of the 'substoichiometric principle' in IDA (Section 3.5) is sometimes viewed as an autonomous IDA method, but even this technique can be considered as a combination of the general IDA approach with a variety of other methods, i.e. with the ability of the analytical chemist to isolate two precisely equal but otherwise arbitrary amounts of substance from two solutions of different concentrations.

A chronological review of progress in IDA is given in Table 3.1

Table 3.1 — Chronological survey of development of IDA.

1923 Hahn [1]	Used ^{231}Pa to find the yield of isolation of ^{234}Pa
1926 Ehrenberg [2]	Used ^{212}Pb to measure the fraction of PbCO$_3$ precipitated by known amounts of CO$_3^{2-}$. It is a radio-reagent method, not IDA
1932 Hevesy and Hobbe [3]	Used ^{210}Pb to find the yield of separation of Pb from minerals
1933 Starik [4]	Independently used ^{210}Pb for the same purpose
1934 Hevesy and Hofer [5]	Used ^{2}H$_2$O for the determination of the water content of the human body
1940 Rittenberg and Foster [6]	First applied the term 'isotope dilution', analysed amino- and fatty acids in biological media with tracers containing ^{2}H
1946 Keston et al. [7]	Introduced derivative IDA
1946 Henriques and Margnetti [8]	Reverse IDA in biology
1948 Bloch and Anker [9]	Introduced the method of double dilution
1950 Keston et al. [10]	First application of double isotope derivative dilution
1957 Bojesen [11]	The first application of superior double isotope derivative dilution
1958–1960 Růžička and Beneš [12], Suzuki [13], Zimakov and Rozhavskii [4]	Introduced independently the principle of substoichiometry in IDA
1961 Růžička and Starý [15]	Developed the theory of substoichiometry
1967 Obrusník and Adámek [16], Braun and Ladányi [17]	Independently introduced displacement substoichiometric analysis
1974 Klas, Tölgyessy and Klehr (see Section 3.6)	Introduced the principle of sub- and super-equivalence IDA

3.1.2 Classification of the methods of IDA

In the literature there is much controversy over the question of IDA classification. This is due mainly to the inconsistency of applying different criteria to the various procedures.

In our opinion, the main criteria to be employed for this classification are:
(a) the manner of introducing the radioactivity into the system;
(b) the determination of the amount of the isolated fraction;
(c) the number of dilution steps during IDA;
(d) the relative weights of the unknown sample and the diluent.

According to the first criterion we can distinguish IDA with stable isotopes; IDA where the non-active sample is diluted with a radioactive substance ('direct IDA', 'single IDA'); 'reverse IDA' (dilution of a radioactive sample with stable isotopes); 'derivative dilution analysis', where the substance to be determined, originally non-radioactive, is made radioactive by a quantitative irreversible chemical reaction with a radioactive substance; IDA where two different radioactive isotopes of one element are mutually diluted; 'IDA after activation', where radioactivity of the element to be determined is induced by a suitable nuclear reaction and subsequently diluted with non-radioactive material; 'pseudoisotopic dilution analysis' where the diluent substance is not an isotope of the analyte, but has sufficiently similar chemical behaviour to it.

According to the second criterion, we distinguish between 'classic IDA'; 'substoichiometric IDA'; 'mixed IDA'.

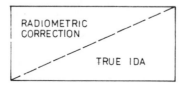

ACCORDING TO RADIOACTIVITY INTRODUCTION

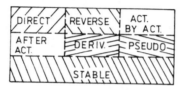

ACCORDING TO MASS DETERMINATION

ACCORDING TO NUMBER OF DILUTION STEPS

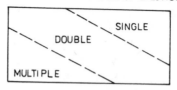

Fig. 3.2 — Classification of IDA.

In classic IDA, the amount of isolated fraction is determined by classical analytical methods such as gravimetry, titrimetry or colorimetry. In substoichiometric IDA this operation is replaced by isolation of equal amounts of material from two solutions of different concentration. In 'mixed IDA' the amount isolated is determined by a radioanalytical method such as activation analysis, or by a radio-reagent method.

According to the number of dilution steps we distinguish single IDA (or 'simple IDA'), for which one dilution step is sufficient; double IDA, in which two dilution steps are involved; multiple IDA, where three or more dilution steps are needed.

Using the fourth criterion we divide IDA into 'radiometric correction', where the weight of the radioactive diluent is negligible compared to the weight of the sample (vice versa in reverse IDA), and 'true IDA', where the two weights are comparable.

A graphic representation of the proposed classification is given in Fig. 3.2.

Theoretically, the total number of combinations based on the number of types within each criterion would amount to $7 \times 3 \times 3 = 63$. In practice, however, some of the 63 theoretical combinations are 'forbidden' for various reasons, and some could be further subdivided by use of additional criteria important for that particular combination. A full description of all the combinations used in the abundant literature is beyond the scope of this book.

As a result, it is useful to choose only several selected types, and treat them in some detail so that a deeper understanding of them can be obtained. This treatment is given in the following sections.

3.2 DILUTION WITH RADIOACTIVE ISOTOPES

This category of IDA has also been called direct IDA, simple IDA and single IDA. It is a method of improving quantitative analysis by using radioisotopes of the substance to be determined; in this way losses during purification or other steps are taken into account. Direct IDA is the type most frequently used in practical analytical work [18]. Whenever the problem of non-quantitative isolation before determination is encountered, direct IDA may be considered as a potential solution. Naturally, the availability of a radioisotope of the analyte is one of the most important prerequisites for using direct IDA.

If the amount of radionuclide (W_0) introduced is negligible compared to the amount of analyte (W_x) the result is calculated not from Eq. (3.2) but from the obvious, simpler relationship

$$W_x = W_0 A_0 / A_2 \qquad (3.3)$$

Let us discuss the influence of the statistical character of radioactive decay on the precision of direct (and also reverse) IDA. First, it is generally assumed that the error arising from any uncertainty in W_0 and

W_2 [Eq. (3.2)] is negligible compared to the error arising from the uncertainty in A_0 and A_2. This assumption is certainly justifiable for the gravimetric determination of W_0 and W_2, where an error of 0.1% or better can fairly easily be achieved. On the other hand, a 1% error in counting rates is considered normal for use of isotopes in chemical analysis. The influence of any error inherent in the measuring equipment is generally neglected.

From Eq. (3.3), the conclusion is reached that there must generally be an optimal ratio of W_0 to W_x. High W_0/W_x ratios result in only a very small change in the specific activity and hence a large uncertainty in W_x, whereas small W_0/W_x ratios lead to small count rates A_0 and A_2 and correspondingly large uncertainties, since the standard deviation of a total number of counts N (given by $N = At$, where t is the counting time, and A the count rate) is $\sqrt{N}$.

It can be shown that the optimal ratio W_0/W_x for an analysis depends on several factors, such as the specific activity S_0 (associated with W_0), the time available for measuring the count rates, the yield of the isolation procedure, etc. Six different methods of direct (or reverse) IDA can be formulated. The relative error $\varepsilon_R = \Delta W_x/W_x$ for each of them can be calculated, and the recommended ratio W_0/W_x for each of them evaluated. In all instances, a further assumption is made, namely, that the error in A is due only to the uncertainty in the count rate A, and that the error in the background measurement can be neglected. All the conditions for the count rate to be proportional to the number of radioactive atoms present must be fulfilled.

Equation (3.2) can be rewritten in the form $W_x = W_0(i - 1)$, where $i = S_0/S_2$. Differentiating this equation and rearranging gives:

$$\frac{dW_x}{W_x} = \left(1 + \frac{W_0}{W_x}\right)\frac{di}{i}$$

In this equation the differentials can be replaced by absolute errors (Δ) and the expression for the relative error of a product or ratio ($z = xy$ or $z = x/y$)

$$\frac{\Delta z}{z} = \left[\left(\frac{\Delta x}{x}\right)^2 + \left(\frac{\Delta y}{y}\right)^2\right]^{1/2}$$

introduced. Thus

$$\varepsilon_R = \frac{\Delta W_x}{W_x} = \left(1 + \frac{\Delta W_0}{W_x}\right)\left[\left(\frac{\Delta S_0}{S_0}\right)^2 + \left(\frac{\Delta S_2}{S_2}\right)^2\right]^{1/2} \tag{3.4}$$

In this expression the square root term must express a relationship corresponding to the type of experiment. For example, consider the following situation: the specific activity S_0 is known before the experiment, and has a fixed relative error $\Delta S_0/S_0 = \delta$. The specific activity S_2 is always measured for a fixed time t_2, and k % of $(W_0 + W_x)$ is always isolated for measurement $[k = 100\, W_0/(W_0 + W_x)]$. What amount W_0 should be taken, then, for determining W_x, if ε_R is to be a minimum? It can be shown that in the example of IDA under consideration, Eq. (3.4) can be transformed into

$$\varepsilon_R = \left(\frac{W_x + W_0}{W_x}\right)\left(\delta^2 + \frac{1}{W_0 k S_0 t_2 p}\right)^{1/2} \tag{3.5}$$

because the relative error in the specific activity is the same as the relative error in the number of counts registered; p is a fixed proportionality constant which depends on the units in which S, W and t are expressed. The only variable in Eq. (3.5) is W_0. An increase in W_0 makes the first term increase and the second term decrease. For ε_R to be at a minimum

$$(\partial \varepsilon_R/\partial W_0)_{W_x,\, S_0,\, t,\, p,\, k} = 0$$

It can be shown that this holds when

$$W_0/W_x = [(1 + 4A)^{1/2} - 1]/2A,$$

where $A = 2W_x S_0 k t_0 \delta^2$.

Such an analysis has been made for several cases of IDA of practical importance. The results are given below. Table 3.2 summarizes the procedures considered, and Table 3.3 gives the formulae obtained for ε_R and for minimum ε_R. Figure 3.3 shows the dependence of ε_R on W_0/W_x for the different types.

Klas has analysed two cases where the error due to the background count rate is also taken into account [19]. The examples included equal amounts of substance isolated and either equal times t_2, t_b, t_0 (t_b = time of background measurement) or equal number of counts N, N_b, N_0. In all cases a value of W_0/W_x in the range 0.25–1 can be recommended.

If a direct or reverse IDA experiment has been carried out and a time τ is available for the measurement of N_0, N and N_b, the influence of the division of τ between t_0, t_2 and t_b on the error from the decay fluctuations can be found in Table 3.4. In this table the total time (τ) available is 60 min. The optimal division of τ, given by Klas, under certain simplified conditions is obtained if the following relationships are obeyed:

Table 3.2   Characteristics of the type of measurement (for Table 3.3).

Type	t_2	t_0	$\tau = t_2 + t_0$	$\dfrac{\Delta S}{S}$	$\dfrac{\Delta S_0}{S_0}$	Yield (k)	Isolated amount (W)
1	—	—	—	const.	const.	—	—
2	$t_2 = t_0$ const.	const.	$2t_0$ const.	var.	var.	var.	const.
3	var.	var.	const.	$\dfrac{\Delta S}{S} = \dfrac{\Delta S_0}{S_0}$ var.	var.	var.	const.
4	var.	const.	—	$\dfrac{\Delta S}{S} = \dfrac{\Delta S_0}{S_0}$ var.	var.	const.	var.
5	t_2 const.	—	—	var.	δ const.	const.	var.
6	t_2 const.	—	—	var.	δ const.	var.	const.

t_2 (t_0) — time of measuring the sample after (before) dilution; S (S_0) — specific activity after (before) dilution; const. — constant; var. — variable.

Table 3.3 — Formulae for relative errors. (Reproduced from [18] by permission of the copyright holders, Pergamon Press, Oxford.)

Type*	Formula for relative standard deviation $\Delta W_x/W_x$	Condition for minimum relative standard deviation for fixed W_x; $\mathrm{d}\,(\Delta W_x/W_x)/\mathrm{d}\,W_0 = 0$
1	$(1+\varphi)\left[\left(\dfrac{\Delta S_c}{S_0}\right)^2 + \left(\dfrac{\Delta S}{S}\right)^2\right]^{1/2}$	$\varphi = 0$ (as low as possible)
2	$(1+\varphi)(2+1/\ )^{1/2}\left(\dfrac{1}{W'S_0 t_0 p}\right)^{1/2}$	$\varphi = 0.39$
3	$(1+\varphi)(2+1/\ )^{1/2}\left(\dfrac{2}{W'S_0 \tau p}\right)^{1/2}$	$\varphi = 0.39$
4	$(1+\varphi)\,W_0^{-1/2}\left(\dfrac{2}{S_0 k t_0 p}\right)^{1/2}$	$\varphi = 1$
5	$(1+\varphi)\left(\delta^2 + \dfrac{1}{W_0 S_0 k t_0 p}\right)^{1/2}$	$\varphi = [(1+4A)^{1/2}-1]/2A$ $(0 < \varphi < 1)$
6	$(1+\varphi)\left(\delta^2 + \dfrac{1+1/\ }{W'S_0 t p}\right)^{1/2}$	$\varphi = [(1+4B)^{1/2}-1]/2B$ $(0 < \varphi < 0.5)$

$A = 2W_x S_0 k t_0 p \delta^2$; $B = 2(1 + W'S_0 t_2 p \delta^2)$; $\varphi = W_0/W_x$.
* See Table 3.2.

$$t_b = \tau\,(k + ki^2)^{1/2}/[1 + i^{1/2} + (k + ki^2)^{1/2}]$$
$$t_0 = (\tau - t_b)/(1 + i^{1/2});   t_2 = t_0 i^{1/2}$$
(3.6)

Here $i = 1 + (W_x/W_0)$; $k = A_b/A_0$.

As an example, after isolation of equal amounts, the samples gave 500 and 200 cpm with a background of 50 cpm.

$$i = (500 - 50)/(200 - 50) = 3$$
$$k = 50/(500 - 50) = 0.111.$$

The values of the times of measurement calculated by Eqs. (3.6) are included in Table 3.4 as the optimal division of time.

In direct IDA of the type corresponding to Eq. (3.3) (radiometric

Table 3.4 — Dependence of the error on the division of total time available for measurement of counting rate. (Reproduced from [19] by permission of the copyright holders, Artia, Prague.)

Division of τ	N_0	N_2	N_b	t_0 (min)	t_2 (min)	t_b (min)	$\tau = t_2 + t_0 + t_b$ (min)	ε_r (%)
So that								
$N_0 = N_2 = N_b$	2222	2222	2222	4.44	11.1	44.4	60	5.6
$t_0 = t_2 = t_b$	10000	4000	1000	20	20	20	60	4.0
Optimal	8000	5600	800	16	28	16	60	3.8
Arbitrary	17500	2000	750	35	10	15	60	5.0

N — number of counts; t — measurement time; subscripts: 0 — original radioactive sample; b — background measurement; 2 — sample after dilution.

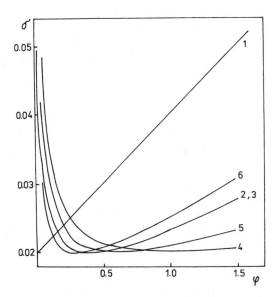

Fig. 3.3 — Errors in direct IDA caused by the statistical character of radioactive decay. δ — Relative standard deviation given by corresponding equation in Table 3.3, $\varphi = W_0/W_x$. Conditions chosen so that the standard deviation at the minimum of every curve is 0.02. The numbers on the curves correspond to the classes indicated in Tables 3.3 and 3.4. For curves 5 and 6 $\varphi_{min} = 0.6$ and $\varphi_{min} = 0.3$, respectively.

correction) the accuracy of the analysis may be improved if the following conditions are fulfilled [20, 21]:
(a) the yield of isolation is greater than 50 %;
(b) the means by which losses occur are not too numerous, all of them are known, and the losses are amenable to correction by radiometric measurement;

(c) the time available for the measurement is not prohibitively short.

In such instances, the count rate associated with the unrecovered fraction of the substance to be determined can be measured, and the correction for non-quantitative yield calculated on this basis. For instance, if 80 % of a labelled substance is isolated, the error in the radioactivity measurement of the isolated fraction is 2 %. If the loss is measured with an accuracy of 4 %, the error in the calculated yield will only be $4 \times 20/80 = 1\%$. This gain in accuracy is due to the fact that the error in the radioactivity measurement is associated with only a small part of the total quantity determined.

3.3 REVERSE IDA, TWO-RADIONUCLIDE IDA, PSEUDOISOTOPIC IDA

3.3.1 Reverse IDA
The terms 'inverse IDA' and 'dilution with inactive isotopes' are also used to describe this technique. Reverse IDA is the inverse of direct IDA. The same principle, formulae and accuracy considerations are involved, but in the reverse technique the amount of the radioactive substance is unknown and it is determined by adding a known amount of inactive substance.

If the specific activity (cpm/g) of the radioactive substance to be determined is known, the weight of the original sample can be found. Very often, however, the specific activity is not known and the analyst will record the results in cpm. If, for example, a radioactive organic substance is taken up by plants growing in an atmosphere of radioactive carbon dioxide of known specific activity, the specific activity of the compound gradually increases to the equilibrium value. After equilibrium has been reached, reverse IDA may be applied to determine weights.

Reverse IDA has several wide areas of application: these include organic analysis, biochemistry [22], biology [23, 24] and activation analysis [25, 26].

A biologist studying the degradation of a compound in an organism can introduce the compound in a labelled form into the organism. The specific activity of the compound (S_0) is known. After any time desired he can take an excretion or tissue of interest. This sample contains the original labelled compound in a mixture with various degradation products (some of them radioactive). Before isolating the original compound (of unknown weight W_0) the analyst adds a known amount (W_x) of the compound in inactive form, ensures homogenization so that the radioactive and inactive molecules behave identically and carries out the separation and purification. After that the specific activity (S_2) of the pure product is found by measuring both the radioactivity (A_2) and the weight (W_2). Equation (3.2) is then valid, except that W_0 is sought and W_x is known.

Basically, the same compound could have been introduced in an inactive form and direct IDA used for its determination in the complex mixture, but with direct IDA the smallest amount that can be applied to the organism is controlled by the sensitivity of the analytical method of determining the quantity of material isolated, whereas in reverse IDA the least amount which can be introduced into the organism is controlled by the specific activity of the labelled compound. Thus an amount smaller by several orders of magnitude can be administered for reverse IDA than for direct IDA if compounds of reasonably high specific activity are available. Moreover, in reverse IDA the purification can be carried out to as high a degree as is desired. Only practical considerations of the determination of S_2 limit the quantity of inactive carrier (W_x) added. Unlike direct IDA, reverse IDA is therefore always suitable for microanalysis. The accuracy of reverse IDA increases with larger amounts of the diluting inactive substance.

In activation analysis it is often necessary to determine the exact counting rate A_t for one radioisotope present in a complex mixture of radioisotopes, so that it can be compared with that of a standard irradiated and measured under identical conditions. If this cannot be done without chemical separation, a known amount of an inactive carrier $(W_c$, several mg) is usually added, isotopic homogenization is ensured, various purification steps are usually performed, and finally the radioactivity A_f of the pure product is measured and its weight (W_f) determined. The desired counting rate is calculated by the formula:

$$A_t = \frac{A_f W_c}{W_f} \tag{3.7}$$

The problem of purification before measuring the specific activity of the diluted sample is usually more serious in reverse IDA than in direct IDA, because the starting mixture may contain radioactivity in many different forms.

Reverse IDA is also used as an effective means for characterizing the radiochemical purity and stability of inorganic and organic preparations (adding inactive species identical to the suspected radiochemical impurities or the main component [27]), of determining the yield of nuclear reactions, such as neutron fission [28], and of determining the amount of inactive carrier in radioactive isotope preparation. It is also used in the radiochemical method of determining the burn-up of fuel in nuclear reactors [29], for determination of fall-out nuclides in water [30, 31], and for solving similar analytical problems.

3.3.2 Dilution of radioactive substance with a second isotopic radionuclide

This is a method, in a sense, between direct and reverse IDA. It is historically the first type that was used [1]. The recovery of a radioactive substance is found by adding a known amount of a different (individually measurable) isotopic radionuclide before separation, and measuring the radioactivity of each radionuclide after the isolation.

The procedure is recommended for the following cases:

(a) where no stable isotope of the radioactive substance is available;
(b) where the reagents used in the purification cannot be sufficiently purified from inactive isotopes;
(c) where the purification of the final product for weighing is too difficult and there is the danger of non-isotopic impurities causing errors in gravimetric yields;
(d) where the sample contains a mixture of radioactive and inactive isotopes and only the radioactive nuclides are of interest; it is not possible to add, for example, 3 mg of inactive strontium to determine the separation yield of ^{90}Sr in cereals if the content of inactive strontium in the material is 1–3 mg;
(e) where the mass of the final product must be kept to a minimum (e.g. to eliminate alpha or beta self-absorption);
(f) where the separation method chosen works well with amounts of the substance too small to be weighed, but cannot be used for weighable amounts. This would occur, for example, where polymerization at higher concentrations interferes with the separation step;
(g) where the separation method involves heterogeneous isotopic exchange (e.g. separation of ^{90}Sr by passing it through a column of strontium sulphate). Several examples of this type of IDA are given in Table 3.5.

The main difficulty with this type of IDA is the measurement of the two radioactive species in a mixture.

The theory of this type of IDA has been given by Currie *et al.* [33]. Let A be the disintegration rates, R the counting rates referring to one radionuclide, r the observed net counting rates and E the counting efficiencies ($AE = R$); β and γ refer to two well-defined methods of counting, j and k refer to the radionuclide to be determined and the diluent, respectively; the zero superscript refers to the initial state. Let method β be suitable for measuring the radionuclide j and method γ for the radionuclide k. The interference factors $\varepsilon_\beta = E_{\beta k}/E_{jk}$ and $\varepsilon_\gamma = E_{\gamma j}/E_{\beta j}$ should be as low as possible.

The assumption of isotope ratio conservation during the isolation step can be expressed as

$$A_j^0/A_k^0 = A_j/A_k$$

As $R_\beta = r_\beta - \varepsilon_\beta r_\gamma$ and $R_\gamma = r_\gamma - \varepsilon_\gamma r_\beta$, the value to be determined can be expressed as

$$A_j^0 = (R_{\gamma k}^0/E_{\beta j})\,(r_\beta - \varepsilon_\beta r_\gamma)/(r_\gamma - \varepsilon_\gamma r_\beta) = (R_{\gamma k}^0/E_{\beta j})\,\varrho,$$

where all the values on the right-hand side of the equation can be determined experimentally.

The uncertainty in A_j^0 arising from the uncertainty in ϱ is of interest. Rather than giving the formula for f_ϱ (the relative standard deviation of ϱ), Table 3.6 gives calculated values of f_ϱ.

Table 3.5 — Examples of diluting radionuclides with isotopic radionuclides.

Species determined	Diluent	Comments
^{239}Np	^{237}Np	Neutron activation [32]
^{144}Ce	^{133}Ce	IDA of vegetable sample [33]
^{90}Sr(^{90}Y)	^{85}Sr	Calcium is an important non-isotopic impurity [34]
^{94m}Nb	^{95}Nb	[33]
^{137}Cs	^{132}Cs	Fission product determination [35]
^{233}Pa	^{231}Pa	Discussion of corrections; neutron activation [36]
^{234}Pa	^{231}Pa	First use of IDA [1]
^{60m}Co	^{60}Co	Activation analysis, ^{60}Co may be added before irradiation [37]
^{227}Ac	^{228}Ac	Alpha emitter + beta emitter [38]
^{210}Po	^{209}Po	For the determination of ^{210}Pb [39]

Table 3.6 — Values of f_ϱ/f for IDA with a second radioisotope. (Reproduced from [33] by permission of the copyright holders, Pergamon Press, Oxford.)

x/ε	0.1	0.2	0.4	0.5	1.0	2.0	4.0	6.0	10.0
0.0	1.50	1.58	1.84	1.97	2.67	4.27	7.74	11.0	17.9
0.2	—	1.83	1.99	2.12	2.83	4.49	8.07	12.0	18.9
0.4	—	—	2.70	2.77	3.44	5.30	9.42	13.7	21.4
0.5	—	—	—	3.73	4.04	6.09	10.80	15.6	24.5

$\varepsilon = \sqrt{\varepsilon_\beta \varepsilon_\gamma}$ (the geometric mean interference factor); $\varrho_0 = \sqrt{\varepsilon_\beta/\varepsilon_\gamma}$; f is the relative standard deviation of the individual measurements of activities; $x = \varepsilon_\beta/\varrho$ if $\varrho < \varrho_0$ and $x = \varepsilon_\gamma \varrho$ if $\varrho > \varrho_0$.

The table should be used as follows. The values of ε_β, ε_γ, f and ϱ must be known. Calculate ε and ϱ_0. Determine x, and for the given x and ε values find f_ϱ/f from the table, and hence f_ϱ.

It can be shown that for small ε and for $f = 1\%$ x must not exceed 28.5. At this value, the calculated initial disintegration rate of j is greater than zero by only twice its standard deviation. Thus for a given counting rate R_k, there exists a minimum detectable rate $R^{min} = \varepsilon_\beta R_{\gamma k}/28.5$, and a maximum detectable rate $R_{\beta j}^{max} = 28.5 R_{\gamma k}/\varepsilon_\gamma$.

3.3.3 Pseudoisotopic dilution analysis

In a limited number of instances the tracer substance and the substance to be determined are ions or molecules differing in nuclear properties, and also having only a limited chemical similarity. However, their identical behaviour in the system studied and/or in the separation steps is a prerequisite for a successful use of pseudoisotopic dilution analysis. Several examples are summarized in Table 3.7.

Various considerations have led to the use of this technique. For instance, it is easier to measure the radioactivity of ^{131}I than that of radioactive carbon or hydrogen, and the synthesis of radioactive insulin may be more complicated than iodination. For routine analysis, the lower cost of using potassium instead of caesium as a carrier for ^{137}Cs to be determined might be important. No suitable radionuclides of an element might be available other than the ones which are being deter-

Table 3.7 — Examples of pseudoisotopic dilution analysis.

Species determined	Diluent	Comments
Insulin	^{131}I-iodinated insulin	^{131}I is in an inert position in the molecule [40]
^{137}Cs	K (natural)	Cs not weighable, dipicrylaminate (DPA) precipitation, KDPA weighed [41]
Fission products, rare earths	$^{231}Am^{3+}$	Precipitation of fluorides and oxalates [28]
^{250}Fm, ^{246}Cf	$^{241}Am^{3+}$	La (0.4 mg) accompanying the transuranium elements [42]
H_2O	N-Acetyl-4-aminoanti-pyrene*	Measurement of body water in living pigs [43]
Extent of a liquid reservoir	Dyes*	Must be no absorption of dye on walls [44]

* May be non-radioactive.

mined, and the use of reverse IDA might require separation methods less efficient than those suitable for the tracer concentration level. The isotopic substance (e.g. ^3H_2O) might undergo undesirable isotope exchange (e.g. exchange of hydrogen between water and active hydrogen in a tissue).

3.4 DERIVATIVE ISOTOPE DILUTION ANALYSIS

3.4.1 Principle
The method of derivative isotope dilution analysis represents a combination of the principles of the radioactive-reagent and isotope dilution methods and is used mainly for the determination of complex organic compounds. Three basic variants of this method are briefly described below (see Fig. 3.4).

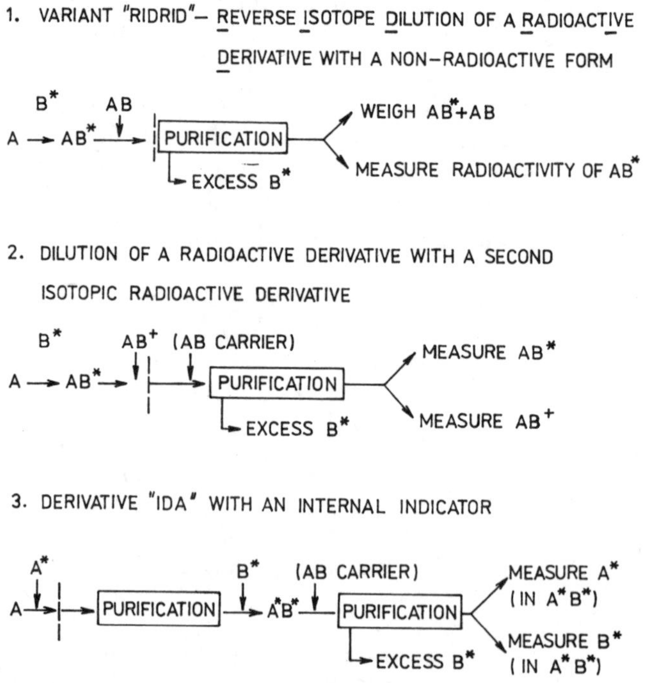

Fig. 3.4 — The graphical representation of three variants of derivative IDA.

The first variant (RIDRID) [45] is used, as a rule, when difficulties arise in the preparation of labelled substances of A, or when several chemically similar substances (A_1, A_2, ..., A_n) reacting with a substance

B are to be determined. The reaction A + B → AB should be quantitative, or its relative yield should at least be constant. The specific activity of B*(S_{B*}) should be known. Provided that the surplus of B is removed, losses of the substance AB are allowed, since this part of the method represents a reverse isotope dilution. An important requirement is to prevent an uncontrolled isotopic exchange between the substances AB* and AB. In the case of inorganic (ionic) compounds (for example precipitates) such an exchange is quite possible. Thus, the application of this variant for systems of this or similar types is impossible. The calculation of an unknown amount of the substance A consists of the following stages.

(1) Calculation of the specific activity of AB according to the formula $S_{AB*} = S_{B*}M_B/M_{AB}$, where M is the molecular weight.
(2) Calculation of the amount of the isolated substance AB* from the value of B*.
(3) Calculation of the initial amount of AB* by using the known amount of inactive AB added, the weight of the isolated precipitate of AB + AB* and the amount of AB* isolated.
(4) Calculation of the initial amount of A, based on the stoichiometry of AB and the initial amount of AB*.

The number of published papers dealing with this variant (characterized by the use of only one radionuclide) is relatively limited [46].

The second variant is a modification of the first [47]. In this case, to determine the yield of the substance AB*, the radioactive substance AB** of known activity (labelled with a radionuclide which is different from that present in the radio-reagent B*) is added, instead of a weighed amount of the substance AB and a subsequent weighing. The nuclear characteristics of the two radionuclides should be sufficiently different to allow measurement of the radioactivities of both radionuclides separately. Undesirable isotopic exchange (for example, the surplus of B* should not undergo exchange with AB**) should also be prevented in this variant (dilution with a derivative compound in a radioactive form). Purification methods suitable for small (tracer) amounts of substances can be used in this variant. However, the yield can be increased by addition of a carrier and purification methods suitable for weighable amounts of substances can also be used.

The determination of an amino-acid with two radionuclides can be mentioned as an example: A = amino-acid (glycine); B* = p-iodo-(131)benzosulphonyl chloride (pipsylchloride); AB* = pipsylglycine labelled with ^{35}S.

The third variant (dilution with the labelled initial substance) is most widely used. The main reason for its importance is the fact that the operation of dilution (the addition of an 'inner indicator' A*) is carried out at the beginning of the whole analytical procedure, so that the following stages can be performed with an incomplete yield. Rather low yields (10–50 %) are not unusual. The substance A* serves as an 'inner indicator' for the determination of the yield of the substance A*B, whereas the radio-reagent B* is used for the determination of the absolute amount of the substance A*B*. The procedure of the third variant can also be modified by an addition of an inactive carrier AB, provided that undesirable isotopic exchange does not occur. The variant is also suitable for the determination of ionic (inorganic) compounds, since no requirements for the absence of an isotopic exchange are necessary, as was the case in the preceding two variants.

This method has occupied a prominent position among routine clinical analytical methods in biochemistry and physiological chemistry. It has recently been applied in pharmacological chemistry. Reviews on the determination of steroids by methods using two different radionuclides have been published by Maier-Hueser [47] and on the determination of their metabolites by Hardouin [48].

The main advantages of the methods described are high sensitivity, potential selectivity, and the possibility of using purification procedures with an incomplete yield.

However, the method exhibits some disadvantages.

(1) The measurement of the radioactivity of two different radionuclides present in the mixture requires a relatively more complex measuring apparatus than that used for the measurement of one radionuclide. Furthermore, additional, often complicated operations of chemical separation of substances containing different radionuclides (for example $^{14}CO_2$ and ^3H_2O), are necessary. Thus, these methods are applied in specialized laboratories rather than in routine analytical (biochemical clinical) practice.

(2) To obtain the required accuracy it is necessary to keep within the determined optimum range of the activity ratio of both radionuclides.

(3) The calculation of the concentration of the analyte from radiometric data is, as a rule, rather complicated, so computers may need to be used.

(4) Radioactive organic substances are often expensive, and undergo autoradiolytic decomposition during storage. The specific activity of some commercially available substances is lower than required. The assortment of commercially available labelled compounds is still expanding but does not seem unlimited. The preparation of these compounds in a laboratory is very often troublesome.

(5) An 'isotopic effect' occurs in some cases, i.e. there is a difference in the chemical behaviour of molecules of the same compound labelled with different radionuclides. In these cases, it is impossible to reach a constant activity ratio of the two radionuclides (for example, $^{14}C/^{3}H$) during the purification, even if the product has been highly purified [49].

(6) The radio-reagents applied (acetic anhydride, thiosemicarbazide, fluorodinitrobenzene) can often react with a series of substances, especially with those which are chemically similar to the compound to be determined. Thus, the selectivity of these methods can be ensured only by a high degree of purification. However, the purification is very often a prolonged and troublesome operation, including many chromatographic procedures. Furthermore, the purification operations are not always fully successful.

However, because of the low selectivity of some radio-reagents the method sometimes assumes the character of a multicomponent analysis. Thus, for example [50], five radioactive preparations (^{35}S) of corresponding amino-acids were added to a mixture of five amino-acids. After the reaction with p-iodobenzosulphonyl chloride, a mixture of derivatives was obtained which was then separated by paper chromatography.

(7) Decisions as to the suitability of the methods described is sometimes difficult. Reproducibility is low in some cases and the analysis is sometimes rather expensive for its purpose [51].

The complications mentioned above are the main reason why the simpler methods of 'radioimmunological analysis' (see Chapter 6) have been more popular during the last two decades.

The application of three different radionuclides (for example, ^{32}P, ^{14}C, ^{3}H) has also been described [52, 53]. Such a method permits radiochemical analysis when a substance formed during chemical purification (because of decomposition of certain components of the mixture) is identical to the one being purified. The third radionuclide is added in the form of the corresponding compound before the purification. In this way the errors caused by the decomposition can be corrected.

3.4.2 Calculations of the results, errors and measurement of two radionuclide activities

In using derivative IDA with two radioactive nuclides the calculation of the amount to be determined may be complicated. Different authors use different methods of calculation. We mention here three typical ways to calculate the result:

— the basic procedure [54],
— method based on the proportionality of the amount of substance determined to the ratio of count-rates of the two radionuclides [55],
— the general procedure proposed by Riess [56].

The basic procedure can be illustrated by an example taken from the work by Hoffmann *et al.* [54], who determined testosterone (concentration $= \mu g/100$ ml of plasma), using M μg of labelled 4-^{14}C-testosterone as an internal standard with total radioactivity A cpm. As the radioactive reagent they used acetic acid anhydride, labelled with ^{3}H, with specific activity of testosterone acetate $= f$ cpm and that of ^{14}C $= a$ cpm. The value of x found in the blank (no testosterone in the plasma) was x'. The initial volume of the plasma was V ml.

If printing errors in the paper quoted are corrected, the relationship

$$x = [bAF/af - (M + x')](100/V) \tag{3.8}$$

can be derived, where F is the stoichiometric factor testosterone/testosterone acetate. The ratio b/f denotes the total amount of both radioactive and non-radioactive testosterone found by measuring the radioactivity of ^{3}H; the ratio a/A is the purification yield found from the count rate of ^{14}C by using the 'internal standard'; M is the correction for the amount of the internal standard, which is combined with the correction for the blank.

The method based on the proportionality of the amount of the substance determined to the ratio of count-rates of the two radionuclides is similar.

The basic formula is

$$A_2/A_1 = ka_2^0/a_1^0 + ka_2^0 x/ma_1^0 \tag{3.9}$$

where x is the amount of substance A to be determined, m the amount of labelled substance added (with specific activity a_1^0), the radio-reagent B has specific activity a_2^0, k is the ratio of stoichiometric factors, A_1 the count-rate of substance isolated corresponding to radioactive A, and A_2 the count-rate corresponding to radioactive B [55].

In a given series of determinations, the values a_1^0, a_2^0, m and k remain constant, so the formula above can be simplified to

$$x = \text{const}_1 A_2/A_1 + \text{const}_2$$

Values for const_1 and const_2 can be determined for a given series of determinations, if the ratio A_2/A_1 is measured for two known x values.

In the calculation of the ratio A_2/A_1, it is necessary to apply a correction to the count-rates for the contribution of B to A_1 and A to A_2.

The general procedure by Riess [56] is sufficiently described in the original paper and also in a recent book by Kyrš and Bilimovich [49]. The method is far more complicated than the two procedures above. Its characteristic feature is the use of formalized tables for the calculation of the amount to be determined, and the fact that the pertinent calculations can be done with a desk calculator (Hewlett-Packard 9100 B).

The main advantage is the inclusion of precise corrections for the blank, and for the mutual interference of the radioactivity of A and B. In this method all calculations, including auxiliary steps, are clearly defined and their sequence is determined.

It is recommended that analysts who use derivative isotope dilution only occasionally should check the calculation based on one of the three methods, by application of one of the other two, even if only in a quick and approximate procedure.

The problem of measuring separately the radioactivity of two radionuclides in a mixture remains to be discussed.

Currently, gamma, alpha and beta spectroscopy are the first choice, with solid scintillators for ^3H and ^{14}C, and more frequently, liquid scintillators for beta emitters ($^{45}Ca + ^{141}Ce$; ^3H, ^{14}C, ^{131}I).

In the past, radioactivities were measured with two different absorbers. As an example, an aluminium absorber with a thickness of 0.056 mm absorbs 50 % of the radiation of ^{131}I and 97 % of that emitted by ^{35}S. Knowing the total count-rate ($A_S + A_I$) without the absorber, and the rate ($A_{S,a} + A_{I,a}$) with the absorber, the individual values of A_S and A_I can easily be calculated [57].

In special cases, the two radionuclides can be separated chemically, and the radioactivity of each of them measured separately [58–60]. As an example, the ignition of an organic sample can be mentioned, with the formation of $^{14}CO_2$ and ^3H_2O followed by separation (freezing out of water at $-80\,°C$, that of CO_2 at $-200\,°C$).

The chemical separation approach is preferable if the two energies of the radionuclides used are too similar to be adequately resolved by the spectroscopic procedures available.

The precision of the results depends mainly on the ratio of the amounts of the two radioactivities. A rigorous treatment of this problem was given by Davies and Deterding [61]. The main recommendation can be summarized as follows: the higher the efficiency of counting for the radionuclide and the greater its interference with the measurement of the other radionuclide, the lower the amount of the given radionuclide that should be introduced into the sample.

One of the sources of systematic errors can be insufficient radiochemical purity of radioactive organic reagents. Therefore, the importance of checking the radiochemical purity (per cent of radioactivity present in the given chemical form) of the starting substances cannot be overemphasized. A very useful list of ways to check the radiochemical purity of radio-reagents was given by Burger *et al.* [62]. Among the procedures recommended, chromatography plays a very important role [63].

3.4.3 Application of computers for the evaluation of results

Conventional evaluations of the results of a given type of analysis are time-consuming and can lead to errors. It is therefore desirable to use computers for these purposes. The application of desk-top computers or pocket calculators seems advantageous for individual analyses. However, for routine analyses, the application of external computers is recommended. The application of 'in-line' computers is more expensive. More than 10 programs for such evaluations have been published since 1963. A complete program (SCINT) in Fortran-IV has been published by Boeckx *et al.* [64] for the combination ^{14}C and ^3H, and it can evaluate 75–100 samples in 1 min. The price of such evaluations is $\$9.5$ per 1000 samples. The program begins with summing the counting rates measured, and ends with the determination of the errors. A program has also been published by Kliman [65] for the analysis of steroids by using two radionuclides. The program is based on the use of a PDP-4 digital computer.

The application of computers has aided in finding an unusual method for the measurement of samples containing two radionuclides. The method [66] is based on the use of three measuring channels. A partial overlapping of energies (which is the disadvantage of the usual method of two channels) allows (using the three-channel method) work in a rather broad range of the activity ratio of the two radionuclides.

The application of computers is even more advantageous in the case of three radionuclides.

3.5 SUBSTOICHIOMETRIC IDA

3.5.1 Introduction

The main disadvantage of classic IDA is the fact that the concentration range in which it can be applied is dictated by the possibility of determining the amounts of the substance isolated by a 'non-radioactive' analytical method, e.g. gravimetry, spectrophotometry, etc. It follows from Eq. (3.2) that unless the mass of substance isolated (m) is known, the corresponding specific activity cannot be determined. In the equation for IDA, however, only the ratio of two specific activities occurs.

A very important question then arises. Is it possible to find the ratio of two specific activities (S_1/S_2) of the same substance, labelled with the same radionuclide, without knowing the values of each? The answer is yes, if the corresponding masses are exactly equal, even though unknown, since the ratio of the count-rates (A) is then exactly the same as the ratio of the specific activities.

This in turn produces a further question: which is easier for an analyst dealing with low concentrations of substances — to determine both m_1 and m_2 exactly or to isolate an arbitrary but identical amount of a substance from two solutions of different concentrations of the substance?

The current answer is isolation of equal and low amounts of the substance and this gives one of the basic ideas of substoichiometric IDA.

The next question is obviously how to perform this isolation. The answer indicates the next basic principle of substoichiometric IDA. It can be done by treatment of the substance with identical and stoichiometrically insufficient amounts of a reagent which fulfils the following requirements:

— the reagent is effectively completely consumed in the reaction and gives a single reaction product,

— the reaction product and the excess of test substance can easily be transferred into different phases or different physical locations so that the radioactivities A_1 and A_2 can easily be measured separately.

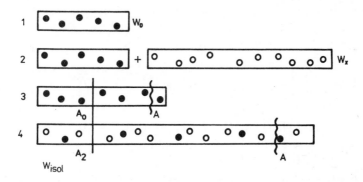

Fig. 3.5 — The graphical representation of the principle of substoichiometric IDA.

A schematic presentation of the principle of substoichiometric IDA is given in Fig. 3.5, where full circles represent the radioactive nuclides, open circles the non-radioactive ones. Line 1 shows the amount of radioactive tracer $W_0 = 5$, and line 2 shows its addition to the unknown amount of non-radioactive substance to be determined ($W_x = 10$). In line 4 the stage of full isotopic exchange after adding W_0 to W_x is shown. The lines 3 and 4 show the parallel isolation of an equal amount W_{is} from two solutions of different concentrations. The cutting curves A indicate that before the isolation a certain part of the substance to be determined after mixing with W_0 can be lost without introducing error into the analysis. The calculation of the result

$$W_x = W_0 \left(\frac{A_0}{A_2} - 1 \right) = 5 \left(\frac{3}{1} - 1 \right) = 10$$

can be illustrated by the figure; e.g. $A_0 = 3$ follows from the number of full circles in the left-hand part of line 3. The equation $A_2 = 1$ is illustrated in the left-hand part of line 4.

The idea of using separation of equal amounts of a substance from solutions of different concentrations in IDA was probably first proposed by Zimakov and Rozhavskii [14], and independently by Suzuki [13], and was applied by Růžička and Beneš for determining silver [12].

In our opinion Růžička and Starý [15, 25] were the first to fully recognize the outstanding significance of substoichiometric analysis for trace determination, develop the necessary theory, and propose the most rational procedure and the corresponding formula, which is still the one most frequently used. They also applied this type of analysis to a wide variety of metals, and coined the term 'substoichiometric IDA', which is more convenient than 'radioactivation analysis based on the quantitative isotope dilution principle' [67].

It is evident that the main factor in substoichiometric IDA is the choice of reagent. The main requirements can be summarized as follows:

— sufficiently high formation constant of the reaction product,
— sufficient difference between formation constants of different elements (selectivity),
— chemical stability in solutions (thermal, photochemical) at extremely low concentrations,
— formation of only one type of complex under the conditions suitable for the analysis (this condition concerns only the final state of the separation procedure, not the reaction mechanism),
— favourable kinetics of the reaction and separation,
— no tendency towards adsorption on reaction vessel walls.

To ascertain whether a given reagent is suitable for substoichiometric isolation and under what conditions, various experiments are performed, which can be summarized as follows.

(1) Investigation of the pH range in which the substoichiometric separation is feasible. Usually the radioactivity of the phase containing the reaction product is plotted *vs.* pH; the plateau obtained corresponds to the acceptable pH range.

(2) Investigation of the range of concentrations allowing the isolation of identical amounts. A fixed relative deficit (e.g. 50 %) of reagent is used, but the concentrations of metal and reagent are varied. Here again a plateau should exist, and the lowest concentrations allowing use of the substoichiometric principle can be found.

(3) Two-phase titrations give valuable information on the degree of consumption of the reagent. The ideal titration curve has two linear branches with no rounding in the neighbourhood of the point of intersection.

(4) The efficiency of substoichiometric isolation is usually checked by using constant reagent concentrations and increasing amounts of labelled element, of constant specific activity. The activity of the phase containing the reaction product is plotted *vs.* the metal concentration (see Fig. 3.6). A two-branch curve is obtained, cor-

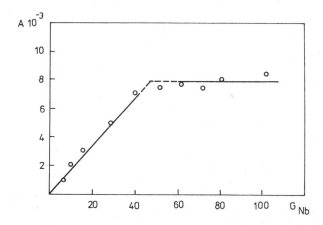

Fig. 3.6 — The dependence of the radioactivity of the organic phase on the concentration of labelled niobium in substoichiometric extraction. Solvent extraction with 5-sulphonitrophenol in butyl alcohol, reagent concentration constant, specific activity of Nb constant, A — activity in cpm, G_{Nb} — concentration of niobium, in µg/5 ml. (Reproduced from [68] by permission of the copyright holders, Akadémiai Kiadó, Budapest.)

responding to a radiometric titration of the reagent with the labelled metal. The activity should be constant when the metal is added
in excess.

(5) The kinetics of substoichiometric isolation, from which the necessary time for the reaction to be completed can be deduced (Fig. 3.7).
Sometimes (e.g. reaction of niobium with sulphonitrophenol S) the
kinetics of complex formation in the aqueous phase must be studied
separately from the kinetics of transfer of the product to the other
phase.

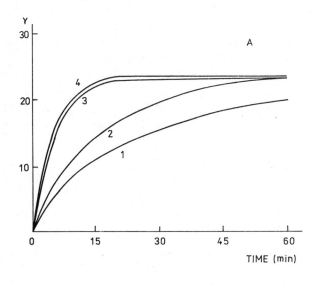

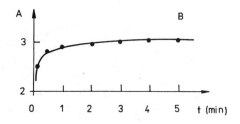

Fig. 3.7 — Kinetics of substoichiometric solvent extraction. A — effect of Cl^- concentra-
tion on extraction rate of $10^{-6}M$ Pd with $10^{-7}M$ dithizone from $2.5M$ H_2SO_4. [HCl]:
$1 - 5 \times 10^{-3}M$, $2 - 1 \times 10^{-2}M$, $3 - 6 \times 10^{-2}M$, $4 - 0.1M$; $Y = \%$ Pd extracted.
(Reproduced from [69] by permission of the copyright holders, Pergamon Press, Oxford.)
B — rate of extraction of the Fe(III)–oxine complex; A — activity of organic phase (10^3
cpm), t — shaking time (min); $2 \times 10^{-4}M$ Fe(III), $2 \times 10^{-4}M$ oxine, $6 \times 10^{-3}M$ tartrate,
pH 9.4 (Reproduced from [72] by permission of the copyright holders, Akadémiai Kiadó,
Budapest.)

(6) The reproducibility (standard deviation) is checked by repeated parallel experiments, as for other analytical procedures.

(7) Investigation of interference by other elements.

3.5.2 Classification of substoichiometric methods of analysis and formulae used

3.5.2.1 *Classification*

Substoichiometric analysis can be classified according to several criteria.

From the number of dilution steps we distinguish between simple and multiple (see p. 73) substoichiometric isotope dilution analysis. In multiple dilution analysis two or more dilution steps are involved.

Substoichiometric IDA can also be classified according to whether the exact amount of substoichiometric reagent used need be known. In most procedures it need not.

The way in which the radioactivity is introduced can also serve for classification as follows:

(1) Direct substoichiometric IDA.

(1.1) Direct substoichiometric IDA with the addition of labelled element.

(1.2) Direct substoichiometric IDA with induced radioactivity.

(2) Reverse substoichiometric IDA.

(2.1) Activation analysis with substoichiometric separation.

(2.2) Substoichiometric determination of the radioactivity of radioactive elements.

(2.3) Substoichiometric determination of the concentration of radioactive elements.

Methods (1.1), (1.2) and (2.3) require the use of extremely low concentrations of reagents, whereas categories (2.1) and (2.2) work with inactive carriers, so the separation can be done at a higher concentration level. The order of importance of the individual categories is $(1.1) > (2.1) \gg (2.2) > (2.3) \gg (1.2)$.

The difference between categories (1.2) and (2.1) must be emphasized. Both use activation, but in (2.1) it is a principal feature; relatively high concentrations of carrier are added and the purity of reagents is not of paramount importance. In (1.2) the opposite is true; here the activation is only an auxiliary procedure, and the reagents used after the activation must be free from traces of the inactive element to be determined.

Category (2.3) involves the determination of inactive carrier concentrations (or total element concentrations) in the preparations of radioactive nuclides. In this case neither special addition of a labelling substance nor induction of radioactivity is needed because the object of the analysis is itself radioactive. Therefore this type of procedure is relatively frequently used.

3.5.2.2 *The formulae used*

The most frequently used formula was given on p. 68. The formula preferred by Kudo and Suzuki [70] requires the exact amount of sub-stoichiometric reagent to be known. In this variant an amount W of a radioisotope of known specific activity A/W is added to the test solution containing an amount W_x of the element in question. A known amount of the element W_{is} ($< W + W_x$) is separated from the mixture and its radioactivity A_{is} is measured. Then

$$(W_x + W)/A = W_{is}/A_{is}$$

and

$$W_x = W_{is}A/A_{is} - W$$

W_{is} must be calculated from the amount of substoichiometric reagent used and the known stoichiometry of the reaction. If the reagent partly decomposes during the separation steps, the value of W_{is} is uncertain.

The superiority of the variant of substoichiometric IDA which does not need knowledge of the amount of reagent used was clearly illustrated by Briscoe and Humphries [69], in experiments in which palladium was extracted with dithizone at various concentration levels, with a 75 % deficit of reagent. In no case was 25 % of the palladium extracted. With $10^{-6}M$ dithizone the Pd extraction was $\sim 24\%$, at the $10^{-7}M$ level $\sim 23\%$, and at the $10^{-8}M$ level only $\sim 21\%$. These results indicate that some portion of the reagent is lost by chemical decomposition (oxidation by impurities), or sorption on the reaction vessel walls or some other process, and that the relative loss increases, as expected, with decreasing reagent concentration. Whereas in the classic substoichiometric variant this loss does not alter the results as long as it is equal in both the isotopically diluted and the undiluted solution, in the variant requiring knowledge of the amount of reagent used, this loss would cause erroneous results.

The next variant yields a very simple calculation formula.

If the tracer is available with sufficiently high specific activity, so that the concentration of it required is negligible compared to that of the substoichiometric reagent and of the element determined (W_x) the classical procedure can be modified [71, 72]. The tracer, in minute but exactly equal amounts, is added to both the test solution and the standard (W_0) and the following equation is used for calculation of the results:

$$W_x = W_0 A_0 / A_x \qquad (3.10)$$

where A_0 is the activity of the portion isolated from the standard, and A_x that of the portion isolated from the test solution.

Two typical variants of multiple substoichiometric IDA will now be briefly treated. First the original procedure of Zimakov and Rozhavskii [14] is considered. It requires the addition of W_1 µg of labelled element to one aliquot of the solution to be analysed, and W_2 µg to a second aliquot ($W_2 \gg W_1$). The aliquots each contain W_x µg of the element to be determined. From each aliquot equal amounts of the elements are isolated and their radioactivities are measured (A_1, A_2). The formula

$$W_x = \frac{W_1 W_2 (A_2 - A_1)}{W_2 A_1 - W_1 A_2} \qquad (3.11)$$

is used for the calculation of the result.

The method has been used for determination of mercury in grain, by the solvent extraction of mercury dithizonate (labelled with ^{203}Hg). The sensitivity was 0.14 µg or better and coefficient of variation < 3 % [73].

This procedure involves the use of two different amounts of the labelled element. There is another type of multiple substoichiometric IDA in which two different amounts of non-radioactive substance are added [74]. It has two variants (A and B).

Variant A. To two aliquots of the sample solution analysed (each containing W_x µg of the element to be determined) exactly equal amounts of carrier-free radionuclide and different amounts (W_1 and W_2) of the non-radioactive element are added. The solutions are diluted to the same volume, equal amounts of the labelled element are isolated from each solution and the count-rates are measured (A_1, A_2).

The result is calculated by using the formula

$$W_x = \frac{(A_2 W_2 - A_1 W_1)}{(A_1 - A_2)} \qquad (3.12)$$

Variant B. To an aliquot of sample solution (containing W_x µg of the element to be determined) an amount kW of non-radioactive substance is added ($0 < k < 1$); to a standard solution an amount W µg of non-radioactive substance is added. Both solutions are spiked with the same amount of carrier-free radioisotope and the procedure is continued as in variant A.

The calculation formula is

$$W_x = W\left(\frac{A_{st}}{A_x} - k\right) \qquad (3.13)$$

The procedures [74] were applied to the determination of caesium in various glasses, by substoichiometric extraction with $10^{-5}M$ magnesium dipicrylaminate or sodium tetraphenylborate in nitrobenzene. The error in this case can be expressed as

$$\Delta W_x / W_x = \left(\frac{k \Delta W}{W_x} + 1 \right) \left(\frac{\Delta A_{st}}{A_{st}} + \frac{\Delta A_x}{A_x} \right)$$

Consequently, if $k = 0$, the error is minimum if the value of the second factor is constant; the calculation formula is then simplified to $W_x = W A_{st} / A_x$, which is identical to the simple formula (3.10).

The most important advantage of multiple substoichiometric IDA emphasized in the literature is that the matrix is present in all samples used for the analysis, so the matrix effect can be expected to be the same in all extractions, and thus not cause erroneous results. The same effect can sometimes be achieved in classic substoichimetric IDA by removing the trace elements of interest from the sample solution before isotopic dilution. However, we have been unable to find in the available literature convincing experimental examples of cases where the matrix effect altered the results of classical substoichiometric IDA but the same reagent gave accurate results in the multiple dilution procedure. Therefore the usefulness of multiple substoichiometric IDA still remains to be unambiguously proven.

3.5.2.3 Calculations based on extrapolation and calibrations
We shall now describe several modifications of substoichiometric IDA in which the result is obtained from several experimental points on a straight line, which may be either an extrapolation or a calibration graph.

An example of the first case is substoichiometric IDA by variation of the amount of carrier [75]. In this method the substoichiometric separation is performed on a series of aliquot samples which contain increasing known amounts of added inactive carrier besides the amount to be determined, which is found by linear extrapolation.

An example is the determination of antimony in metallic tin with potassium dichromate as substoichiometric reagent [75]. An accurately weighed amount of the tin (~ 1 g) is dissolved in $HCl + H_2O_2$. Sb(V) is reduced to Sb(III) by passage of SO_2 through the solution ($\sim 3M$ HCl). Arsenic(III) is removed by solvent extraction with benzene. The solution is partly evaporated, and then diluted to volume in a 50-ml standard flask with $6M$ HCl.

Aliquots (0.5 ml) of this test solution (containing a weight W_x of Sb) are placed in 7 test-tubes, to which are added increasing amounts

(0.1–0.6 ml) of Sb(III) carrier solution (Sb 7.83 µg/ml; amount of Sb added $= x$), and 0.2 ml of ^{125}Sb(III) tracer solution (Sb, W_t) and the contents are diluted to 8.3 ml with $1M$ HCl. To each solution 0.1 ml of $4.0 \times 10^{-4} N$ potassium dichromate is added. After oxidation for 120 min, each solution is shaken with 6 ml of $0.05M$ N-benzoyl-N-phenylhydroxylamine and a 7-ml volume of the aqueous phase [containing non-extractable Sb(V)] is removed for activity measurement (A_x). Then $x = f(1/A_x)$ is plotted. The graph is linear and extrapolation to $1/A_x = 0$ gives the value of $W_t + W_x$ (Fig. 3.8).

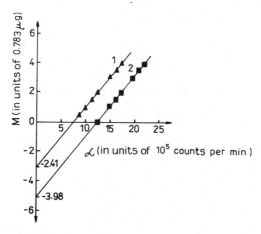

Fig. 3.8 — Redox–substoichiometric determination of antimony content in metallic tin according to the method of variation of carrier amount. M — amounts of Sb(III) carrier added, α — activity of Sb(V) isolated; 1 — tin sample, 2 — 1.57 µg of Sb(III) added to (A). Result — 1.22 ± 0.05 µg of Sb in 10 mg of Sn; 1.20 µg of Sb carrier added to ^{125}Sb. (Reproduced from [75] by permission of the copyright holders, Akadémiai Kiadó, Budapest.)

This extrapolation procedure is based on the relationship

$$(x + W_t + W_x)/W_{is} = A/A_x$$

where W_{is} the amount of Sb oxidized and remaining in the aqueous phase, and A the total radioactivity added. This relationship can be transformed into

$$x = -(W_t + W_x) + AW_{is}/A_x$$

where W_t, W_x and A are constant for the whole series in the procedure outlined, and W_{is} is constant because all the oxidant is consumed. Therefore $x = f(1/A_x)$ is linear and for $1/A_x = 0$ we obtain $-x = W_t + W_x$.

The procedure described is naturally more complicated than classical substoichiometric IDA [15, 25]. The advantage of this variation is not stated by the authors, but it can be taken that any matrix effect should be eliminated, and that the graphical method will minimize the overall error. The choice between this type of procedure and classical substoichiometric analysis using the simple IDA formula should be thoroughly investigated for each individual case. Only after evaluation of the results for several systems can a general conclusion be drawn as to which approach is the better.

An interesting feature of the method [75] is that the radioactive antimony is added late in the procedure. In our opinion this is only partly advantageous. Any losses of Sb prior to the isotope dilution step cause deviations of the A_x values from the linear plot. Early addition of the ^{125}Sb would of course lengthen the time of work with radioactivity but would also eliminate the source of reduced precision.

Some authors maintain that instead of one of the calculation formulae a 'pseudosubstoichiometric' calibration curve can be used [76].

For example, in determination of Ru with dithizone the following procedure was used [77].

To 1 ml of ^{106}Ru standard solution ($1 \times 10^{-5}M$), volume a ml of inactive $1 \times 10^{-5}M$ Ru standard solution was added ($a = 1, 2, 4, 6, 8, 10$) and after the addition of ethanol, buffer solution and adjustment of pH with $HClO_4$ or NaOH, 1 ml of $5 \times 10^{-5}M$ dithizone was added, and the aqueous phase was diluted to 25 ml and extracted with 5 ml of chloroform. The activity of 1 ml of the organic phase was counted for 1 min (count-rate A). The equation of the linear calibration line was $Ru = 7.16 \times 10^3/A - 10.98$ (nmole).

There have been attempts to use the method of sub-superequivalence when the conditions for substoichiometry are fulfilled [78]. It was found that the method of sub-superequivalence (p. 117) gives significantly higher precision. The precision of the determination of 1 μg/l. Hg in water with thio-oxine was 14–26 % by the substoichiometric calibration curve method and 4.9–8.4 % by the sub-superequivalence method. Whether this conclusion about the superior precision of the latter method is generally valid remains, in our opinion, still open to question.

3.5.2.4 Rarely used modifications
In direct substoichiometric IDA with induced radioactivity, the sample is irradiated without parallel irradiation of a standard [79–82]. It is then divided into two equal parts, each containing a weight W_x of the element in question. A known amount W of carrier is added to one part. From both samples an equal amount (W_{is}) of substance is isolated

$(W_{is} < W_x)$ and the radioactivities of the isolated fractions are measured $(A_x$ for W_x; A for W, $W + W_x)$.

The unknown amount is calculated from

$$W_x = A_x W/(A_x - A)$$

It is evident that this method is essentially substoichiometric IDA, not activation analysis, because the relationship between the induced radioactivity and the amount of substance present is not relevant. The activation only replaces the addition of a labelled substance. The method requires pure reagents, unlike destructive activation analysis. Kudo and co-workers [70, 82] used it for the determination of traces of metals in semiconductor materials (e.g. Zn in GaAs) and called it 'radioactive analysis based on the quantitative isotope dilution principle'.

Some authors [83] measure the radioactivity of both the complexed and the unbound fractions, and call this substoichiometric analysis. Strictly speaking, the method is not IDA, and belongs to the radio-reagent class. An example is the determination of thulium (W_{Tm}) with substoichiometric amounts (W_{DCTA}) of 1,2-diaminocyclohexanetetraacetic acid (DCTA) and use of ^{170}Tm, according to the formula

$$W_{Tm} = W_{DCTA}(1 + A_p/A_s)$$

where A_p is the radioactivity of a precipitate of thulium oxalate (free Tm), and A_s the radioactivity of the Tm–DCTA complex remaining in solution [83]. This procedure belongs obviously to the group requiring knowledge of the exact amount of substoichiometric reagent used.

In some cases authors have modified the substoichiometric method by introducing measurement of the non-complexed fraction instead of the complex. In this case again the main advantage of isotope dilution analysis, the tolerance of losses of the determinand element in an early purification step, is sacrificed and the procedure passes into the realm of radio-reagent methods.

An example of this type of procedure is the determination of In in zinc phosphide by complexation with substoichiometric amounts of a complexone and separation of the complexed fraction by sorption on an anion exchanger [84].

The anion exchanger is also used for elimination of the zinc as its anionic chloride complexes. ^{115m}In tracer is added before isolation of the Zn but only to check the total recovery of In and not to provide a correction for possible loss. The authors measure the radioactivity of the free indium eluted from the column.

The result is calculated from

$$W/(W - W_{is}) = A_{st}/A'_{st}; \; (W + W_x)/(W + W_x - W_{is}) = A_x/A'_x$$

where W_x is the unknown amount in the sample, W the known amount of In added to both sample and standard, W_{is} the amount of In bound by the complexone, A_x and A_{st} are the total radioactivities of carrier-free ^{115m}In added to the sample and standard, respectively, and A'_x and A'_{st} are the radioactivities of the free In in the eluate. The value of W_{is} can be eliminated from the two equations and the value of W_x calculated from W, A_{st}, A_x, A'_{st}, and A'_x [84].

Finally we describe a semiempirical method of finding the weight of the element in the sample (W_x). It was used in a case where it was observed that the classical substoichiometric equation was not quite valid [85].

The basic relationship $W_x + W = WA/A'$ (where W is the weight of standard, A the activity isolated from test solution, A' the activity isolated from standard only) leads to the expectation that a plot $W_x + W$ vs. $1/A$ is a line passing through the origin. In some cases a straight line is obtained but does not pass through the origin. In that case another system is prepared with the same amount of radioactivity but W_a of inactive substance instead of W_x. The modified equation is then

$$W_x/W_a = \left(\frac{A - A'}{A'}\right)\left(\frac{A - A''}{A''}\right)$$

where A'' corresponds to the activity isolated in the second system.

The derivation of this relationship follows from the semiempirical relationship found valid, $W_i = k(1/A_i - 1/A)$ as described above. (A_i is a variable corresponding to A', A'', etc., W_i a variable corresponding to W_x, W_a, etc.) By insertion of $i = a$ for the first system and $i = x$ for the second the expression for W_x/W_a is derived. It is to be noted that W (the weight of substance in the radioactive standard solution) is not involved in the calculation [85].

This approach was used for the determination of µg amounts of mercury in water by substoichiometric extraction with silver diethyldithiocarbamate.

This survey of methods for calculating the results reflects the many approaches to the essence of substoichiometric IDA. In our opinion there is no need to invent new ones. Practically all cases can be treated by one of the earlier procedures. What is needed is a judicious choice of method procedure for each case. The procedures involving measurement of the uncomplexed determinand should be avoided unless there is good reason to use them.

3.5.3 Substoichiometric separation by solvent extraction

The substoichiometric separation step involves two main processes: (i) the chemical reaction in which the substoichiometric reagent is consumed; (ii) the separation of the complexed (reacted) and non-complexed (non-reacted) portion into different phases (or 'spots' in paper chromatography) for subsequent radioactivity measurement of the complexed (reacted) portion.

Most of the combinations of these two steps met in the literature are given in Table 3.8.

Table 3.8 — Chemical reactions and separations in substoichiometric analysis.

Reaction, reagent	Distribution, separation	Remarks
With an extractant	Between aqueous and organic phases	Solvent extraction; reaction + separation = one operation [25]
With a complexone	(a) Solvent extraction (b) Anion exchange (c) Cation exchange (d) Electrophoresis (e) Precipitation (f) Co-precipitation	Separations on exchangers under static or dynamic conditions, kinetics of complexonate dissociation important [25]
With an oxidant or reductant	Solvent extraction, etc.	'Redox substoichiometry' [86]
Complex of test substance reacts with a metal	Between aqueous and organic phases	'Replacement substoichiometry' [17]
With an adsorbent	Between aqueous phase and adsorbent	For reaction and separation the same adsorbent is used [25, 87]
With a precipitant	Between aqueous phase and the precipitate	For reaction and separation the same precipitant is used [88]
With electrons	Between electrode and solution	Substoichiometry based on electrolysis [12, 25]

In the next sections attention will be paid to the important aspects of these reactions and separations, the choice of a suitable reagent, its concentration, reaction conditions, kinetics, thermodynamic aspects, etc.

Table 3.9 — Substoichiometric reagents and substances determined. An asterisk indicates use in reverse IDA (neutron activation). Compiled mainly from [25, 87, 88].

Reagent	Substance
$AgNO_3$	Cl^-, I^-
Acetylacetone	Be^{2+}
Arsenazo III	Th^{4+}
Ba^{2+}	SO_4^{2-}, Hf*, Zr*
α-Benzoinoxime	Mo*
N-Benzoyl-N-phenylhydroxylamine	La*, Sb(III)*, Cu*, Fe*, Mn*, Sc*, Sn*
Bindschedler's Green	Hg*
Copper diethyldithiocarbamate	Au*, Au, Pd*, Pt(II)*, As*, As, Hg
Cryptand I[a]	Na
Cryptand II[b]	K
Crystal Violet	Mo*
Cupferron	Fe(III)*, Sb(III), Mo*, Sb(III)*, Zr*
1,2-Diaminocyclohexanetetra-acetic acid	Tm
Diaminobenzidine	Se(IV)
Dicyclohexyl-18-crown-6	Na*
Diethylammonium diethyldithiocarbamate	Tl
Diethyldithiophosphoric acid	Pd*
Diethylenetriaminepenta-acetic acid	Zr
Dimethylglyoxime	Ni*
Dithizone	Bi*, Cu*, Hg*, Zn*, Pt*, Ag*, Cd, Cu, Pb, Hg, Ag, Zn, Ir, $(C_6H_5)_2Hg$, Tl
Electrons	I, Ag
Ethylenediaminetetra-acetic acid	In, In*, Fe, Ho, Tm, Y, Nd, Sc, Eu, Yb, Tb, Cr, Tl(III), Co, Mn, Sr, Zr, Pb, Zn, La
Ethyl α-isonitrosoacetoacetate	Pt*, Pd*
Glass	F^-
Hydrazine	Tl(III)
8-Hydroxyquinoline	Ga*, Mo*, La*, Sr, La, Fe
Isonitrosoacetophenone	Pd*, Cd*, Ni*
$KBrO_3$	Sb(III)
$K_2Cr_2O_7$	Sb(III)
$KMnO_4$	Sb(III)
La(III)	F^-*, SiF_6^{2-}
Lithium dipicrylaminate	Cs
Malachite Green	Ga
8-Mercaptoquinoline	Re(VII)*, Pd*
Methyl Violet	Mo(IV)*, Ta*, Sb
Mono- and dialkylphosphoric acids	Hf
Monolaurylphosphoric acid	Sn, Sb, Ti
MoO_4^{2-}	P*, P
Neocuproine	Cu*, Hg*, Sb*
1-Nitroso-2-naphthol	Co*, Pd*, Co
2-Nitroso-1-naphthol	Pd*, Co

Table 3.9 (continued)

Reagent	Substance
1-Phenyl-3-methyl-4-benzoylpyrazol-5-one	Co*, Th
1-(2-Pyridylazo)-2-naphthol	Tl(III), Co
Rhodamine B	Au*, Re*, Cu
Silver diethyldithiocarbamate	Hg
$SnO_2 . xH_2O$	F^-*, SiF_6^{2-}
Sodium diethyldithiocarbamate	Cu*, Ag*, Ni*, Se, Bi, Cr*
Sodium dipicrylaminate	K
Sodium tetraphenylborate	Cs
Tetra-n-pentylammonium bromide	Ta*
Tetraphenylarsonium chloride	Cr(VI)*, Mn(VII)*, P*, Re(VII)*, Au*, Ga*, Fe*, I*, Sb(V)*, Cr(VI), Ta
Tetraphenylphosphonium chloride	Ta
Tetraphenylphosphonium sulphate	F^-
2-Thenoyltrifluoroacetone	Mn*, Sc*
2-Thenoyltrifluoroacetone + excess of TOPO	Ca
Thionalide	Tl(I)*, Sb(III), As(III)
Thio-oxine	Pd*
Thiourea (+ reduction)	Tl(III)
Toluene-3,4-dithiol	W*, W
Trialkylbenzylammonium chloride	Zn, Cd, Hg
Trimethylsilane	F^-
Tri-n-octylamine	Cr(VI)*, Mo*
Tri-n-octylphosphine oxide	Au*, Ga*, Tl(III)
Tri-n-octylphosphine oxide with excess of hexafluoroacetylacetone	Zn*, Cu*, Eu*
Tri-n-octylphosphine sulphide	Au*
Triphenylguanidine	Au*, Ta*, Ta
Zinc dibenzyldithiocarbamate	Ag
Zinc diethyldithiocarbamate	Au*, Cd*, Se*

a 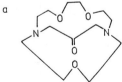 b

A rather comprehensive list of substoichiometric reagents used and the substances determined with them is given in Table 3.9. Representative examples of reagents and additional information on substoichiometric IDA covering the period 1975–1983 are given in Table 3.10.

Table 3.10 — Some reagents used in the period 1975–1983 in substoichiometric IDA or activation analysis.

Reagent	Metal determined	Separation	Matrix, sensitivity, precision	References
Complexones				
EDTA	Pb	Dithizone extracts excess of Pb	Pb in rocks, ± 10%	[89]
	Fe	Acetylacetone extracts excess of Fe	0.03 µg/ml	[90]
	In	AV-17 anion exchanger retains Fe—EDTA complex	In zinc phosphide	[84]
	Tl(III), Co	—	—	[92]
	Y, Nd	TTA extracts excess of Y, Nd		[93]
	Zr	TTA or *N*-benzoylhydroxylamine or 1-phenyl-3-methyl-4-benzoylpyrazol-5-one extracts excess of Zr	Zr in metallic Ti	[93, 94]
	Zr	Excess of Zr co-pptd. with Cu(OH)$_2$ or Fe(OH)$_3$	In steel, 10^{-6} – 10^{-4} g/ml	[95]
	Bi	Extract uncomplexed Bi as iodide complex	0.01–0.6 ppm in rocks, calibration curve used	[96]
Diethylenetriaminepenta-acetic acid	Zr	See above case with EDTA;	Zr in metallic Ti	[93]
	Y, Nd	TTA extracts excess metal	—	[94]
1,2-Diaminocyclohexanetetra-acetic acid	Tm	Pptn. or co-pptn. of excess Tm with La oxalate	Carrier in ^{170}Tm	[83], see also [97]
Chelate extractants				
Dithizone	Tl	Displacement with Hg^{2+}, chloroform	0.01 µg/ml	[98]
	Bi	—	0.1% Bi in alloys, ± 3–5%	[99]
	Pd	—	In rocks	[69]
	Cu	—	0.44 ppm in ZnSe; 0.018 ppm in distilled nitric acid; In ^{65}Cu carrier	[100]

Reagent	Element	Extraction	Remarks	Reference
	Ag	Into CCl_4	In Pt	[101]
	Zn	Into CCl_4	100 µg	[102]
	Ru(III)	Into chloroform	20 ng/ml Ru	[77]
	Cd	Into chloroform	In tap and natural waters	[91]
Diethyldithiocarbamate	Se, Tl(I), Tl(III)	—	—	[91]
Cu diethyldithiocarbamate	Au(III), Pd	—	20–200 ppm	[103]
Zn diethyldithiocarbamate	Ag	—	± 4 %	[91]
Ag diethyldithiocarbamate	Hg	—	—	[85]
Thio-oxine and its alkyl compounds (8-mercaptoquinoline)	Hg, Cu, Cd, Co, Zn	—	In H_2O	[78]
1-Phenyl-3-methyl-4-benzoylpyra-zol-5-one	Th	Into xylene	~ 70 µg ± 1.2 %	[104]
Monolaurylphosphoric acid	Ti, Sn, Sb	Into hexane	Ti in metallic Zr and Hf, ± 2 %	[105]
Tri-n-octylphosphine oxide	Tl(III)	Into benzene	20–100 µg Tl, ± 1 %	[106]
1-(2-Pyridylazo)-2-naphthol	Zn	Into chloroform	10–90 µg Zn	[107]
	Co(III)	Into chloroform	2–15 µg Co	
	Tl(III)	Into chloroform	~ 15 µg Calibration curve	
Resacetophenone oxime	Co	Into butanol	—	[108]
Ferroin	Tl(I), Tl(III)	—	—	[108]
Dimethylglyoxime	Pd	—	Low extn. constant	[108]

Table 3.10 (continued)

Reagent	Metal determined	Separation	Matrix, sensitivity, precision	References
TTA + TBP	Ca	—	In spinach, in pine needles	[109]
8-Quinolinol	Fe	Into $CHCl_3$	550 ± 9 ppm in spinach; 193 ± 4 ppm in pepperbush	[75]
Thionalide	As(III), Sb(III)	Into 1,2-dichloroethane	In soil extracts	[110]
Salicylideneamino-2-thiophenol	Sn(IV)	Complexation in benzene phase	In coal fly ash; μg ± 0.5–2.4 %	[111]
Zn dibenzyldithiocarbamate	Ag	Work at μl volume level	± 2–4 %, calibration curve, ~ 0.1 μg/100 μl	[112]
Cupferron	Bi	Into toluene, methyl isobutyl ketone	$10^{-5} M$ Bi	[99]
1-Nitroso-2-naphthol	Co	Into chloroform	1.6 pCi of ^{60}Co	[31]
Ion-association extractants Tetraphenylborate	Cs	Into nitrobenzene	22.2 μg in Cs + K + Na mixture	[113]
Brilliant Green	Tl(III)	Into benzene	Sub-μg amounts	[108]
Trimethylchlorosilane	F$^-$	Into benzene	0.05 μg/ml ± 5 % in urine, tap water, dry plants	[91]
Molybdic acid	PO_4^{3-}	Into methyl isobutyl ketone	Carrier in ^{32}P	[114]
Cryptates	Na, K, Cs	—	Proposition only	[115]
Polyethylene glycols	Ba, Sr, Pb	Into nitrobenzene with hydrophobic anions	$10^{-6} M$ Pb	[116]

Rhodamine B	Tl, In	Extraction of $InBr_4^-$ into benzene + acetone (5:1 v/v) from 2M HBr	10–60 µg In ± 0.7%; no interference from Al, Fe, As, Bi, Pd, Cd, Hg, etc.	[117]
Tetraphenylarsonium	Au	Replacement substoichiometry with Ga	$5 \times 10^{-3} M$ Au	[91]
	Ga	Replacement substoichiometry with Fe(III)	$7 \times 10^{-7} M$ Ga	
	MnO_4^-, Ta, PO_4^{3-}			
Precipitants				
Lanthanum salts	F^-	pH 2–8	—	[118]
Oxalates	Ca^{2+}	—	In soil and plants	[108]
Silver salts	I^-, Br^-	Dialysis separation	0.5 µg/ml I, 1.6 µg/ml Br^-	[119, 120]
Redox reagents				
$K_2Cr_2O_7$	Sb(III)	Extract with benzoylphenylhydroxylamine Sb(III) into $CHCl_3$	Sb in metallic Sn, 1.22 ± 0.05 µg Sb in 10 mg Sn	[121, 122]
Permanganate	Sb(III)	Sb(III) is extd. with N-benzoyl-N-phenylhydroxylamine	In metallic Zn, carrier in ^{124}Sb, 20 ± 0.8 ppm in Zn, 0.48 ± 0.02 µg in ^{124}Sb	[123—125]
Thiourea	Tl(III)	Thiourea reduces Tl(III) to Tl(I); Tl(III) is extd. into amyl acetate	2–25 µg Tl ± 4%	[108]
Hydrazine	Tl(III)	$2Tl^{3+} + N_2H_4 \rightleftharpoons 2Tl^+ + N_2 + 4H^+$, Tl^+ is extd. with isoamyl acetate	5–10 µg Tl, Cr(VI) interferes	[126]
Other Resin loaded with 8-quinolinol-5-sulphonic acid	Zn	Anion exchanger Diaion SA-100 (Cl-form, 100—200 mesh)	± 4%, ~ 100 µg Zn	[127]

3.5.3.1 Choice of reagent

The most common substoichiometric extractants are chelating agents which form extractable inner complexes; ion-association systems are also used but the reagent concentration level must generally be somewhat higher.

One of the most frequently used substoichiometric extractants is dithizone [diphenylthiocarbazone, $S = C(NH—NH—C_6H_5)_2$]. Table 3.11 indicates the range of extraction constants of dithizonates.

Table 3.11 — Extraction constants (K) of dithizonates (compiled from [98] and [128]).

Metal ion	log K^a	log K^b	Metal ion	log K^a	log K^b
Hg(II)	25	26.7	Sn(II)	− 2	− 2
Pd		27	Ga	1.3	− 1.3
Bi		10.7	Tl(I)	− 3.8	− 3.8
Cu(II)	6.5	10.5			

Metal ion	log K^a	log K^b	Metal ion	Conditions for quantitative extraction (acidity)	
Ag	5.8	8.9			
In	0.6	7			
Zn	0.6	2.6	Au(III)	0.5M H$_2$SO$_4$	
Co(II)	−15	2.3	Fe(II)	pH 7–9	
Cd	0.5	2.0	Mn	pH 10	
Pb	−0.9	0.85	Po	pH 0.6–9	
Ni	−2.9	−0.6	Pt(II)	0.5–5M H$_2$SO$_4$	
			Te(IV)	1–3M HCl	

[a] Solvent CHCl$_3$.
[b] $5 \times 10^{-5}M$ dithizone in CCl$_4$.

The stability constants of extractable chelates are important because their values determine which metals can be selectively determined. If for two metal ions with the same charge the ratio of the stability constants is $\geqslant 200$ for their complexes with the same reagent, the more stable complex can be separated under the usual substoichiometric conditions with sufficient completeness (with excess of reagent the required ratio would be $\geqslant 10^4$).

Furthermore, knowledge of the extraction constant allows calculation of the minimum pH value needed for $\geqslant 99.9\%$ consumption of the extractant; for a given concentration of extractant and metal it is $pH \geqslant - (1/n)\log K - \log 0.01\, C_{HA}$, where n is the charge on the metal ion and C_{HA} the initial molar concentration of the extractant in the organic phase.

For instance $10^{-7}M$ Cu can be determined by solvent extraction of its dithizonate into CCl_4 at pH $\geqslant -(1/2)10.5 - (-2-7) = 3.8$. For Zn the pH would be > 8. On the other hand, the upper pH value is limited by the dissociation and distribution constants of the reagent and the formation of hydroxo-complexes of the metal.

The practical reliability of formulae of this type should not be overestimated. The calculated pH range must always be checked experimentally, and discrepancies are often encountered. Moreover, the dependence of percentage extraction on pH obtained with excess of reagent obviously cannot usually be used to determine the range of pH suitable for substoichiometric determination. This can be illustrated by the extraction of $10^{-5}M$ Th with 1-phenyl-3-methyl-4-benzoylpyrazol-5-one [104]. With excess of reagent 100 % extraction occurs between pH 3.5 and 5.5 (below and above this range the extraction yield falls), but with substoichiometric amounts of the reagent extraction is complete only between pH 4.6 and 5.5.

It is somewhat unexpected that useful and reproducible substoichiometric determinations can be made with reagents which are not capable of extracting more than ~ 90 % of the element in question when present in excess. As an example the extraction of iridium with dithizone at pH 6, on heating, can be mentioned [76].

In investigating suitable conditions for substoichiometric separation it is possible, of course, to use spectrophotometric methods, instead of radiometric experiments, if the compound possesses suitable spectral properties. An example is of the substoichiometric separation of Pd for activation analysis. Pošta and Kukula [129] studied the dependence of extraction on pH by extraction titration of Pd with the $0.01M$ reagent, monitored at 550 nm.

When very low concentrations of extractants are used, their chemical stability has to be taken into account. Some practical advice can be drawn from the work of Lo *et al.* [85]. The following aspects should be considered.

(1) The stability of the stock solution during storage; the effect of temperature, daylight and type of any accompanying reagent ion on the stability [e.g. the Ag salt of diethyldithiocarbamate (AgDDC) is more stable than HDDC or NaDDC].

(2) The decomposition of the reagent during the extraction (isolation) procedure; e.g. $2M$ HCl and HNO_3 decompose AgDDC rapidly whereas $HClO_4$ and H_2SO_4 have no effect even at 5–$6M$ concentration.

(3) The stability of stock solutions of the labelled element and inactive carrier should be ensured, so that effects of some form of degradation of the latter are not erroneously ascribed to factor (1) or (2).

(4) The criterion of stability is usually constancy of the percentage of labelled substance isolated under standard conditions. Other criteria, such as the presence of decomposition products of the substoichiometric reagent can also be used.

3.5.3.2 Selectivity

The selectivity of substoichiometric solvent extraction analysis is frequently enhanced by masking the interfering elements in the aqueous phase. For instance, zinc can be determined very selectively by substoichiometric IDA with dithizone in the presence of diethyldithiocarbamate, which masks all other metals extractable with dithizone. A suitable masking reagent can enable determination of an element in the presence of another that forms more extractable complexes. For instance, silver can be determined with dithizone in the presence of copper if the latter is masked with EDTA [101].

Another important factor can be the kinetics of the process (see Fig. 3.7). An example is the extraction of Fe(III) with $2 \times 10^{-4}M$ oxine in chloroform at pH 9.4, which requires 5 hr of shaking [72]. Kinetic factors can also cause interference. Mercury, for instance, lowers the rate of solvent extraction of palladium by the exchange reaction with copper diethyldithiocarbamate because the formation of the mercury complex is fast and that of the palladium complex is slow [103].

On the other hand, kinetic factors can also lead to a substantial increase in selectivity. Thus the hydrolysis of $SbCl_6^-$ in $1M$ HCl in the first 5 min after dilution of its solution in $8M$ HCl is negligible, and $SbCl_6^-$ can be extracted under these conditions with $\sim 0.002M$ tetraphenylarsonium chloride in dichloroethane. At the same time $FeCl_4^-$ and $GaCl_4^-$ decompose rapidly in $1M$ HCl so that even 50-fold ratio of Fe(III) and Ga(III) to Sb(V) does not interfere with the substoichiometric isolation of Sb(V) under these conditions [130]. (This system is a typical example of the use of ion-association extraction systems in substoichiometric analysis.)

Another example of selectivity being increased by kinetics is the substoichiometric determination of ruthenium in the presence of some 30 metals. Excess of dithizone is added at room temperature and the excess together with all the metals except Ru is extracted. To the remaining aqueous phase substoichiometric amounts of dithizone are added and ruthenium is extracted at elevated temperature [77].

The repeated use of the same extractant in substoichiometric analysis is a rather common procedure. In a procedure by Shigematsu and Kudo [131], besides other purification steps dithizone was used three

times during the determination of Pt in optical glasses by substoichiometric neutron-activation analysis (p. 98):
(a) solvent extraction separation of the interfering elements from Pt
(IV) in $2M$ HCl after addition of 0.3 mg of inactive Pt carrier and
excess of dithizone;
(b) separation of Pt(II) with excess of dithizone, followed by mineralization of the extract;
(c) substoichiometric extraction of Pt(II) with $10^{-6}M$ dithizone followed by measurement of the extract in a Ge(Li) detector with a
4096-channel pulse-height analyser.

Another example is the determination of cadmium in commercial
aluminium [82] by activation analysis. After the addition of 400 µg of
Cd carrier the first separation is performed with excess of dithizone in
carbon tetrachloride in the presence of sodium tartrate, sodium hydroxide, hydroxylamine and potassium cyanide in the aqueous phase.
After stripping with $0.1M$ HNO$_3$ the second extraction is performed
with 1 ml of $10^{-3}M$ dithizone. Copper has been similarly determined in
semiconductor gallium arsenide.

3.5.3.3 Modifications and pitfalls
Substoichiometric separation of several metals can be performed under
dynamic conditions on a column. The column is originally loaded with
an organic solution of the complex of the extractant with a suitable
element (Me) retained on an appropriate support (fluoroplast). Passage
of the test solution through the column results in the elements with
lower extraction constants than that of Me passing through the column
into the eluate, those with higher constants being retained and forming
a sharp zone in the upper part of the column. For Me, the conditions
for its substoichiometric isolation by replacement must hold. The method has been used for the determination by activation analysis of Zn,
Cu, Cd, Ag, Hg, Mn, Co and Fe in Y, Mo, Nb and Be, with diethyldithiocarbamate and for the activation determination of Fe, Ga and Au in
Al, with tetraphenylarsonium (TPA) salts. In the latter case the column
was loaded with Fe–TPA complex. Iron was replaced by gallium, and
gallium by gold. After the successive substoichiometric replacement the
column was fully saturated with gold, the organic phase was washed
from the column with acetone, and gold was determined in the solution.
Similar results could be obtained with quaternary ammonium compounds [132].

The aqueous/organic volume ratio used in substoichiometric extraction is usually 1 : 1 but other ratios can usually be used without complications. An exception is the determination of 25–150 µg of fluoride

by solvent extraction with $\sim 8 \times 10^{-5}M$ tetraphenylstibonium sulphate at pH 1.5 into chloroform. The recommended volume ratio is about 1.35. If it is lowered to $\leqslant 1.0$ the concentration of the ion pair becomes such that a dimeric form is extracted into the organic phase, and this leads to serious error [133].

In some cases the extracts obtained in successive substoichiometric extractions are combined, and this can increase the reproducibility of the determination. This procedure was used by Jedináková and Sladkovská [101] in the determination of Ag (in Pt) with dithizone. The first extraction is performed with 10 ml of $1 \times 10^{-6}M$ reagent solution in CCl_4, and the second with 5 ml of the same solution. For the measurement of radioactivity 1 ml of the combined extract is used.

Zimakov [73] observed that in substoichiometric extraction of Sr with 8-hydroxyquinoline at pH 10–12, the extraction should be performed immediately after the addition of the buffer to the initial solution of Sr (pH $\sim$ 1–3), otherwise the reaction is far from giving complete consumption of reagent and hence irreproducible. The authors ascribe this phenomenon to the formation of strontium hydroxide and hydroxo-complexes at high pH. The discovery and investigation of this phenomenon was one of the most important stages in development of the substoichiometric determination of strontium.

The presence, in the aqueous phase, of high concentrations of 'inert' salts (such as NaCl) which presumably do not react with the reagent used for substoichiometric separation, may influence the determination in various and hardly foreseeable ways. One may be the increase in the permittivity of the aqueous phase and the concurrent change in the equilibrium constants. The influence of 'inert' salts may depend upon the reagent concentration and the type of diluent used for a particular extractant, and should be investigated for each case of substoichiometric separation. Salting-in and salting-out effects are well known in solvent extraction. An unsuspected source of error may be the combined effect of a very high concentration of an 'inert' anion forming a weak complex with the determinand element, and the very low concentration of substoichiometric complexing agent.

3.5.3.4 Unusual extractants
An interesting attempt to use the synergic effect in solvent extraction for substoichiometric separation was made by Mitchell and Ganges [134]. They studied the extraction of mg amounts of Zn, Cu, Eu, Co, Fe(II), Mn and Lu with mixtures of hexafluoroacetylacetone and tri-n-octylphosphine oxide in cyclohexane. They envisaged practical use of this system in neutron-activation analysis for Mn(II).

As already mentioned, ion-association systems can also be used for substoichiometric isolation. A cation can be extracted with a large hydrophobic anion (e.g. Cs^+ with dipicrylaminate) or an anion with a large hydrophobic organic cation (e.g. I^- with tetraphenylarsonium).

It can easily be shown that for 99 % consumption of the reagent the extraction constant should obey the relationship log $K_{ex} \geqslant 2 - $ log [Me], where [Me] is the concentration of unreacted metal. Therefore, for the determination of $10^{-7}M$ concentrations the value of log K_{ex} should exceed 9. This value is seldom reached for ion-association systems and consequently the procedure can be used only for higher concentration levels. An exception is the extraction of iodide with the phenylmercury cation, for which K_{ex} is $> 10^{10}$ [91].

An interesting case of possible substoichiometric isolation is the determination of phosphorus by solvent extraction as the ion-association compound of molybdophosphate with a suitable hydrophobic cation (tetraphenylarsonium, quinolinium, lutidinium, ammonium, etc.) [114]. In most cases the composition of the compound extracted corresponds to the stoichiometric ratio P : M : reagent = 1 : 12 : 3. The question is whether adding substoichiometric amounts of molybdate is superior to adding substoichiometric amounts of the base. The answer is as could be expected. The co-ordination reaction between phosphate and molybdate is preferred for substoichiometry. Of the bases added in excess, ammonium is the best choice because it does not form extractable ion pairs with molybdate in the absence of phosphate. Only in samples with excess of molybdate can phosphorus be determined to advantage by using solvent extraction with substoichiometric amounts of the organic cation. Phosphorus can be determined by this type of procedure at concentrations of the order of $10^{-4}M$.

An unusual substoichiometric separation has recently been proposed for the determination of μg amounts of Sn [111]. Sn(IV) iodide is extracted into benzene, then Sn(IV) is complexed with a substoichiometric amount of salicylideneamino-2-thiophenol (SATP) in the benzene phase, and the excess of tin is back-extracted. The first extraction of SnI_4 serves a twofold purpose: to separate tin from certain interfering elements and to facilitate the substoichiometric complexation with SATP. After the first extraction the benzene extract (5 ml) is treated with 1 ml of $2.5 \times 10^{-5}M$ SATP in ethanol and 0.1 ml of 10 % pyridine in ethanol. After mixing and 10 min standing, the excess of Sn is stripped with 5 ml of buffer solution (pH $\sim$ 4, $2.5 \times 10^{-5}M$ tartrate). The authors give the following comments on the procedure. The addition of pyridine to the non-aqueous medium promotes the complexation of Sn with SATP. The 1 : 1 complex is formed in the organic phase under substoichiometric conditions, then the washing with an appropriate buffer strips the tin and SATP and causes formation of the $Sn(SATP)_2$ complex, which is finally again transferred into the organic phase. Tartrate may somewhat accelerate the formation of the $1:2$ SATP complex by complexing in the aqueous phase with Sn(IV) released from the $1:1$ SATP chelate. This relatively complicated separation procedure is reproducible within 0.5–2.4 % and has been applied to the analysis of coal fly-ash.

An interesting question is the possibility of modifying the substoichiometric reagent to increase its stability or render it more suitable for the determination of a given substance. Thio-oxine (8-mercaptoquinoline) is a typical example [78]. For Hg it is recommended to use the thio-oxinates of Zn or Cu as reagent because they are more stable than thio-oxine. For copper it is recommended to use 8,8′-diquinolyl disulphide

which is reduced by Cu(I) to form Cu(II)-thio-oxinate. If a modified reagent, 8,8′-(2,2′-dimethyl)diquinolyl disulphide or 8,8′-(4,4′-dimethyl)diquinolyl disulphide, is used, the extract does not exhibit any changes of radioactivity or colour with time, unlike the case for the non-methylated reagent.

For cadmium it is recommended to use 2,7-dimethylthio-oxine because the Cd chelate is more soluble than the thio-oxinate in chloroform. For cobalt the best reagents of this family are thio-oxine and 4-methylthio-oxine. Both reagents react with Co^{3+} to form $Co(C_9H_6NS)_3$ or the analogous methylated compound. Microamounts of zinc are more successfully extracted with substoichiometric amounts of 2,7-dimethylthio-oxine than with thio-oxine. 2-Methylthio-oxine and 4-methylthio-oxine cannot be used for substoichiometric isolation of Zn.

3.5.4 Complexones in substoichiometry
The use of complexones in substoichiometric determinations is very promising. The reagents form stable complexes with most metals (Table 3.12), are chemically stable, rather cheap and usually give rapid reaction. The main difference from chelating extractants is that those ensure both substoichiometric complex formation and separation into the second (organic) phase, whereas the complexones ensure only sub-

Table 3.12 — Logarithmic stability constants of complexonates of some metals. EDTA — ethylenediaminotetra-acetic acid; DTPA — diethylenetriaminopenta-acetic acid; DCTA — diaminocyclohexanetetra-acetic acid; NTA — nitrilotriacetic acid. (From [135] by permission of the copyright holders, Izdat. Nauka, Moscow.)

Cation	log K			
	EDTA	DTPA	DCTA	NTA
Zr^{4+}	29.5	35.0	—	20.0
Hg^{4+}	29.0	35.0	—	20.0
Bi^{3+}	27.9	—	—	—
V^{+}	25.9	—	—	—
Pu^{4+}	25.8	—	—	—
U^{4+}	25.5	—	—	—
Fe^{3+}	25.1	27.9	29.3	15.9
In^{3+}	24.9	—	—	—
Th^{4+}	23.3	—	25.1	—
Tl^{3+}	23.2	—	—	—
Sc^{3+}	23.1	—	25.4	—
Hg^{2+}	21.8	26.7	25.0	—
Ga^{3+}	20.3	—	23.6	—
Pd^{2+}	19.5	—	—	—
Lu^{3+}	19.1	22.4	21.5	12.5
Er^{3+}	19.0	22.7	20.6	12.0
Cu^{2+}	18.8	21.5	22.0	13.0
VO^{2+}	18.8	—	20.1	—
Tm^{3+}	18.6	22.7	21.0	12.0
Ni^{2+}	18.6	20.2	19.4	11.5
Yb^{3+}	18.7	22.6	21.1	12.4
VO_2^{+}	18.1	—	—	—
Pu^{3+}	18.1	—	—	—
Ho^{3+}	18.1	22.8	20.0	11.85
Dy^{3+}	18.0	22.8	20.0	11.7
Pb^{2+}	18.0	18.9	19.7	11.4
Cf^{3+}	17.9	—	—	11.3
Y^{3+}	17.8	22.1	19.2	11.5
Tb^{3+}	17.6	22.7	19.5	11.6
TiO^{2+}	17.5	—	19.9	12.3
Eu^{3+}	17.3	22.4	18.6	11.5
Gd^{3+}	17.1	22.5	18.8	11.5
Cm^{3+}	17.1	—	18.4	—
Sm^{3+}	16.9	22.9	18.4	11.5
Am^{3+}	16..9	—	18.3	11.0
Zn^{2+}	16.5	18.6	18.7	11.6
Cd^{2+}	16.5	19.3	19.9	9.8
Nd^{3+}	16.4	21.6	17.7	11.2
Co^{2+}	16.3	19.3	19.6	10.4
Pr^{3+}	16.2	21.1	17.3	11.1
Al^{3+}	16.1	—	17.6	9.5
Ce^{3+}	15.8	20.5	16.8	10.8
La^{3+}	15.1	19.5	16.3	10.5

Table 3.12 (continued)

Cation	log K			
	EDTA	DTPA	DCTA	NTA
Fe^{2+}	14.3	16.0	—	8.8
Mn^{2+}	14.04	15.6	16.8	7.4
V^{2+}	12.7	—	—	—
Ca^{2+}	10.7	10.9	13.2	6.4
UO^{2+}	10.0	—	—	—
Be^{2+}	9.3	—	10.8	7.1
Mg^{2+}	8.7	9.05	10.3	5.4
Sr^{2+}	8.63	9.7	10.7	4.98
Ba^{2+}	7.8	8.6	8.0	4.8
Ag^+	7.1	—	8.2	5.2
Tl^+	6.6	—	—	—

stoichiometric complex formation in the aqueous phase, and the 'free' cation has to be separated from its complex with complexone by independent means. The separation is usually performed with ion exchangers under static or dynamic conditions. The complexed portion is sorbed on an anion exchanger or the unbound metal is retained on a cation exchanger. Paper chromatography, electrophoresis, extraction with weaker extractants, and precipitation or co-precipitation of the unbound metal have also been used. The separation process used must be a 'mild' one, so that removal of the unreacted metal does not significantly shift the equilibrium of the reaction of the metal with the complexone. This can be ensured in two ways.

(1) If dissociation of the complex is relatively slow the 'mildness' of the separation procedure can be due to a short time of contact.
(2) If the dissociation of the complex is rapid, the reagent (precipitant, extractant, ion exchanger) used for separation must remove, say, 99.9 % of the unbound metal from the aqueous phase, but not cause a significant shift of the metal–complexonate equilibrium, i.e. the equilibrium constant for its reaction with the metal must be within certain limits.

The maximum permissible concentration of the extractant can be calculated from the relationship

$$C_{max}^N = 10^{-3} \beta \, [Y^{n-}][H^+]^N / K$$

where K is the extraction constant of the chelate MA_N, β the stability constant of the complexonate MY, and $[Y^{n-}]$ the equilibrium concentration of the anion of the complexing agent.

This equation indicates that the extractant concentration should be higher if masking conditions are favourable ($[Y^{n-}]$ high) and extraction conditions unfavourable ($K/[H^+]^N$ low).

Examples of suitable extractants in substoichiometric determination of zirconium with EDTA are thenoyltrifluoroacetone, benzoyl-phenylhydroxylamine and phenylmethylbenzoylpyrazolone. In the substoichiometric determination of Bi (down to 0.01 ppm in rocks) with EDTA, excess of Bi is extracted as iodide into methyl isobutyl ketone [96].

To find a suitable concentration of an extractant (C_e) for the separation of uncomplexed metal, the dependence of the radioactivity of the extract is measured as a function of the concentration of the extractant at constant aqueous pH, substoichiometric concentration of complexone, and labelled metal concentration.

In the region of low C_e, the extraction increases with the increase in C_e because the separation of unbound metal becomes more complete. Above a certain C_e value a plateau should occur, the radioactivity of the extract being independent of C_e because all unbound metal is separated and dissociation of the complexonate is still negligible. The C_e values corresponding to this plateau should be used in the analytical procedure. For higher C_e values the extraction increases again because of increasing ligand exchange with the complexonate. A similar procedure would be useful in development of a method for separating the unbound portion of metal by batch or column anion exchange. The effect of pH can likewise be investigated at constant complexone and extractant concentration (Fig. 3.9).

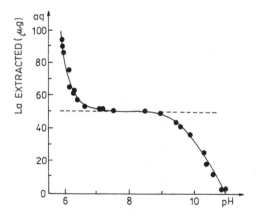

Fig. 3.9 — Effect of pH on the separation of La with EDTA. [La]$_{aq}$ — amount of La in the aqueous phase (μg), 50% substoichiometry, $10^{-4}M$ EDTA, $0.25M$ 8-hydroxy-quinoline in chloroform (for the extraction of uncomplexed La). (Reproduced from [136] by permission of the copyright holders, Akadémiai Kiadó, Budapest.)

To illustrate the competition between complexone and extractant for the metal let us calculate a hypothetical case, assuming $V_{aq} = V_{org}$, characterized by the relationships:

$[Me]_{aq} + [MeY]_{aq} + [Me]_{org} = 10^{-6}M$ (total concentration of metal)

$[MeY]_{aq} + [Y]_{aq} = 5 \times 10^{-7}M$ (total concentration of complexone)

$[Me]_{org}/[Me]_{aq} = D$ (distribution ratio in the absence of complexone)

$K = [MeY]_{aq}/[Me]_{aq}[Y]_{aq}$ (stability constant of complexonate)

We have four unknown values ($[Me]$, $[MeY]$, $[Me]_{org}$, and $[Y]$) and the four equations can be solved if we know the values of K and D. The results for two values of K and three of D are summarized in Table 3.13. The ideal case would be $[MeY]_{aq} = [Me]_{org} = 5 \times 10^{-7}M$, δ very low, $[Me]$ very low.

Table 3.13 — Theoretical calculations of a substoichiometric isolation in the presence of complexone and extractant.

K	Case	D	$[MeY]$	$[Me]$	$[Me]_{org}$	$[Y]/(5 \times 10^{7})$ $= \delta$
10^{10}	a	4.005	4.995×10^{-7}	1×10^{-7}	4.005×10^{-7}	0.001
	b	49.5	4.95×10^{-7}	0.1×10^{-7}	4.95×10^{-7}	0.0099
	c	454	4.545×10^{-7}	0.01×10^{-7}	5.455×10^{-7}	0.09
10^{15}	d	499	5.000×10^{-7}	1×10^{-9}	4.99×10^{-7}	10^{-6}
	e	4990	5.000×10^{-7}	1×10^{-10}	4.999×10^{-7}	10^{-5}
	f	5×10^{4}	4.9995×10^{-7}	1×10^{-11}	5.0005×10^{-7}	10^{-4}

It is evident that if $K = 10^{10}$ no choice of D renders the determination feasible. The values of $[Me]$ are too high. A best compromise involving errors $\sim 1\%$ would be case b, in case a only 80% of metal unbound is extracted, in case c 9% of the complexone remains unbound. If K is equal to 10^{15} the situation becomes much more favourable. In case d 0.2% of unbound metal remains in the aqueous phase but both case e and f could be used in practice. Identical calculations would be valid for batch ion-exchange separation of the free metal.

Instead of solvent extraction or ion exchange other separation procedures have been used. For example, paper electrophoresis (4 V/cm, 60 cm) has been used for the separation of the Zr complex with diethylenetriaminepenta-acetic acid from unreacted zirconium. The separation took 2 hr and the two peaks were 5–6 cm apart, which is adequate for identification of the anionic Zr–DTPA complex. The non-complexed Zr remains at the start, probably because of formation of hydroxo-compounds with isoelectric properties. The electrolyte solution was $0.2M$ $NaClO_4$ at pH 1. Amounts of zirconium as low as 10^{-10} g in 0.01 ml could be determined with a relative standard deviation of $\pm 10\%$ [95].

The role of kinetic factors in substoichiometric determinations based on reactions of metals with complexones, and subsequent separation by electromigration on silica gel, was elucidated by Stepanov and Gedeonov [83, 97], who concluded that under the conditions used, substoichiometric analysis was feasible only if the average life time of the complex was much longer than the time needed for separation by electromigration. The corresponding electromigration curves are given in Fig. 3.10.

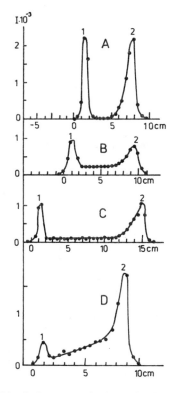

Fig. 3.10 — Influence of kinetic factors on the determination of Tm with substoichiometric amounts of DCTA or EDTA by electromigration. I = radioactivity of the spot on the paper strip. 1 — complexed Tm; 2 — unbound Tm; 1.75×10^{-9} mole DCTA. A — 25°C, $E = 50$ V/cm, 5 min; B — 25°C, $E = 1.56$ V/cm, 167 min; C — 60°C, $E = 50$ V/cm, 5 min; D — EDTA, 25°C, $E = 50$ V/cm, 5 min. (Reproduced from [97] by permission of the copyright holders, Izdat. Nauka, Moscow.)

The half-life of the Tm^{3+} complex with DCTA at pH 2.6 and 20 °C is 5.7×10^3 sec and that for the EDTA complex only 150 sec. It can be seen that the separation becomes unsatisfactory if the migration time is too long (B), the temperature is high so that the half-life of the complex is shortened (C), and the ligand does not form a complex with suffi-ciently long half-life (D).

Since higher stability constants usually correspond to lower dissocia-tion rates, it is evident that the thermodynamic and kinetic conditions required for this type of substoichiometric analysis can be fulfilled simultaneously by using highly stable complexes.

The kinetic inertness of the Tm–DCTA complex at around pH 3.2 allows the separation of the unbound Tm by adding 1 mg of Tm carrier (or La carrier) and (co-)precipitating the radioactively labelled thulium with oxalic acid. The period of half-dissociation of the Tm–DCTA complex is 1.4×10^4 sec (at 25 °C, ionic strength 0.1, pH 3). The method was used for determination of the amount of carrier in a radioactive solution of ^{170}Tm.

In the determination of Fe(III) and Tl(I) with substoichiometric amounts of EDTA the unbound metal is also separated by co-precipita-tion [e.g. Tl(I) with $BaCrO_4$].

The difference between the stability constants of the Y and Nd complexes with EDTA and DTPA has been used [92] for determination of the two metals in a model mixture. Only Y was labelled and the fraction of complexed Y was measured radiometrically (Φ_Y for EDTA and Ψ_Y for DTPA). It was found experimentally that for various combinations of the initial yttrium and neodymium concentrations (C_Y, C_{Nd}) the following linear relationships were valid:

$$C_Y + a_1 a_{Nd} = C_{EDTA}/\Phi_Y \text{ and } C_Y + a_2 C_{Nd} = C_{DTPA}/\Psi_Y$$

The empirical parameters a_1, a_2 were determined experimentally. From C_{DTPA}, C_{EDTA} and the measured values of Φ_Y and Ψ_Y for two sets of experiments, the unknown values C_Y and C_{Nd} can be calculated. Similar two-system procedures are theoretically possible with other sub-stoichiometric reagents but without the validity of the empirical linear relationships above. The theory has been developed by Kyrš and Hálo-vá [137]. The practical usefulness of the method still remains to be demonstrated.

3.5.5 Reverse substoichiometric isotope dilution,
including substoichiometry in neutron-activation analysis
Reverse substoichiometric IDA can be essentially used in three ways: as the final step in destructive activation analysis; for determination of

the *radioactivity* of radionuclides in a mixture; for determination of the *total concentration* of a substance already in radioactive form. The common feature of all these procedures is that the sample to be analysed is isotopically diluted with a non-radioactive form of the determinand, unlike direct IDA. In the first two cases the amount of non-radioactive carrier is theoretically unlimited (in practice mg amounts are used) since only the radioactivity of the test element is needed. In the last case the result is given in mass/volume units and the concentration level of the separation procedure is practically predetermined as in direct substoichiometric IDA.

3.5.5.1 Substoichiometry in destructive neutron-activation analysis

The main steps of this type of determination include irradiation of sample (amount W_x) and standard (W_{st}) under identical conditions; addition of exactly equal amounts of non-radioactive isotopic carrier to both sample and standard, and ensuring complete isotope exchange; purification of the test element in the sample solution, 100 % chemical yield not being required; isolation of equal amounts of test substance from both purified sample solution and standard solution and measurement of their radioactivities (A_x, A_{st}); calculation of the result from the relationship:

$$W_x = W_{st} A_x / A_{st}$$

The advantages of the procedure are obvious: no necessity to determine the yield by classical analytical procedures; enhanced selectivity of separation.

The procedure is nowadays used relatively rarely because of the progress of non-destructive activation analysis. Some examples of recent use are given in Table 3.14.

An additional difference between substoichiometric IDA and activation analysis is the method of measuring activity. In substoichiometric IDA simple non-discriminating scintillation or Geiger–Müller counters are generally used, whereas in activation analysis even the high selectivity of the substoichiometric separation does not always render the use of gamma spectroscopy with Ge(Li) detectors superfluous. The requirement of selective extraction in substoichiometric separation is strictly necessary only if the interfering element consumes a significant portion of the reagent or it cannot be discriminated spectroscopically from the element determined.

Table 3.14 — Substoichiometric procedures in activation analysis and reverse IDA.

Reagent	Element determined	Comment	Matrix, precision	References
		Chelate extraction		
N-Benzoyl-N-phenyl-hydroxyl-amine	Sb Sn	—	In Ge In In	[138] [139]
Ethyl α-isonitrosoacetoacetate	Pd(II)	Into chloroform	In noble metal mixture	[140]
		Ion-association extraction		
Tetraphenylarsonium	Sb(V)	Into dichloroethane, chloroform	—	[130]
Ternary amines, quaternary ammonium salts	Anionic complexes of metals	—	Impurities in Al	[141]
		Oxidation		
$KBrO_3$ $KMnO_4$	Sb(III) As(III)	Sb(III) extd. with N-benzoyl-N-phenylhydroxylamine	Sb in Sn and in Zn	[110, 124]
		Sorption		
$SnO_2 \cdot xH_2O$ sorbent	F^- or SiF_6^{2-}		O in Si crystals	[25, 87]
		Precipitation		
H_2PtCl_6	^{137}Cs	Reverse IDA	In environ. samples In K salts	[18] [142]
		Co-precipitation		
Molybdophosphoric acid	^{137}Cs	Reverse IDA	± 5 % error	[18, 142]

The use, in activation analysis, of substoichiometric separation with incomplete consumption of reagent is met rather rarely, because of the relatively high concentrations involved. An exception is the procedure by Jaskólska [139] in which activation analysis for tin in indium is combined with substoichiometric extraction ($> 80\%$) of tin with $0.015M$ N-benzoyl-N-phenylhydroxylamine in $CHCl_3$. It is claimed that the result is reproducible although the amount of tin extracted was not linearly related to the BPHA concentration (this is similar to the case for substoichiometric extraction of weak complexes). In this work the radiochemical purity of the fraction of Sn isolated was not sufficient for the ^{121}Sn beta radiation to be counted (presumably ^{82}Br and ^{187}W interfered). Hence the determination was done on the basis of the gamma radiation of ^{117}Sn, measured with a Ge(Li) detector and a 512-channel pulse-height analyser. A similar unexplained anomaly (but not preventing substoichiometric isolation after activation) was observed by the same author in the determination of trace amounts of Sb in germanium, with the same extractant.

3.5.5.2 Substoichiometry in analysis for radionuclides in a mixture

The procedure differs from the preceding case only insignificantly: no special activation step is involved, and the origin of the mixture is irrelevant; no standard is involved; the separation yield is known from the amount of substoichiometric reagent used and consumed.

Two examples dealing with the analysis of fission products illustrate the procedure. Wertenbach and von Beckmann claim that their procedure (Table 3.15) achieved appreciable simplification of the analysis and improved reproducibility ($\pm$ 1–2.5 %). They recommend always checking that the amount of precipitant is really substoichiometric [143].

The other example concerns the determination of the radioactivity of a mixture of ^{95}Zr and ^{97}Zr in a solution of fission products [144]. The separation is done by solvent extraction with cupferron. The extract is measured with a Ge(Li) detector and a 4000-channel gamma analyser. The gamma spectrum allows differentiation between the peaks of ^{95}Zr (724 keV and 756 keV) and ^{97}Zr (744 keV). The precision is $\pm$ 5 % relative. If the substoichiometric isolation is omitted the spectrum of the mixture of fission products is much more complicated and can be deciphered only by using a computer. Consequently, the advantage of the substoichiometric isolation is the financial economy (no computer calculations and less sophisticated gamma-spectrum equipment). The good agreement of the results obtained with substoichiometric isolation and with a conventional procedure is illustrated in Table 3.16. Similar substoichiometric methods have been developed for ^{99}Mo, ^{89}Sr and ^{140}Ba.

Table 3.15 — Final* substoichiometric determination of activities of fission products in irradiated fuel elements, based on precipitation [143].

Element determined	Amount of isotopic carrier added (mmole)	Precipitant, accurate amount	Conditions (at boiling point)
Sr	0.25	5 ml of 0.02M oxalic acid	In ammoniacal solution, 10 ml
Zr	0.25	5 ml of 0.04M BaCl$_2$	In 0.6M HF
Mo	0.4	5 ml of 0.03M Pb(CH$_3$COO)$_2$	With ammonium acetate buffer
Te	0.3	5 ml of 0.08M TiCl$_3$	Reduction by Ti(III) In 3M HCl
I	0.3	5 ml of 0.01M PdCl$_2$	In 0.5M HCl
Cs	0.5	10 ml of 0.0146M sodium tetraphenylborate	In acetic acid solution
Ba	0.25	5 ml of 0.02M H$_2$SO$_4$	In dilute HCl solution
Ce	0.25	5 ml of 0.025M oxalic acid	In 0.02M HNO$_3$

* Purification by repeated precipitation or extraction steps after addition of the carriers and prior to the final substoichiometric isolation.

Table 3.16 — Number of fission events (n) in fissile material, calculated from ^{97}Zr radioactivity. (Reproduced from [144] by permission of the copyright holders, Akadémiai Kiadó, Budapest.)

Method	$n \times 10^{-11}$			
	Sample 1	Sample 2	Sample 3	Sample 4
I	2.77	17.5	35.3	73.3
II	2.86	16.3	35.2	71.7

I — from gamma spectrum of fission products without chemical separation; computer evaluation.
II — from simplified gamma spectrum after substoichiometric separation.

3.5.5.3 Determination of carrier concentration in radioactive nuclide preparations

This procedure is probably used very frequently, since for a typical substoichiometric IDA the concentration of the radioactive tracer must naturally be known exactly. The authors of most papers do not mention how this value was determined. In our opinion most of them used reverse substoichiometric IDA for this purpose, based on the same procedure as that of the determination itself.

In Yuzawa and Suzuki's paper [109] on substoichiometric IDA of calcium in biological materials the use of reverse substoichiometric IDA for the purpose is explicitly stated. As an alternative the titration of Ca with $0.01M$ EDTA was used.

Determination of the amount of carrier involves the following steps. The radioactive sample solution containing $2W_x$ mg of carrier (unknown) is divided into two equal portions. To the second portion an exactly known amount (W) of non-radioactive sample substance is added and an ordinary substoichiometric separation is applied to both portions (radioactivities A_1 and A_2). Then

$$W_x = W \bigg/ \left(\frac{A_1}{A_2} - 1\right) \qquad\qquad (3.14)$$

3.5.6 Other types of substoichiometric separation

In this section we will deal with replacement (displacement) substoichiometry, which is suitable for substances forming relatively weak complexes with the substoichiometric reagent, redox substoichiometry, in which oxidants or reductants are used in substoichiometric amounts, and with separations based on precipitation, adsorption and electrolysis. Substoichiometric concentration is also mentioned.

3.5.6.1 Displacement substoichiometry

Displacement substoichiometry [145] is a variant which is not characterized by use of a certain type of substoichiometric reagent (theoretically any type of reagent can be used) but is based on partial decomposition of a complex by a competitive reaction with an auxiliary reagent which forms a more stable complex and is used in substoichiometric amount [146, 147].

Let us use the example of solvent extraction. The metal to be determined is extracted into the organic phase with excess of reagent, unbound reagent is removed from the organic phase, and a substoichiometric amount of another metal (giving a higher extraction constant with the reagent) is added in a new aqueous phase which is equilibrated with the extract. The equilibrated aqueous phase afterwards contains the substoichiometrically isolated portion of the metal to be determined.

The inherent selectivity of the procedure is illustrated in Fig. 3.11. The position of the metals on the vertical line reflects the logarithmic value of the corresponding extraction constant.

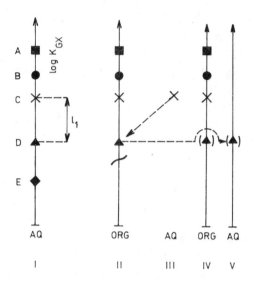

Fig. 3.11 — Schematic representation of displacement substoichiometric determination. The positions of the elements A–E on the line correspond to increasing values of the extraction constants. For other details see the text.

In stage I we have 5 metals A, B ... E in aqueous solution. Stage II represents the extract (after extraction with excess of reagent). Element E is not extracted, owing to its low extraction constant. Stage III is the aqueous phase with a substoichiometric amount of element C. Stage IV is the organic phase after the corresponding equilibration. The elements A and B cannot be displaced, so their concentration in the extract is the same as in II. A portion of D equivalent to the amount of C added is transferred to the aqueous phase V, which is measured for radioactivity. The distance l_1 should theoretically correspond to a ratio of at least 200.

Prediction of the possibility of replacing one element by another in a complex, from the extraction (stability) constants, seems all that is needed, but it has to be experimentally tested (because of possible kinetic effects, formation of mixed complexes, etc.).

Generally the prediction will work, and as a rule the elements with higher two-phase stability constants displace those with lower values. It should be borne in mind, however, that such series of constants are only of real use for predictions under certain conditions: the position of the elements in the series may change when the pH, the organic solvent or the composition of the aqueous phase is altered.

In displacement substoichiometry with diethyldithiocarbamates [145] the displacement reaction is usually rather rapid if the difference in the extraction constants of the two metals is large, but slow for metals close to each other in the series, and to ensure complete displacement a longer shaking time is then required. This is characteristic of the pairs Zn–Fe, Fe–Cd and As–Fe. The displacement also occurs faster in carbon tetrachloride than in chloroform.

There may be other reasons besides improved selectivity for using displacement substoichiometry, for example, the complex anion TaF_6^- is extracted with tetraphenylarsonium chloride into chloroform [68] only with a high excess of reagent, indicating that the association constant of the compound formed is low. Therefore replacement substoichiometry was successfully tried, based on the reaction

$$\overline{(C_6H_5)_4AsTaF_6} + AuCl_4^- \;\rightleftharpoons\; \overline{(C_6H_5)_4AsAuCl_4} + TaF_6^-$$

where the bars indicate the organic phase. Here the substoichiometric reagent is non-radioactive $AuCl_4^-$ ($10^{-4}M$) and the tantalum is labelled with ^{182}Ta. The reaction is considered suitable for the substoichiometric separation of tantalum in activation analysis.

Let us give some examples of complexes and displacing elements suitable for displacement substoichiometry. With dithizone, suitable pairs of metals are In–Hg, Tl–Hg, Sn–Hg (the first metal is the one determined); with diethyldithiocarbamate Hg–Cu, Pb–Cu, As–Cu, Cd–Zn, Cu–Zn, Ag–Zn, Hg–Zn; with tetraphenylarsonium Ta–Au, Ga–Fe, Au–Ga [145].

An example of a sensitive determination of Tl(I) by displacement substoichiometry from its dithizonate by Hg(II) will be given in some detail to illustrate the procedure, as follows.

(1) Isotope dilution with carrier-free ^{202}Tl.
(2) Complete extraction of Tl(I) with dithizone into chloroform from an EDTA, CN^-, tartaric acid, hydroxylamine medium at pH > 10.5 (concentration level $10^{-7}M$ dithizone).
(3) Washing the organic phase several times with ammonia solution at the appropriate pH to remove the dithizonates of co-extracted elements and free dithizone; this procedure leads to 1–2 % loss of Tl.
(4) Shaking the organic phase with substoichiometric amounts of Hg^{2+} (pH 11.5, tartaric acid).
(5) Measuring the activity of ^{202}Tl in the aqueous phase.

The reproducibility of the determination is ∼ 1 % and lead does not interfere even if present in 80-fold ratio to Tl. In this way Tl has been determined in minerals, sediments, ion exchangers and hydrothermal water in concentrations down to 0.01 µg/ml.

The procedure is very rapid because the isotopic exchange Tl(I)/Tl*(I) is obtained very quickly [98]. It has likewise been shown theoretically that displacement substoichiometry in ion-association systems can been used if the ratio of the stability constants of the halide complexes of the displacing and displaced metal is greater than 100.

Displacement substoichiometry is relatively frequently used in activation analysis and was originally designed for this type of analysis.

3.5.6.2 Redox substoichiometry

In redox substoichiometry, introduced by Polak in 1971 [148, 149] (see also [86, 150]), the basis is partial oxidation or reduction of the element to be determined, followed by the transfer of either the reacted or the unreacted fraction into a second phase. In this respect redox substoichiometry is similar to that using complexones. The inherent selectivity of this type of procedure is based on the fact that many metals exhibit only one oxidation state and therefore do not interfere. Oxidation reactions are much more frequently used than reductions.

One of the elements that can be determined by redox substoichiometry is arsenic [110]. Concentrations of potassium bromate or ceric sulphate even as low as $5 \times 10^{-4}N$ are completely consumed in the oxidation of As(III) in $5N$ H_2SO_4. Unoxidized arsenic is extracted with thionalide. It is noteworthy that small amounts of osmium tetroxide or potassium bromide accelerate the oxidation of As(III) to As(V) so a reaction time of 10 min is sufficient. In this way 9.7 ± 0.3 ppm As was found in 'Orchard leaves' reference material by using neutron activation, the National Bureau of Standards certified value being 10 ± 1 ppm.

Potassium permanganate has been used as a substoichiometric oxidant in concentrations as low as $5 \,\mu M$ (brown test-tubes recommended!) for the determination of antimony [125] (see Table 3.17). The oxidation of Sb(III) in ∼ $0.4M$ HCl takes 1 hr. Tervalent antimony can be separated by extraction with excess of N-benzoyl-N-phenylhydroxylamine. Potassium dichromate can also be used as the substoichiometric oxidant.

An example of the use of substoichiometric amounts of a reductant is the reduction of Tl(III) with thiourea (2–25 µg of Tl can be determined with a precision of 4 %) [151].

Another reductant used is iron(II). Chromate is partially reduced to

Cr(III) which is then adsorbed on TiO_2. Hydrazine has been proposed for the determination of Tl(III). It is interesting to note that the presence of free halogens and halides strongly reduces the rate of reduction in this system, so these compounds must be absent [126].

Table 3.17 — Comparison of the results obtained by redox substoichiometry in IDA and other methods (determination of Sb in metallic Zn with $KMnO_4$). (Reproduced from [125] by permission of the copyright holders, Akadémiai Kiadó, Budapest.)

Amount of Sb found (ppm)	Method
19.2 ± 1.3	Neutron activation
19.5 ± 1.1	Colorimetric analysis
19.8 ± 0.1	Substoichiometry (A)
19.7 ± 0.8	Substoichiometry (B)

(A) — oxidation in $1.2M$ HCl + $1.8N$ H_2SO_4 medium.
(B) — oxidation in $0.4M$ HCl medium.

3.5.6.3 Other substoichiometric separations

Precipitation is rarely used for substoichiometric IDA because the separation of small amounts of a solid phase from a solution is difficult. One way round this problem is formation of the insoluble compound as a colloid and its separation from the solution by dialysis or electro-dialysis with semipermeable membranes. This method was proposed [119] for the determination of iodide with ^{131}I in the form of AgI. The sensitivity found was $0.2\,\mu g$ which is only slightly inferior to the theoretically calculated value.

The theory of substoichiometric precipitation has been developed rather thoroughly but very often discrepancies between theoretical expectations and experiments are found.

The lower pH threshold for the substoichiometric precipitation of fluoride with lanthanum is ascribed to kinetic effects [118].

Precipitation substoichiometric procedures are sometimes used in activation analysis or determination of radioactivities of individual radionuclides in mixtures (see p. 101). The advantage is that the co-precipitation of foreign non-radioactive substances can be tolerated unless it consumes significant amounts of the precipitant.

For the determination of smaller amounts in IDA the separation of the precipitate can also be facilitated by forming it on paper or using the 'ring-oven' technique.

Adsorption is also rarely used for substoichiometric determination, mainly because of the low reproducibility and the inhomogeneity of the active sites of the sorbent with respect to sorption affinity. Therefore the amount adsorbed depends markedly on the concentration of the substance to be adsorbed, and full saturation of the surface is difficult to achieve. Attempts have been made to determine fluoride by adsorption on glass (10^{-10} g of F^- in 1 mg of tooth enamel) [152].

SnO_2 has been proposed for isolating equal amounts of F^- and SiF_6^{2-}. It is possible to separate ~ 9 mmole of fluoride and ~ 1.5 mmole of fluorosilicate. The reaction ratio between F^- and $SnO_2 . xH_2O$ was estimated to be $1 : 5.9$, that for fluorosilicate $1 : 56$. The method was applied to the activation analysis of silicon crystals for oxygen, by using a packed column containing 2 g of $SnO_2 . xH_2O$ [87].

Ion exchange on a resin loaded with a suitable functional group could be a promising substoichiometric technique [127]. A chelating reagent is adsorbed on an ion-exchange material, which is then stirred with the test and standard solutions for 2 hr, and the radioactivity of the resin phase is finally measured. This type of procedure has been tested for the determination of $\sim 100 \, \mu g$ of Zn with 100 mg of Diaion SA 100 anion exchanger containing $\sim 1 \, \mu$mole of 8-quinolinol-5-sulphonic acid. Interfering metals can be removed beforehand by dithizone extraction.

Electrolytic separation requires the parallel connection of two electrolytic cells, one containing the standard and the other the isotopically diluted sample. This procedure has been applied with mediocre success for the determination of silver and copper [12]. The irreproducible adsorption of radionuclides on the electrodes lowers the sensitivity of the method. Attempts to determine traces of copper in highly pure zinc were made by using spontaneous deposition of Cu from aqueous solutions on a cathodically polarized platinum surface.

The high selectivity of substoichiometric separations can be used to advantage without isolation of equal amounts. In this case the substoichiometric separation (sometimes repeated) is a preliminary step in which the element to be determined is enriched with regard to accompanying and interfering elements [153–156]. If the element is radioactively labelled prior to this type of substoichiometric enrichment (or concentration) procedure, the yield of each substoichiometric separation can be followed. Table 3.18 gives a hypothetical example of such a procedure, with three substoichiometric extraction steps. The procedure was successfully tested by Shamaev [155] on mixtures of Cs, Rb, K and Na by solvent extraction of the tetraphenylborates into nitrobenzene. In a typical procedure the initial and final amounts (mg) were as follows: Na initial 9140, final 0; K 2184, 0.025; Rb 135.3, 0.12; Cs 1.53, 0.041.

Table 3.18 — Hypothetical substoichiometric concentration procedure. Two-phase reactions: $A + L \rightleftharpoons AL(K_A)$; $B + L \rightleftharpoons BL(K_B)$; $Me + L \rightleftharpoons MeL(K_{Me})$; $K_{Me} = 10$; $K_B = 50\,K_A$. (After [155].)

	A	B	Me	% Me in mixture
1 Originally present (μmole)	1000	100	1	0.091
2 Isolated after applying 50 μmole of L	33	16	0.67	1.34
3 Isolated after applying 5 μmole of L after step 2	0.37	4.16	0.46	9.2
Final yield	0.037 %	4.16 %	46 %	

3.5.7 Sensitivity of substoichiometric IDA

3.5.7.1 Factors affecting sensitivity

Starý [91] has evaluated the sensitivity (absolute detectable amount) of good substoichiometric IDA procedures as 10^{-9}–10^{-8} g, corresponding to the use of $10^{-7}M$ concentration of substoichiometric reagent. In some cases this sensitivity exceeds [49] that of activation analysis. An example is the determination of palladium [69]. The smallest amount of Pd determinable by activation analysis is 10 ng, compared with 3 ng by the substoichiometric method. Moreover, the latter method is expected to be applicable to any type of matrix, whereas self-shielding effects may cause low results in activation analysis.

The sensitivity of substoichiometric IDA is limited usually by one of the following factors,
— the specific activity of the diluting radionuclide,
— the chemical stability of the reagent or the reaction product (e.g. chemical instability of 1-phenyl-3-methyl-4-benzoylpyrazol-5-one becomes a limiting factor at a concentration of $\sim 10^{-4}M$!),
— the stability constant of the product of the substoichiometric reaction, which determines the minimum concentration of the test substance able to consume the reagent completely,
— kinetic factors: very low concentrations of metals and some reagents can exhibit a prohibitively low reaction rate [e.g. the extraction of Au(III) with Cu diethyldithiocarbamate into $CHCl_3$].
— the blank value.

A low stability constant is the most important factor for elements or compounds which form weak complexes (alkali and alkaline-earth metals, anions, organic substances). Below a certain concentration level the reagent added is, to an increasing degree, unconsumed; the amount of substance isolated is then the larger the greater is its total concentration in the system used for isolation. This is the point where typical IDA ends but radio-reagent methods (radiometric methods) can still be employed. Theoretical calculations show that if ideal conditions could be attained (exactly constant pH, complete equilibration, etc.) concentrations lower by some two or three orders of magnitude could be determined than those required for isolating exactly equal amounts.

If kinetics is the limiting factor, the addition of a suitable catalyst can prove helpful.

The dependence of sensitivity on the concentration ratio of reagent to element was investigated by Zimakov and Vinokurova [157], who came to the conclusion that when the substoichiometric reagent concentration is increased,

— in solvent extraction and in precipitation the threshold pH and the sensitivity are both increased,
— in ion-exchange separation of ions not bound in a complex the threshold pH and the sensitivity both decrease,
— the interference of other elements increases in all cases investigated.

Attempts have been undertaken to increase the sensitivity of substoichiometric IDA by reducing the volume of the two phases in the isolation step. Ag can be determined by solvent extraction with $50\,\mu l$ of $5 \times 10^{-10} M$ Zn-dibenzyldithiocarbamate in CCl_4 from $100\,\mu l$ of aqueous phase. It was found that the error was between 0.4 and 8 % and the detection limit near $0.1\,\mu g/100\,\mu l$ [112].

The blank value (rather than instrumental factors) controls, in many instances, the sensitivity limit in substoichiometric IDA and can be lowered by previous careful purification of the reagents. Many practical instructions for lowering the blank are given in the book by Růžička and Starý [25].

Sorption on the electrodes, competing with electrolysis, limits the sensitivity of the electrolytic substoichiometric determination of silver [12]. High sorption of traces of metals on glassware at the high pH required by substoichiometry theory might sometimes occur.

3.5.7.2 Sensitivity of cation determination by extraction with hydrophobic anions

This will be examined in some detail since it has not been treated in the literature up till now, although similar systems have been proposed for substoichiometric determinations. The following treatment clearly illustrates some pitfalls of the calculations and show that it is dangerous to apply formulae found in the literature, without thorough analysis of their underlying assumptions.

Let as calculate for what caesium concentrations and pH values the substoichiometric isolation of caesium is feasible, by extraction with a nitrobenzene solution of dicarbollide acid $[H(C_2B_9H_5Cl_6)_2Co]$.

The dicarbollide acid is fully ionized in both nitrobenzene and the aqueous phase. The anion is so hydrophobic that its concentration in the aqueous phase can be neglected. Consequently, the solvent extraction of caesium obeys the equation:

$$Cs^+_{aq} + H^+_{org} \rightleftharpoons Cs^+_{org} + H^+_{aq}$$

It is known from the literature that the equilibrium constant $K_{ex} = [Cs^+]_{org}[H^+]_{aq}/[Cs^+]_{aq}[H^+]_{org}$ is 10^3 for nitrobenzene and water [158]. If B is the total concentration of the reagent in the organic phase, x the value of $[H^+]_{org}$, h the concentration of mineral acid in the aqueous phase, used to adjust the pH value, and C the total concentration of caesium, the expression for K_{ex} assumes the form:

$$K_{ex} = \frac{(B-x)(h+B-x)}{(C-B+x)x}$$

This involves the assumption that the formation of protons due to the dissociation of water molecules is negligible. Let us introduce two simplifications:

$C = 2B$, the usual condition of substoichiometry;

$x/B = z$, the fraction of unreacted extractant.

Rearranging the expression for K_{ex}, we obtain:

$$z^2(K_{ex} - 1) + z\left(K_{ex} + 2 + \frac{h}{B}\right) - 1 - \frac{h}{B} = 0$$

Remembering that $K_{ex} = 10^3$ and z should be between 10^{-2} and 10^{-3}, we can certainly neglect the first term in z^2. Hence $B = (1-z)h/[z(K+2)-1]$ which is approximately equal to

$$B = \frac{h}{10^3 z}$$

The result is as expected. The reagent concentration can be lower if
— the corresponding equilibrium constant is higher,
— the acidity of the aqueous phase is lower,
— the condition of substoichiometry is fulfilled less rigidly.

The somewhat surprising fact is the simplicity of the final expression. For our case, if $h = 10^{-4}M$ (then $[OH^-] = 10^{-10}M$) and $z = 10^{-3}$, the acceptable concentration B corresponds to $10^{-4}M$ so $2 \times 10^{-4}M$ Cs^+ could be isolated. This has been verified in practice. It would, of course, be safer to use $h = 10^{-3}M$ so that an accidental decrease in acidity due to presence of impurities does not cause erroneous results. This would cause a corresponding increase in the concentration of caesium that could be isolated.

3.5.8 Precision and accuracy of substoichiometry

The possible sources of systematic errors in substoichiometric IDA are:
— insufficient purity of reagents or cleanliness of apparatus,
— contamination of the sample by impure laboratory air,
— incomplete isotope exchange,
— sorption on reaction vessel walls or pipettes (e.g. Ag in acid solutions),
— inadvertent addition of superstoichiometric amounts instead of substoichiometric,
— non-negligible difference between amounts isolated from solutions of unequal concentrations,
— evaporation of organic solvent during the separation procedure.
  The precision of the result can be influenced by:
— insufficient pipetting precision,
— non-reproducible decomposition of the reagent at low concentration,
— non-reproducible heating of the solution by intensive mixing (provide for constant temperature by adequate cooling),
— the statistical character of radioactive decay.
  The methods for minimizing such error are obvious from what has already been said on this type of analysis. Adsorption (e.g. of $\sim 10^{-5}M$ Th) on glassware can be prevented by silylating with trimethylchlorosilane.

  Most of the factors are naturally valid not only for substoichiometric IDA but also for other trace analysis methods.

  The accuracy can be affected by the choice of several arbitrary values inherent in the procedure, such as
— the amount of radioactively labelled substance added,
— the substoichiometric ratio, i.e. the ratio between the concentration of reagent actually added and that required theoretically by the stoichiometry of the reaction,
— the division of measurement time between that for the activity of the isolated portion of standard, the isotopically diluted sample, and the background.
  An exact mathematical solution of the combination of these three aspects has not yet been given in general form, but separate optimiza-

tions of the three factors have been made, with various assumptions, and can be found in the literature [159].

The problem of choice of amount of radioactively labelled substance is not specific to substoichiometric IDA and has been dealt with in the book on IDA by Tölgyessy *et al.* [18]. If the amount of labelled substance added is equal to that of the substance to be determined, no severe effect on the precision should arise.

An illustration of the influence of the substoichiometric ratio on the error of the determination under certain conditions is given in Fig. 3.12. Although the validity of the curves given is certainly not universal, it can be assumed that use of $h = 0.5$ will not be too unfavourable [160].

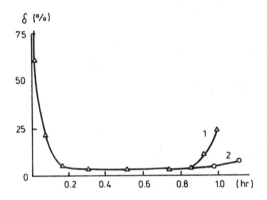

Fig. 3.12 — The dependence of the error in the determination of Sb(III) on the substoichiometric ratio h. Solvent extraction with monolaurylphosphoric acid in hexane; h = concentration of reagent used/equivalent concentration of reagent. Curve 1 — result calculated assuming full consumption of reagent; curve 2 — result calculated with respect to violation of the conditions of substoichiometric isolation, from known values of reagent concentration and extraction constant. Sensitivity: 2.5 µg Sb/ml. (Reproduced from [160] by permission of the copyright holders, Izdat. Nauka, Moscow.)

The division of time between various measurements might become important when working with very low count-rates; otherwise it is sufficient, as a rule, to determine the required count-rates with the same error, which should generally be in the range 0.1–1 % (see also p. 51).

In work with segmented flow analysers, attention should be paid to errors due to 'memory' effects (cross-contamination of samples caused by adsorption/desorption of analyte during processing).

A survey of the literature shows that the errors given by the authors of various substoichiometric procedures are rarely larger than 10 % or lower than 1 %. Some typical values can be found in the tables on pages 82—85, 100 and 123.

Most authors claim that substoichiometric IDA is precise and accurate. This should be taken 'cum grano salis', being valid only when a suitable reagent and proper conditions have been chosen, and skilled operators employed. Even in such a typical and well developed procedure as the determination of mercury with Zn-dithizonate, the average of six single determinations was $1.07 \pm 0.03 \, \mu g$ when $1.007 \, \mu g$ of Hg was taken, i.e. the precision was $\sim 3 \%$ and the error not better than $+ 6 \%$.

An example of a more accurate substoichiometric determination is analysis for iron by solvent extraction with substoichiometric amounts of oxine, with which Suzuki *et al.* [72] found $548 \pm 9 \, ppm$ of Fe in NBS spinach, the certified value being $550 \pm 20 \, ppm$.

3.5.9 Other advantages and drawbacks, problems of classification, and trends

One of the less known advantages of substoichiometric determination is the possibility of repeating the isolation (e.g. solvent extraction) from the same aqueous solution. It is evident that the number of possible repetitions increases as the amount of reagent relative to that of substance isolated decreases.

The objectives of such repetition are
— to satisfy the operator that the first isolation was really substoichiometric, i.e. that some of the substance was left in the solution,
— to check the reproducibility of the count-rates obtained in repeated isolation and use the average value in calculating the result [109] (see Table 3.19),
— to increase the selectivity of the determination in certain cases: if another element (present in minor concentration) accompanies the substance to be determined into the phase of the substoichiometric reagent, the first extract may contain less analyte than required, because of the competitive extraction, but the second or further extracts can be sufficiently pure.

The advantage of tolerance for incomplete separation in substoichiometric analysis naturally concerns only the part of the analytical procedure starting from the isotope dilution step. In a typical example, mercury is determined in biological material (bovine liver), after combustion of the sample, by solvent extraction with substoichiometric amounts of zinc dithizonate. ^{203}Hg is added to the combustion product, so any loss of mercury in the combustion process will affect the result.

Another less known advantage of substoichiometric IDA in some cases (over activation analysis) is the possibility of using it without preconcentration procedures [78].

Table 3.19 — Reproducibility of successive extractions in substoichiometric reverse isotope dilution analysis.* (Reproduced from [109] by permission of the copyright holders, Akadémiai Kiadó, Budapest.)

No. of extraction	Activity of extract (cpm)	Ca determined (mg/ml)
1st	24911	1.35
2nd	24650	1.32
3rd	24992	1.36
4th	24621	1.32
5th	24854	1.35
6th	24660	1.32
Average	24782 ± 158	1.34 ± 0.02
Rel. std. dev.	0.6 %	1.3 %

* The determination of Ca carrier in ^{47}Ca, solvent extraction with excess of TBP and $3 \times 10^{-3}M$ TTA in CCl_4, pH 8.6.

A typical example of a serious problem is the storage, sample preparation and concentration in determination of traces of mercury, because of its volatility and that of its compounds. Concentration of aqueous solutions of mercury by either hot or cold evaporation is not suitable, because of losses. Owing to the high stability constants of the chelates of mercury with thio-oxine and its methyl derivatives, however, it is possible to determine Hg in trace amounts by applying IDA directly to aqueous samples without any previous treatment (except filtration and acidification to prevent adsorption of Hg on the walls of the containers during storage of the samples) [78].

Although substoichiometric IDA is considered most useful at trace element levels it can also be used for the determination of a major component (such as the total amount of iron in silicate rocks, amounting to 3–13 % Fe_2O_3). It may not be suitable for routine analyses in this case but could be a useful supplement when the results obtained by other methods are to be checked. For example, with EDTA as the substoichiometric reagent, and uncomplexed iron (^{59}Fe) retained on a cation-exchange column, the precision and error can be 1.5 % relative or better [90].

One of the drawbacks arises from self-irradiation effects. In some cases the original specific activity of the labelled element is too high for use without dilution with inactive element, because of radiation effects on the tracer substance itself. For example, if $1\,\mu Ci$ of $^{125}Sb(III)$ is present in 10 ml of concentrated hydrochloric acid, self-oxidation occurs owing to its own radiation. (It can be eliminated by adding $30\,\mu g$ of inactive antimony(III) [125].)

The pronounced advantages of substoichiometric analysis have led some authors to new ideas about the place of these procedures among other radioanalytical methods from the point of view of classification [161]. An interesting new approach to the classification of a certain area of nuclear analytical methods was proposed by Perezhogin [162]. In his opinion the radiochemical methods featuring substoichiometric isolation and radiometric determination of the substoichiometrically isolated portion, should not be considered as variants of IDA or radio-reagent methods, etc., but as an autonomous group of 'radiosubstoichiometric analysis'.

This proposal has the advantage of removing some difficulties arising from the need to decide whether a determination using the substoichiometric principle is a radio-reagent method, IDA or perhaps activation analysis. Another advantage is that the theory of equilibria of two-phase reactions and of some one-phase reactions would form a common basis for this category of radioanalytical methods.

Unfortunately, Perezhogin's proposal has serious drawbacks.

The links connecting typical substoichiometric analytical procedures with the realms of IDA and neutron activation, both in the minds of most analysts working in the field and in their underlying principles, are strong indeed and soundly founded, and the new proposal tends to abolish them. Furthermore, a classification is considered satisfactory if the division into individual categories occurs according to only one criterion or at most to a set of related and fairly equivalent criteria. The criterion of whether the amount of the reagent is less or more than equivalence requires is felt to lie on a somewhat different level from that applying when the analysis is based, for example, on the change of specific activity due to isotope dilution. The question arises as to how the new method of sub- and super-equivalence should be classified. One part of its calibration curve may be substoichiometric, the rest 'super-stoichiometric'.

Therefore, in our opinion, substoichiometry should not be considered as a 'category' but, in accord with the opinion of Růžička and Starý, as a 'principle' penetrating into different fields of radioanalytical chemistry, especially into IDA, activation analysis and the radio-reagent method.

The diversity of opinion on what is IDA and what are radio-reagent methods can be illustrated by the very title of an interesting paper by Ooms *et al.* [163]: "Multielement isotope dilution analysis by means of radiometric titration." At the same time, this paper testifies to the close relationship between substoichiometric IDA and radiometric titration.

Let us conclude this section with a reminder that the substoichiometric principle has been used for the determination of more than 50 trace elements in pure substances, geochemical, biochemical and other materials, as well as in the analysis of radioactive substances, such as fission products, with a sensitivity of 10^{-4}–10^{-6} %, and in special cases even down to 10^{-8} %.

The trends expected in the future development of the method probably include
— efforts to find new highly sensitive, chemically stable and selective reagents,
— development of new effective separation procedures (solvent extraction, chromatography, etc.),
— finding possibilities for simultaneous determinations of several elements,
— automation of the process for routine analysis.

3.6 THE SUB- AND SUPER-EQUIVALENCE METHODS

3.6.1 Principle
This type of IDA (SSE) was proposed by Klas *et al.* in 1974 [164]. The principle consists in the preparation of two series of aliquots of the labelled sample; one series is isotopically diluted, the other is not. A constant amount of a suitable reagent is added to each aliquot and the product is separated and counted. The relative radioactivities of the isolated fractions are plotted versus the extent of isotope dilution and the intersection of this curve with a line corresponding to a special condition is used to calculate the result.

The method is similar to substoichiometric IDA in that only the activities of the portions isolated are measured but differs from it in that the condition of isolating equal amounts need not be fulfilled. In this respect the SSE procedures resemble concentration-dependent distribution, but unlike the latter, do not require the determination of distribution ratios.

Let us now decribe the basic type of SSE. In the following text the amounts (concentrations) y refer to a radioactive substance, those denoted by x to a non-radioactive isotopic substance.

From the solution of the radioactive substance A of unknown concentration, aliquots each containing y mole are transferred into a series of test-tubes. Increasing amounts of non-radioactive isotopic substance A are added to the series of test-tubes so that the first tube holds $y + x$ mole, the second $y + 2x$, and so on up to $y + px$ mole in the pth tube. The contents of these test-tubes (series No. 1) are adjusted to the same volume, pH, etc. The value of x is known (and arbitrarily chosen) and constant within the series.

Into a second set of test-tubes (series No. 2) aliquots containing ξy of the radioactive substance are pipetted, without addition of non-radioactive substance; ξ is a known factor, arbitrarily chosen, and greater than 1 (in several papers $\xi = 2$ was used). The volumes are all adjusted to the same value. Replicates are used in order to achieve a reliable value of the radioactivity distribution, to which the values of the first series will be related.

To both series the same amount of extractant B in the same volume of organic solvent is added, equilibrium is attained, the phases are separated and aliquots of the organic phases are mounted for radio-activity measurement.

The radioactivities of all samples are measured, those from the first series being denoted by $A_1, A_2, A_3 \ldots A_p$, and the average value of the second series by A_f.

The ratios $A_f/A_1, A_f/A_2, A_f/A_p$, denoted by i, are calculated and a plot of i vs. p is prepared. By interpolation on the curve, a hypothetical tube number p^* is found, for which the corresponding i value is equal to ξ.

The value of y is then calculated from the relationship

$$y = p^*x/(\xi - 1)$$

The correctness of the equation is proved as follows.

On the plot of $i = f(p)$ there must be a point (p^*, not generally an integer) which corresponds to the same activity as that for the second series, so the amounts isolated in the two cases are also equal, and

$$(\xi y)_{ser.2} = (y + p^*x)_{ser.1} \quad \text{and} \quad A_f : A_{p^*} = \xi y : y$$

The second relationship expresses the proportionality of the radioac-tivities of equal amounts to the initial radioactivities introduced. It follows that $A_f/A_{p^*} = \xi$, but A_f/A_p is equal to i by definition, so from the first relationship $y = p^*x(\xi - 1)$ is obtained.

If $\xi = 2$ then $y = px = z$ (where z denotes the known amount of inactive substance added in the first series). If i is plotted against z then the value of z corresponding to $i = 2$ is equal to the unknown concentration y.

Let us illustrate the procedure by the determination of selenium with ^{75}Se [165].

Seven aliquots (1.0 ml) of the sample to be analysed [aqueous solution of Se(IV) labelled with ^{75}Se, containing the unknown concentration y] were placed in 20-ml stoppered tubes and then 0.10 ... 60 μl of an inactive Se solution (Se 0.5 μg/10 μl) were added (series 1). Exactly 2 ml of the same labelled Se solution was added to each of 3 additional 20-ml tubes ($\xi = 2$) (series 2). One ml of acetate buffer (pH 3.6) was added to each tube of both series and all were brought to the same volume (3.9 ml) with demineralized water. Next 100 μl of the $10^{-3}M$ extractant solution (diethyldithiocarbamate) and 5.0 ml of CCl_4 were added to each tube and all tubes were vigorously shaken. The relative activity of 2.5 ml of each organic phase was then determined with an NaI-well scintillation counter (activities A from the first series, A_f the average activity from the second). The data were plotted as $i = A_f/A$ vs. z — the known amount of Se(IV) added to the first series. The plot is given in Fig. 3.13. The amount found was $y = 1.97$ μg of Se. When this procedure was repeated 6 times, the amounts found yielded an average of 1.92 ± 0.06 μg. The relative standard deviation was $\sim 3\%$.

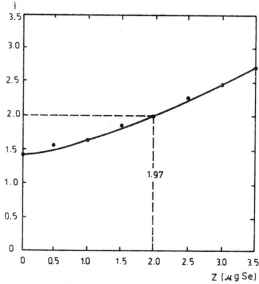

Fig. 3.13 — Determination of Se by the SSE method of IDA. (Reproduced from [165] by permission of the copyright holders, Akadémia Kiadó, Budapest.)

3.6.2 The four basic variants

It can easily be shown that instead of increasing the amounts of the inactive substance in the first series the amount of the radioactive species could be increased. On the other hand, this type of determination is also possible for non-radioactive substances by dilution with a radioactive substance (analogously to reverse and direct IDA).

Altogether we obtain four variants, which can be schematically characterized as in Table 3.20.

Table 3.20 — Variants of the SSE method: y denotes the radioactive substance, x the non-radioactive; $p = 1, 2, 3...$; the bars indicate the unknown quantities.

Variant	Substance to be determined	Diluent	Series 1	Series 2
I	Radioactive (y)	Inactive (x)	$\bar{y} + px$	ξy
II	Radioactive (y)	Radioactive (y)	$x + p\bar{y}$	ξy
III	Inactive (x)	Inactive (x)	$y + p\bar{x}$	ξy
IV	Inactive (x)	Radioactive (y)	$\bar{x} + py$	ξy

In variant II the value of p^* is found from the intersection of the curve $\xi/i = f(p)$ and that for $\xi/i = p$ (a straight line with slope of 1). The desired value is $y = x/(\xi - p^*)$.

For variant III (which is mathematically very similar to I) the unknown concentration is calculated from $x = y/p^* (\xi - 1)$. Variant IV leads to the equation $x = y(\xi - p^*)$ because of the similarity to variant II. The theory of propagation of errors allows us to find conditions that keep the errors of the determination at a reasonable level. Any of the four variants can be the choice under appropriate conditions. Practically, the choice lies between variants I and II, or between III and IV. Variants II and III may be desirable if the amounts of the substance to be determined are very high, as no limit exists in this respect. On the other hand variant IV may seem preferable if the supply of the radioactive auxiliary substance is ample. Variant I requires a sufficient supply of the non-radioactive auxiliary substance, which usually causes no difficulty. For all variants the value of ξ recommended is in the range 2–3 [166].

3.6.3 Comparison with concentration-dependent distribution

Practically the same experiments (extractions) as in the SSE method can be performed by concentration-dependent distribution (CDD) analysis. The only difference is that the radioactivity of both phases must be measured in CDD analysis for each distribution and the distribution ratio plotted *vs.* the known concentration of the diluent.

If we take the analogy of CDD with variant I of SSE, the plot would be D (distribution ratio in series 1) *vs. px*. The distribution ratio of the second series would be denoted by D_f.

The situation is illustrated in Fig. 3.14, showing the calibration curve of D *vs.* $z\,(-px)$. The distribution ratio D that corresponds to the value for the second series (ξy) is denoted by D^* and the corresponding z value by z^*. Again the obvious formula is obtained, from the condition that if concentrations in the two series are equal, the D values must also be equal and vice versa: $z^* + y = \xi y$.

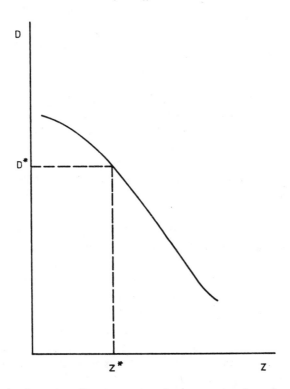

Fig. 3.14 — A schematic calibration curve of concentration-dependent distribution compared to SSE. D — distribution ratio, z — amount or concentration of non-radio-active substance added.

The choice between CDD and SEE boils down to the choice between measuring both phases and plotting $D = D(z)$ or measuring only one phase and plotting $i = i(z)$. In the latter case $i^* = \xi$, in the former case $D^* = D(\xi y)$.

Consequently, it is evident that the same set of distribution data can be treated by either method. SSE is superior to CDD in that it requires less sample preparation for the radioactivity measurements and fewer measurements. On the other hand, the calibration curve $D = f(z)$ is steeper than $i = f(z)$, which would favour CDD.

It is difficult to compare the relative ease with which the errors $\Delta i/i$ in SSE and $\Delta D/D$ in CDD can be made equal. If there is a significant error in the determination of A_f, the whole i-curve is shifted (a systematic error in this sense) but the D curve in CDD would be unaffected, which slightly favours CDD. The precision of the final y-value determined also depends on the errors in ξ or D_f. The error in D_f could probably exceed that in ξ for reasons given above. On the whole, a thorough comparative theoretical analysis of the precision of both CDC and SSE seems to be an insurmountable task at present and an experimental approach is preferred.

3.6.4 Applications of SSE, advantages and trends
Several examples of practical use of the SSE method are given in Table 3.21. It is obvious that the systems studied are of the same type as those used in substoichiometric IDA.

The main advantages are as follows:
— higher sensitivity than substoichiometric IDA in the systems where the stability constants limit the sensitivity of substoichiometric IDA,
— systems with variable complex composition, not allowing isolation of equal amounts at any concentration level, can be used,
— unlike the CDD method, only one phase is measured.

The drawbacks in comparison with substoichiometric IDA are the same as those with the CDD method (see p. 206).

The development of this radioanalytical method will probably involve the use of other systems, not yet described, finding criteria for preference of SSE over CDD analysis or vice versa, applications of automated versions of the procedures, and applications to new types of sample (environmental samples, etc.).

3.7 ACCURACY, PRECISION, SENSITIVITY, ADVANTAGES AND DRAWBACKS

3.7.1 Accuracy
The most common potential sources of systematic errors are described below.

(1) Incomplete isotopic exchange may be a serious cause of error if the substance to be determined and the diluent differ in oxidation state, or chemical and physical state, etc. Complete exchange between col-

Table 3.21 — Determinations using the SSE method.

Element	Reaction	Conditions	Remarks	References
Zn	Extraction with dithizone in CCl_4	Diethyldithiocarbamate masking of interfering elements	Reproducibility better than 15%; detn. in an antioxidant	[167. 168]
Co	Complexation with EDTA	Separation of free Co^{2+} on cation exchanger	1 µg/ml	[169]
Tl(III)	Complexation with EDTA	Excess of Tl(III) pptd. as Tl_2O_3	~ 4 mg/5 ml	[170]
$TlCl_4^-$	Extraction with Brilliant Green into benzene	From 1.8–3.5M HCl	~ 0.2 µg ± 2.5%	[170]
Fe	Complexation with EDTA	Free Fe extracted with acetylacetone	0.03 µg/ml	[171]
Hg	Extraction with Cu diethyldithiocarbamate	Acetate buffer pH 4.6	In corn, fungicides, human hair; 1–25 µg of Hg	[172]

loidal zirconium and its acidic solution requires a long digestion in a strongly acidic medium or exchange as hexafluorozirconate (ZrF_6^{2-}). For elements with several oxidation states a strongly reducing or oxidizing medium may be of assistance, or a redox buffer (a mixture of macroamounts of oxidant and reductant) may be useful. In some instances (iodine, ruthenium) even this may fail and it is necessary to make the element pass successively through all oxidation states in question. Carriers in reverse IDA should be added at as early a stage of the procedure as possible (before dissolution). With organic substances, such as vitamin B_{12}, some complications may arise from retention within cellular matter and bonding with proteins. Heating or treatment with enzymes and/or an acid–nitrite mixture may be helpful in this respect.

It is worth noting that when a certain amount of organic substance is burnt in a definite volume of oxygen enriched in ^{18}O, the abundance of $C^{18}O_2$ in the carbon dioxide can be found and the original amount of oxygen correctly calculated. This means that complete isotopic exchange takes place during the combustion. However, in the determination of nitrogen in steel, unsatisfactory results are obtained if ^{16}N is applied in gaseous form; it is necessary to add ^{15}N as aluminium nitride.

If isotope exchange is slow, the analysis should be performed with different periods of time allowed for isotopic exchange, to ascertain when exchange is complete. In some instances, taking samples at different places in the reservoir can be informative as to whether complete homogenization has been achieved.

(2) Undesirable isotopic exchange may occur if the radioactive atom in a molecule exchanges with the same element in a different molecular species. For instance, in the IDA of bromonaphthalene by use of radioactive bromine, the presence of lithium bromide might cause a decrease in the specific activity of the bromonaphthalene by the undesirable isotopic exchange:

$$*BrC_{10}H_7 + Br^- \rightleftharpoons BrC_{10}H_7 + *Br^-$$

Thus the radioactive atoms should always occupy as inert a position as possible in the molecule. From this point of view an organic molecule labelled with nitrogen is preferable to one labelled with deuterium. However, radiation-induced exchange may also occur. The atoms in positions normally considered to be inert may sometimes undergo exchange because of the action of radicals formed by the interaction of ionizing particles with the medium.

If the reagents used in the analysis contain the substance to be determined, or any substance which undergoes isotopic exchange with it, in amounts comparable with the weight to be determined, the results will be false. Therefore a blank should be run whenever such a danger exists (in trace analysis) and very often the reagents must be thoroughly purified. Indeed, reagents, apparatus and airborne contamination may all affect the precision of IDA.

(3) Erroneous radioactivity measurements obviously cause error. If all the rules for work with radioactive tracers are adhered to, count-rates are strictly proportional to concentrations of radioactive nuclides and no systematic errors of this sort occur. Very briefly, the most commonly used conditions for eliminating errors of this kind are:

(a) the geometrical shape and position of samples counted must be identical;

(b) corrections for dead time and self-absorption must be made;

(c) for short-lived radioisotopes, a correction for decay should be made or, better, simultaneous measurements should be made on the samples and a previously prepared standard; if the daughter element is also radioactive, similar care must be exercised.

These requirements are met more easily if the radionuclides chosen for IDA have suitable characteristics, namely, a medium half-life (from days to several years) and a readily detectable radiation energy ($E_\beta \geqslant 0.4\,\mathrm{MeV}$). A list of suitable beta emitters is given in Table 3.22. Naturally the statistical counting error will remain, and can be minimized in the usual way.

Table 3.22 — Suitable radionuclides for IDA with theoretical sensitivity ranges. (Reprinted from [173] by permission of the copyright holders, Interscience Publishers Inc., New York.)

Sensitivity[a] (g)	Radionuclides
10^{-8}–10^{-10}	^{42}K, ^{86}Rb, ^{99}Mo, ^{115}Cd, ^{114m}In, ^{181}Hf, ^{182}Ta, ^{187}W, ^{191}Os, ^{203}Hg, ^{204}Tl, ^{210}Bi
10^{-10}–10^{-12}	^{46}Sc, ^{59}Fe, ^{64}Cu, ^{72}Ga, ^{82}Br, ^{85}Kr, ^{109}Pd, ^{186}Re
10^{-12}–10^{-14}	^{24}Na, ^{58}Co, ^{125}Sb, ^{137}Cs
$\leqslant 10^{-14}$	^{32}P, ^{45}Ca, ^{77}As, ^{89}Sr, ^{90}Y, ^{95}Zr, ^{95}Nb, ^{103}Ru, ^{111}Ag, ^{131}I, ^{140}Ba, ^{141}Ce, ^{199}Au

[a] Hypothetical sensitivity for the specific activity theoretically achievable.

(4) The effect of isotopic differences in molecular weight is usually negligible, because even in the solution of the radioactive isotope the relative concentration of the radioactive atoms is very low. Only in work with stable or long-lived isotopes must a correction for the difference in molecular weight be applied.

Note: If specific activities are expressed in counts per mole and amounts in g, equation (3.2) must be altered to

$$W_x = W_0 \left(\frac{S_0}{S_x} - 1 \right) \frac{M_x}{M_0} \qquad (3.15)$$

where M is the molecular weight.

(5) Fractionation of the isotopic substances during the purification step is a rare phenomenon but could occur with light molecules containing ^1H, ^2H or ^3H, in processes where the kinetics controls the separation or in multistage processes, especially in the gaseous phase. In recent years this phenomenon has been found to be more common than was expected, so its effect on IDA, especially of organic substances, must not be overlooked.

(6) Radiochemical purity of the tracer used can be important. If the tracer contains radioactivity in a form other than that assumed to be present, the observed decrease in specific radioactivity will be larger than that corresponding to dilution with the inactive substance to be determined. Therefore the radiochemical purity of any tracer used should be checked. A suitable way is to use reverse IDA. Radiochemical purity requires not only the absence of foreign radionuclides but also the absence of the same radionuclide in a different chemical form. For example $H^{32}PO_3$ impurities in $H_3^{32}PO_4$ are inadmissible.

Very serious errors of this type may result if the tracer molecule is synthesized by reacting substance A with a radioactive reagent (B) and the purification from excess of B is performed after dilution with inactive AB. (All these operations are, of course, individual steps in the preparation of the tracer for an actual analysis.)

If 1 mg of A is reacted with 2 mg of B* to form 2 mg of AB* and 1 mg of excess B*, and 100 mg of pure inactive AB is added, and the chemical purity of AB is improved in the purification process from $\sim$ 99.0 % after dilution to, say 99.9 %, 0.1 mg of B* remains in the final sample of AB* + AB. This means that $\sim$ 9.1 % [0.1/(1.0 + 0.1)] of the radioactivity of the preparation is in the form not of AB* but of B*. This may cause a serious error.

There are two ways of preventing this low radiochemical purity. Either the synthesis can be performed on, say, a 10-mg scale, and purification carried out before any dilution, or (if after the synthesis is finished no exchange between AB and B takes place), inactive AB and B can be added with dilution, followed by purification. If the specific activities of AB and B before purification are equal, the chemical and radiochemical purities will be identical.

Non-radioactive impurities can be tolerated if they do not affect the determination of the original specific activity.

(7) The isolated fraction may be insufficiently pure. The purification of the portion used for measuring the specific activity may be checked by making sure that the specific activity does not decrease with further purification. In derivative analyses with two radionuclides the ratio of the two count-rates should not change with additional purification steps.

The purity of the isolated fraction is often critical in derivative analyses where excess of the radio-reagent must be removed, or in reverse IDA where the radioactivity of the substance in question may represent only a few per cent of the total activity of the original mixture. In such instances no excessively large dilution ratios should be used, because chemical purity higher than 99.9 % is sometimes difficult to attain. It can be calculated that if the radioactivity of the substance to be determined in reverse IDA is 2 % of the total radioactivity, then if it is diluted 50 times and the final chemical purity is 99.9 %, the error amounts to about 5 %.

In substoichiometric IDA the chemical purity of the separated portions is irrelevant if the inactive impurities do not consume the reagent and the requisite of isolating equal amounts is met.

(8) Decomposition of radioactively-labelled substances is dangerous, especially if the original specific activity was determined a certain period of time before dilution, so that the real specific activity is lower than the one taken for the calculation. This problem is serious when substances with extremely high specific activities are used and the concentration of the radioactive substance is much higher than $\sim 1\ \mu Ci/ml$. However, even material with a specific activity of $1\ \mu Ci/\mu g$ is reasonably stable if stored in the cold. If decomposition occurs during purification, the result is not affected, because the labelled and unlabelled molecules are decomposed to equal extents.

(9) Other systematic errors, e.g. in determination of the yield of thorium isolation from biological samples, are connected with the detection efficiency.

3.7.2 Precision
The following non-systematic errors may limit the precision of IDA.

(1) The error in determining the quantity isolated. If the quantity isolated is weighed, the error is usually negligible relative to the errors of type (2) and (3) but if colorimetric, spectrographic, polarographic and similar methods of determination are used the errors may become

comparable to or greater than the counting error. In such instances it is useless to try to attain extreme accuracy in counting.

(2) The error in determining the activity of the isolated portion involves the uncertainty arising from the statistical nature of radio-active decay, the variation in the background count, the instability of the measuring device, the irreproducibility of geometrical conditions for counting and similar factors. Inexperienced workers expect that the overall error is given only by the error due to the statistical character of radioactive decay, which is often far from true. The real error is usually larger than the error calculated on this basis. For instance, the error caused by the variation in form of the sample, in the geometry of counting and similar effects is usually about 1–2 % and, of course, cannot be diminished by increasing the number of counts registered. In order to minimize this error in determining the activity, radioactive tracers with high specific activities should be used, the yield of isolation should not be too low, the radioactivity should be measured for a sufficiently long time, the background should be kept at a minimum, reliable and stable measuring apparatus should be used, and the degree of dilution leading to a minimum error should be chosen.

(3) The error in the original specific activity is minimized if the specific radioactivity of the tracer is determined as many times as possible for each series of determinations; this is also recommended for substoichiometric determinations. On the other hand, care must be taken to ensure that the original specific activity and the activity after dilution are strictly comparable.

(4) The error due to inaccurate weighing of radioisotope or inactive carrier added can often be neglected, compared to the other errors.

(5) Fortuitous contamination by airborne activity, contaminated apparatus, etc. plays a role only in trace analysis (and may also cause systematic errors).

3.7.3 Sensitivity

IDA procedures cannot be more sensitive than the following main limiting factors allow:

(a) the least amount amenable to determination or purification in direct IDA;
(b) the original specific activity in reverse IDA;
(c) the specific activity of the diluting radioactive tracer or the specific activity of the radioactive reagent in derivative IDA;
(d) the equilibrium constants of extraction, complexation, hydrolysis, precipitation and similar reactions used in substoichiometric IDA;

(e) the blank value of the determination (reagent contamination),
(f) the stability of reagents at very low concentrations, sorption of the reagent on glass, etc.;
(g) the volume of the solution used for IDA;
(h) the neutron or charged particle flux in IDA after activation;
(i) other interference factors.

Most of these factors have been discussed in the preceding text, or the items in question are self-explanatory. Thus only some additional comments will be made.

Factor (a) is the cause of the statement that IDA does not increase the sensitivity of an analytical method. This statement is essentially true but there are two instances where the introduction of IDA can improve sensitivity. First, if the limit of sensitivity of an analytical procedure (e.g. gravimetry) is due to losses (partial solubility of precipitates) which become appreciable when working at a lower concentration level. Second, if a classic determination is replaced by substoichiometric IDA. Naturally, in the second instance the improvement in sensitivity relates to two completely different methods so that really no improvement in the sensitivity of a particular analytical method is achieved.

The specific radioactivity of a tracer with a given capture cross-section is controlled either by the time of irradiation or by the neutron flux available or, in many instances, by the concentration of the inactive isotopic substance which was present in the sample for irradiation. The theoretically attainable specific activities corresponding to a pure sample of one radionuclide (controlled by the half-life and atomic weight) are only exceptionally obtained in practice.

Most of the important beta-radioactive nuclides, together with approximate sensitivities controlled by the specific activity theoretically attainable, are listed in Table 3.22. It can be seen that the sensitivities are very high.

Factors (c)–(g) are typical for substoichiometric IDA and factor (c) is of prime importance. It must be realized that this factor, in general, includes the role of the counting efficiency and background because minimum specific activity is involved.

The methods of decreasing the background are well known, but the possibility of using back-scattering (platinum) may be mentioned. A sensitivity as low as 0.0 ± 0.7 cpm per sample at 95 % confidence level with a background of 1.0 cpm and a counting efficiency of 50 % could thus be achieved in the IDA of ^{90}Sr in milk and plant material. Also, the time available for counting may limit the sensitivity, as is the case with automated procedures using flow-cells for counting. Such procedures are less sensitive than manual methods.

The fact that carrier-free radioisotopes can be prepared by nuclear processes unsuitable for activation analysis [nuclear fission, ^{7}Li(p,n)^{7}Be, RaD from radon deposits] leads to higher sensitivity of substoichiometric IDA than of activation analysis in such instances. Also, if the same nuclear reaction is involved in both methods, the radioisotope for IDA can be prepared beforehand by a longer irradiation or by using a reactor with a higher neutron flux.

The values of the respective equilibrium constants (see pages 93 and 96) explain why the use of very low concentrations often causes difficulties in isolating precisely equal amounts of substance.

Generally, the most sensitive procedures of IDA are substoichiometry ($\sim 10^{-10}$ g), derivative dilution with double labelling (without weighing), IDA with determination of the amount isolated by activation and of course, reverse IDA or dilution with a second radioisotope.

3.7.4 Advantages, drawbacks and trends

The principal and decisive advantage of IDA is the possibility of using non-quantitative isolation procedures, so that in some instances IDA is the only analytical method for solving a problem. This possibility allows the analyst to perform an isolation quickly, to choose a purification method from a very wide selection of methods and to tolerate a partial decomposition of the analyte during the analysis. A typical example of how the use of IDA can save time is the analysis of a mixture of triphosphate and pyrophosphate. By using IDA the time required for analysing such a mixture can be shortened from 1–3 days to 2–4 hr [174].

Taking losses into account leads to increased precision. In the analysis of albumen, in which 26 amino-acids were determined, the mass-balance agreed with good precision when IDA was used, whereas for classical procedures the difference was several tens per cent [175].

It is a great advantage that the losses may be allowed to be irreproducible in an IDA procedure, unlike analysis with radioactive indicators, in which the whole separation procedure is first tested with use of radioactive tracers, the average yield is determined, and a constant correction is applied to the amounts found in pure form at the end of standardized procedure applied without any tracer.

As the element to be determined is radioactive, its path through the analytical scheme can be followed, to be sure of its identity and to check for adequate purity by the constancy of its specific activity. Also, certain types of IDA are readily amenable to automation.

The isotopic dilution usually leads to a greater similarity (e.g. in weights) of all the samples prepared for measuring specific activity than for the original amounts, so that greater precision can sometimes be achieved. For instance, if samples containing 1 and 3 mg are diluted with 1 mg of isotopic diluent, the original ratio of weights is 3, but the ratio for determining the specific activities is only 2.

In IDA, there is no danger that the determinand element can also be formed by a different nuclear reaction from another component of the irradiated sample (which may be a cause of false results in activation analysis).

The reproducibility of substoichiometric IDA is generally higher than for the concentration-dependent distribution method (saturation analysis).

On the other hand, IDA has several drawbacks. It is not an autonomous analytical method; two pieces of apparatus are needed (e.g. balance + radioactivity meter, balance + mass spectrograph, colorimeter + radioactivity meter, etc.). Therefore, IDA is not suitable for an occasional analysis and the two alternatives should always be carefully weighed:

(a) tedious quantitative separation without radioactivity or mass-spectrographic measurements;

(b) easy and rapid non-quantitative separation, with the necessity of carrying out one of these additional measurements.

To achieve sufficient accuracy the range of the unknown amount must be known beforehand, otherwise it is not possible to choose a suitable amount of diluent. However, in some types of IDA this requirement does not apply.

In some instances an analytical procedure must be somewhat modified before being adapted for IDA, e.g. the volume of the sample at one of the initial stages of the procedure must be known accurately [176].

In working with radioisotopes certain safety precautions must be observed for health reasons. This is usually not a difficult problem, but it may be rather complicated if, for instance, flame-photometric determination is finally used.

Stable isotopes are usually expensive, and so is the apparatus for measuring isotopic compositions; qualified operators are required and the analysis takes much time.

As compared, say, with emission spectrography, IDA is not essentially a multielement analysis. It is not practicable to store radioactive isotopes of a large number of elements or to store or prepare a large assortment of labelled compounds, as would be needed if many different substances were expected to be submitted for analysis. Indeed, the synthesis of labelled organic compounds is a problem in itself. Moreover, there are several elements for which no suitable tracers are available.

Compared to activation analysis, the major drawback in IDA is the necessity of using reagents freed from traces of the substance to be determined and the very low concentration level at which the isolation must be performed. Therefore IDA is not advantageous for the analysis of traces in samples that require large amounts of concentrated reagents for their dissolution, because their purification may be difficult.

In comparison to saturation analysis or the method of concentration-dependent distribution, substoichiometric IDA sometimes has poorer sensitivity. The choice of reagents is also more limited.

One type of trend in the development of IDA can probably be characterized by a quotation from Lyon [177]: "There seems to be no lack of information being passed back and forth within the literature, but most of it continues to be rehashing, refurbishing or re-evaluating of previously reported methods or techniques. We left our literature search with a feeling of déjà vu."

On the other hand, progress can be expected in some directions. A new possibility in isotope dilution is the simple determination of the specific activity by Cherenkov photometry. If the compound of the isotope is coloured or can be converted into a coloured compound, it is possible to measure the activity and the concentration of the compound in one sample. Kulcsár et al. [178] used the yellow phosphovanadomolybdic acid for determination of the specific activity of ^{32}P.

New simple ways of determining specific activities (such as using Cherenkov radiation) are being developed. The determination of the separation yield of one radionuclide from another might be more frequently used owing to the progress of high-resolution measuring techniques. The general opinion on the degree of hazard involved in work with radioactive tracers is gradually changing and the maximum tolerable exposure levels are being increased in some cases, which would render IDA more competitive with classical methods.

For the use of IDA in organic chemistry and biochemistry the increasing assortment of commercially available labelled compounds and their decreasing price may be of great influence.

3.8 EXAMPLES OF APPLICATIONS

Inorganic applications up to about 1970 were reviewed in the book by Tölgyessy et al. [18], totalling some 150 papers. Approximately the same number of applications in organic chemistry and biochemistry were quoted in the same book. The review on derivative IDA by Kyrš [49] dealt with about 100 papers.

Special applications include the use of IDA for solving the following problems. Volume can be determined by IDA if its direct measurement is impossible (e.g. the total amount of water in caves) or when the system in question cannot be destroyed (e.g. total volume of blood in a living organism) (see also 3.9.4.3).

The capacity of ion exchangers can be determined by IDA when their conversion into the H- or OH-form is impossible [179]. IDA has also been used for the determination of important chemical properties of substances such as solubility, dissociation constant [180], reaction rates and equilibria [181], and self-absorption coefficients [182]. It is also possible to measure flow-rates by methods based on the dilution principle [183].

Many typical procedures are thoroughly described in instructive form in the monograph by Tölgyessy et al. [18] and need not be repeated here.

As a typical example of a recent application of IDA, the determination of exchangeable water in coal, charcoal and activated charcoal [184] will be briefly described.

The background of this analysis is the necessity for meaningful and reproducible studies on the ability of coal and coal-derived solids to sorb solutes from solutions. This requires simulating the conditions that occur in coal aquifers and coal slurry pipelines, so only a solid pre-equilibrated with water can be used as the sorbent. Such a sorbent, especially one with approximately equal masses of solid and water, shows an apparently high degree of sorption, which is really only a dilution effect resulting from mixing with the water held on the surface and in the pores of the moist solid.

It is therefore necessary to know the extent of exchange of interstitial and surface water held by the moist solid.

Determination of water by heating the moist solid does not give a value of the total exchangeable water and tells nothing of the rate of exchange of water with a separate liquid phase. The knowledge of this rate is important for comparison with the apparent rate of sorption.

Exchangeable water was determined by using tritiated water (HTO), followed by liquid scintillation counting. The solids studied were in-itially equilibrated (72 hr with frequent shaking) with HTO and the specific activity (dpm per unit volume) was measured. The HTO was then drained from the solids under standard conditions. (The uniform-ity of this operation is the main factor affecting the reproducibility of the results.) The solids were then re-equilibrated with pure water in a slurry. The amount of tritium appearing in the initially pure water provided a measure of the exchange of water with the solid.

The procedure was as follows. The critical operation of drainage included placing the slurried solids on a coarse sintered-glass filter and filtration under suction at a low air flow-rate for 5 min. During filtration the solids were stirred gently several times with a spatula, and water clinging to the sides of the funnel was removed with filter paper. The re-equilibration was done with 2.5-g portions of the moist solids in 15-ml glass vials with 5 ml of distilled water. The slurries were periodically agitated, and 0.2-ml portions of water were periodically withdraw for counting. Counting with a wide-open ^3H window continued until a 20 count-rate error of 1.0 % or less was achieved. The counting efficiency was determined for each sample and was used to compute the absolute activity as disintegrations per minute (dpm). Final dpm values were corrected by subtraction of the background count-rate.

The calculation was performed by using the equation

$$S_c V_c = S^0 (V_w + V_c)$$

where S_c is the known specific activity of the tritiated water associated with the solid, and determined from the radioactivity of the HTO-spiked water equilibrated with solid prior to each run (dpm/ml), V_c the unknown volume of water associated with the solid as interstitial and surface-bound water (ml), S^0 the measured specific activity of tritium in the water phase (both external and sorbed) at 'infinite' re-equilibration time (dpm/ml), and V_w the known volume of the external water phase (ml).

The results were expressed as per cent water exchanged, equal to $100 V_c/W$ %, where W is the weight of moist solid (g). The per cent water exchanged then represents the amount of water initially sorbed (in g) by the solid, exchanged per 100 g of moist solid. Typical values obtained were 13–20 %. The total exchangeable water, V_c, in the unactivated charcoal was 15.2 %, which may be compared with the value of 21.9 % water determined by heating. The highest values obtained were 24.5 % of exchangeable water (29.7 % found by heating).

The time-dependences of the volume of H_2O exchanged were obtained and confirmed that the exchange is slow, equilibrium being reached only after tens of hours.

This example of IDA is a case where probably no other method could be used to obtain the information required.

A similar example is the determination of accessible hydroxyl groups (∼ 30 %) in wood, by using isotope exchange with deuterium oxide. The equilibration takes 20 hr at room temperature. After removal of the phase containing the excess of D_2O the sample of wood is 'rehy-drogenated' with normal water for 20 hr and the fraction of deuterated wood is found by determination of the deuterium oxide content in the rehydrogenated solution by the interferometric method [185].

In our opinion the results above are strictly correct only if it can be assumed that there is no exchange of hydrogen between the water and the coal matrix. Lenskii and Rachinskii [186] noted that the humidity content in various systems can generally be determined by isotope dilution with tritium if the concentration of isotope-exchangeable hy-drogen in the dry matrix is determined beforehand. This concentration is a characteristic of the chemical composition, so an exact correction can be made for this type of isotope exchange. The highest precision is achieved if the mass of tritiated water is approximately equal to that of the moisture in the sample.

The last example is concerned with IDA of one radionuclide by means of another isotopic to it (see p. 57). Urnezis and Holtzman [39] have introduced an isotope dilution step into the determination of ^{210}Pb from the radioactivity of its daughter element ^{210}Po. A known amount of ^{209}Po and ^{210}Po is deposited on a silver planchet, which is counted in an alpha spectrometer to determine the activities of each isotope. The recoveries generally range from 70 to 90 %. The decrease in counting efficiency of the alpha spectrometer compared to that of the propor-tional counter used for measuring only ^{210}Po is more than compensated for by the reduction in background count-rate, which is 20–25 % of that of the proportional counters. The main advantage is the reliable deter-mination of the overall yield of Po isolation, the sensitivity remaining the same as in the original method without ^{209}Po.

3.9 ISOTOPE DILUTION ANALYSIS (IDA) WITH STABLE ISOTOPES*

3.9.1 Introduction
This section deals with IDA in which the change in isotope ratio of isotopes is evaluated instead of the change in specific activity.

It should be emphasized that evaluation of the change in isotope ratio is also sometimes used for determining some radioactive elements with a sufficiently long half-life. As the isotope ratio is evaluated in IDA with stable isotopes, only the elements having at least two isotopes can be determined. Thus, ~ 80 % of all elements (see Table 3.23) can be determined by using IDA.

---

* By J. Moravec and F. Sus.

The limits of detection for individual elements differ according to the analytical techniques used and the physical properties of the element and are 10^{-4}–10^{-6} g or even lower. Milligram to nanogram amounts of the elements are sufficient for their determination.

To determine the content of an organic compound in a substance by IDA, the same compound, containing a different isotope of at least one element, is used. The concentration of the compound is evaluated from the change in isotopic composition of the given element in the mixture after mineralization.

Table 3.23 — Survey of the applicability of isotope dilution analysis with stable or long-lived isotopes.

Elements which can be determined	H	He	Li	Be*	B	C	N
	O	Ne	Mg	Si	S	Cl	Ar
	K	Ca	Ti	V	Cr	Fe	Ni
	Cu	Zn	Ga	Ge	Se	Br	Kr
	Rb	Sr	Zr	Mo	Tc*	Ru	Pd
	Ag	Cd	In	Sn	Sb	Te	I*
	Xe	Cs*	Ba	La	Ce	Nd	Pm*
	Sm	Eu	Gd	Dy	Er	Yb	Lu
	Hf	Ta	W	Re	Os	Ir	Pt
	Hg	Tl	Pb	Th*	U	Np*	Pu*
	Am*	Cm*					
Monoisotopic elements, not suitable for the determination	F	Na	Al	P	Sc	Mn	Co
	As	Y	Nb	Rh	Pr	Tb	Ho
	Tm	Au	Bi				

* Radioactive spike with long-lived isotope or radioactive element.

The development of IDA has been made possible by introduction of methods and devices for measuring the isotopic composition of elements, especially mass spectrometers. Thanks to its specific features, IDA has found widespread application in studies of complicated analytical problems, analyses of geological samples, including the determination of geological age, analysis of nuclear materials, determination of fission yields, investigation of physical problems in research work, etc. The wider use of IDA for determining organic substances has been hindered by the limited availability of compounds labelled by suitable isotopes.

3.9.2 Methods for determining the isotopic composition

Labelling of the sample by one isotope of the element to be determined is the usual condition for application of IDA. An unambiguous correlation then exists between the amount of spike in the mixture with the sample and the physical parameter being measured, e.g. the isotopic composition of the element, density of the substance and its dependence on temperature, refractive index, etc. The dependence of some of these physical parameters on the amount of spike is of limited use. Moreover, the precision and accuracy of determination of the change in properties of a compound and the consequent evaluation of the isotopic composition do not always comply with the needs of IDA. Some methods for determining the isotopic composition of an element on the basis of physical properties have been published by Cabicar [187].

Mass spectrometry is a universal analytical method for estimation of the isotopic composition of elements; it produces highly accurate results, and the consumption of sample for the analysis is minimal.

3.9.2.1 Principle of a mass spectrometer
A mass spectrometer generally consists of the following basic parts:
(a) ion source and optics, in which the sample is ionized to create a beam of ions (usually positively charged) possessing a defined energy;
(b) analyser system, in which the ions are distributed according to mass/charge ratio of the ion (m/z),
(c) detector system, where the ions of a given m/z ratio are registered;
(d) electronic circuits for power supply, control, signal evaluation, etc.
Mass spectrometers can be classified as dynamic or static, with single or double focusing, etc. In principle, the dynamic spectrometers determine the time needed for the ions (moving in a variable electric or magnetic field) to reach the detector. Because of their high sensitivity, high resolution and (usually) small size, dynamic spectrometers are used for studies of chemical reactions, fast reactions, analysis of gases (e.g. even in the upper layers of the atmosphere), etc.

In static spectrometers, homogeneous magnetic fields act as the analysers; the ions move in circular paths with radius

$$r = \frac{1}{B}\sqrt{2Um/z} \tag{1}$$

where B is the magnetic field, and U the accelerating voltage.

Single-focusing mass spectrometers, illustrated schematically in Fig. 3.15, are usually used for measuring the isotopic composition of elements. The ions are focused by a magnetic field (90°) onto M_1, M_2 ... M_x points in the focal plane, and, after passing the slits, reach the detector system which records signals corresponding to m/z.

The following basic characteristics determine the parameters of the spectrometer and its applicability to a particular problem:
(a) resolving power, defined by the ratio $M/\Delta M$, where ΔM is the difference between two consecutive ion masses still resolved by the spectrometer (at a chosen peak height) and M the mean of the two masses;
(b) ion source yield and transmission of the spectrometer, which both determine the approximate amount of sample needed for the analysis;
(c) abundance sensitivity, establishing the signal level for mass $M \pm 1$ caused by scattering of the ions of mass M.

Most modern spectrometers are fully automatized, including sample exchange or filling with gaseous samples, and often also the sample preparation, and a control computer takes over the function of the operator. In this way, high reproducibility and relative errors of $\leqslant 0.01\,\%$ in the isotopic composition measurements have been achieved.

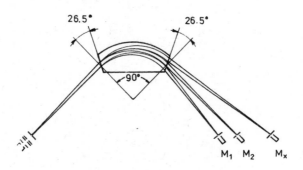

Fig. 3.15 — Ion optical system.

Besides the classical commercially produced instruments, more sophisticated spectrometers are available, especially for measuring the isotopic composition of U, Pu, H, etc. They differ from each other in the arrangement of the ion sources and the electronic and detector systems.

Mass spectrographs form a further group of spectrometers; a photographic plate is used as the detector. The position of a line in the spectrum correspond to the mass/charge ratio of the ion, and the intensity of blackening of the line corresponds to the abundance of the ion. Mass spectrometers are usually used for a complete chemical analysis and determination of a set of elements with the aid of a combined spike containing the isotopes of the elements. The construction of the mass spectrograph and the measurement technique differ significantly from those for measurement of isotopic composition. Spark sources and nowadays laser sources are usually used for sample excitation, the latter giving more possibilities for optimizing the sample excitation conditions. Photographic detection has many disadvantages, resulting from the properties of the emulsion layer (low sensitivity, small dynamic range, dependence of blackening on the ion mass, etc.). A channel multiple diode array, which has a wider dynamic range, is now commonly employed as the detector, and gives a smaller error in IDA by mass spectrometry.

3.9.3 Fundamentals and technique

The usual procedure of IDA with stable isotopes is as follows.

(1) A weighed portion of sample is dissolved.
(2) A measured (known) amount of isotopically enriched component is added to the solution.
(3) The mixture is chemically processed to ensure the proper equilibration of the spike with the sample.
(4) The element to be determined, present as a mixture of normal and diluent isotopes, is treated chemically to yield a pure form. Solvent extraction and ion-exchange chromatography are most frequently used for this purpose. The purification need not be quantitative and may be omitted if no interference by other substances is to be expected.
(5) The isotopic ratio of the extract is measured. Three isotopic ratios need to be known:

$R_{i/j}^p$ — atomic isotopic ratio of the isotopes i, j in the test sample;
$R_{i/j}^s$ — atomic isotopic ratio of the isotopes i, j in the spike;
$R_{i/j}$ — atomic isotopic ratio of the isotopes i, j in the mixture of the sample and the spike.

The concentration of the determinand C (ppm by weight) is calculated from

$$C = \frac{w_s M_p^j (1 - R_{i/j}/R_{i/j}^s)}{w_j w_p M_s^i (R_{i/j} - R_{i/j}^p)}$$

where w_j is the mass fraction of isotope j of the determinand in the sample, w_s the added amount of isotope j of the spike (μg), w_p the weight of the sample (g) and M_p^j and M_s^i are the atomic weights of the element to be determined in the sample and the spike.

It can be seen that the more enriched the spike and the lower the value of the isotopic ratio in the sample, the greater is the accuracy of the analysis.

The maximum accuracy can be achieved with an $R_{i/j}$ ratio of ~ 1 in the mixture; in practice this ratio should not be outside the range 0.25–4.

It is generally inconvenient to measure the exact concentration of the spike (e.g. by gravimetry). Instead, the concentration is determined by calibrating the spike solution with a solution containing an accurately known amount of the element with normal isotopic composition. A typical example of the calibration of a spike is given in Table 3.24.

Table 3.24 — Calibration* of a uranium spike solution.

Notes	Primary standard (natural element)		Spike (isotope-enriched element)
Chemical form	U metal		U_3O_8
Chemical purity	99.975 $\pm$ 0.017 wt. %		$\geqslant 99.9\%$
Isotopic composition	^{233}U	—	99.7021 at. %
	^{234}U	0.0055 at. %	0.2373 at. %
	^{235}U	0.7203 at. %	0.0125 at. %
	^{236}U	—	< 0.0002 at. %
	^{238}U	99.2742 at. %	0.0479 at. %
Weighing and dissolution	~ 100 mg weighed with an error of $< 0.01\%$ relative, dissolved in dilute acid, and diluted to $\sim 500\,\mu g$ U/g solution		~ 100 mg of U_3O_8 Dissolved and diluted to $\sim 500\,\mu g/g$ solution

* Determine the composition of uranium and $^{238}U/^{233}U$ ratios; mix weight aliquots, perform the isotopic exchange, determine the isotopic composition of uranium, and calculate the concentration of ^{233}U in the spike solution.

3.9.4 Practical applications

3.9.4.1 Applications in organic chemistry
IDA with stable isotopes has not found wide application in the analysis of organic substances. In this field, it has been mainly employed for elemental analysis. Compounds labelled with $^2D_2$, ^{13}C, ^{15}N, ^{18}O, ^{34}S, etc. are used or the samples are burnt in oxygen containing ^{18}O. The organic substance is either mineralized or the molecular fragments of the compound, with a chosen m/z ratio, are directly measured.

Milligram amounts are burnt in a platinum boat, in the presence of a catalyst, with a known volume of oxygen enriched with ^{18}O (i.e. a mixture of $^{16}O_2$, $^{16}O^{18}O$ and $^{18}O_2$). In the combustion process, the equilibrium

$$CO_2 + {}^{18}O_2 \; \rightleftharpoons \; C^{16}O^{18}O + {}^{16}O^{18}O$$

is achieved. The carbon dioxide formed is separated and purified and the amount of ^{18}O present determined by mass spectrometry. For 2 % ^{18}O in the oxygen used, the error of the method is 1–2 %. This error can be reduced by an order of magnitude by increasing the ^{18}O content to 20 %.

Nitrogen, carbon and sulphur in organic compounds can be determined similarly. Depending on the element to be determined, gases enriched with $^{15}N_2$, $^{13}CO_2$ or $^{34}SO_2$ are added to the system. With 12–20 % heavy stable isotope in the nitrogen, carbon dioxide or sulphur dioxide, the error of the method is 0.3 %.

To determine nitrogen, the sample is oxidized with CuO instead of oxygen. A correction for the amount of air which may enter the instrument is established by measuring the peak corresponding to the mass of ^{40}Ar. As the ratio of nitrogen to argon in air is constant, the quantity of nitrogen can be calculated.

In other cases, the organic substance is decomposed by the Kjeldahl method to NH_3 and then to N_2, the isotopic composition of which is measured.

The combination of gas chromatography with mass spectrometry (GC–MS system) offers many applications to the analysis of organic substances.

Selected ionized fragments of the molecules of the labelled compound and the non-labelled one to be analysed are directly compared. The following methods for estimating ethyl alcohol in blood and urine can be taken as examples. A 200-µl blood or urine sample is mixed with 90 µl of deuterated ethyl alcohol and after chromatographic separation, the ion currents at m/z 45 ($CH_3CH{=\!=}OH^+$) and 49 ($C_2D_4{=\!=}OH^+$) are recorded. Amounts of ethyl alcohol down to 5 ng can be determined with an RSD of 1.5 %. To determine N^τ-methylhistamine in blood and urine by the GC–MS method, bis(heptafluoro)butyryl derivatives and deuterium-labelled N^τ-methylhistamine are used as internal standards. The pairs of peaks for the fragments m/z 309 and 307 (deuterated derivative) and m/z 517 and 520 (deuterated derivative) are evaluated. The detection limit is 5 pg.

Björkhem et al. discussed the results of an intercomparison between 64 Swedish laboratories for the determinations of cholesterol, glucose, urea, urate, creatinine and cortisol in serum from patients, by IDA and routine methods [188]. IDA gave results with errors lower by a factor of 2–30. The application of the IDA–MS method as a standard method for quantitative determination of steroids was described.

3.9.4.2 Applications in inorganic chemistry

The determination of the gas content of various metals and alloys is an important application of IDA with stable isotopes. A wet chemical procedure has been developed for the determination of nitrogen in various metals. During the dissolution of the sample in acid, ammonium sulphate enriched with ^{15}N is added. All the forms of nitrogen present are converted into elemental nitrogen, which is measured. The method can be applied to samples containing from 2 ppm to 6 % of nitrogen, with an average relative error of $\pm 5 \%$.

IDA based on nitrogen labelled with ^{15}N has been used for the determination of nitrogen in National Bureau of Standards (N.B.S.) standard steel 1044 and stainless steel 347. The results were compared with those obtained by the vacuum fusion procedure. No differences attributable to gas adsorption in the furnace were observed. The concentration of ^{14}N in the gas was close to the natural abundance. For the N.B.S. samples, the analytical results were lower than those obtained by the Kjeldahl method. This reveals that some nitrogen still remained in the melted sample in the crucible.

Determination of oxygen and nitrogen in steels and nickel by IDA with ^{15}N and ^{18}O, covering the concentration range 10^{-4}–$10^{-3} \%$, has been suggested. The gases sorbed on the surface are removed by heating the sample at 1000 °C and 3–5×10^{-4} Pa. The results compare well with those obtained by the reductive fusion method and by activation with ^{3}He ions.

In a method for estimating $5 \times 10^{-3} \%$ C in metals (Ni, Fe, Ti, Mo, W), powdered Zr containing ^{13}C is used for labelling and the isotopic composition of the carbon in the CO_2 released by vacuum fusion at 1500–1600 °C is measured. There is good agreement with the oxidative vacuum fusion method. The factors affecting the blank have been discussed.

Chloride in rocks can be determined by using ^{37}Cl. The samples are decomposed with HF in a Teflon bomb and the chloride is separated as AgCl. A double-filament thermal ion source is used for measuring the isotopic composition and the negative ions are recorded. Chloride (100–200 ppm) can be determined with an error of 1–4 % relative.

A method for the estimation of polyisotopic rare earths, based on use of a triple-filament assembly, without separation of the rare-earth elements from each other, has been introduced. The rare-earth elements can be determined with high precision in 3 hr. The determination of alkaline-earth metals and heavy metals in environmental materials has also been developed. Isotope exchange is ensured by using a mixture of acids under pressure. By use of a combined spike, several elements can be determined simultaneously.

The method makes it possible to determine nanogram amounts of elements with an error of 1 %. Mo (3–9 %) in Ni-based analytical standards has been determined with an error of 0.5–1 %, with ^{96}Mo as a spike. For measurement of the isotopic composition, the carbonization technique with single-filament surface ionization is used. When Zr and/or Nb are absent, the determination can be performed without a separation; otherwise, Mo is separated with α-benzoinoxime.

IDA can be combined with spark-source mass spectrometry in the analysis of geological samples. For labelling, the spikes are added to the sample individually as solutions or mixed with graphite powder. The elements are determined with an error of < 10 %.

3.9.4.3 Application in nuclear chemistry

Because of its specific features, IDA is an essential tool in the analysis of nuclear fuels and materials. Mass spectrometry, alpha spectrometry and gamma spectrometry are used for measurement of the isotope abundance.

IDA has been applied to the determination of fission yields of > 40 isotopes of the elements occurring from fission of uranium and plutonium. Various methods are used for isolating µg and smaller amounts of the individual elements, and optimal conditions for measurement of the isotopic composition have been established. The errors are a few tenths of 1 %.

Evaluation of fuel burn-up can be based on determination of the concentration of ^{148}Nd, with ^{150}Nd or ^{142}Nd as spikes. The blank is determined from the abundance of ^{142}Nd which does not occur by fission. For 0.05–1 µg of ^{148}Nd in the sample, the determination error is 0.1–0.3 % relative.

Procedures are known for evaluating the burn-up by determination of the Mo isotopic concentration, with ^{92}Mo, ^{94}Mo, or a natural isotope mixture for spiking. The presence of ^{92}Mo, ^{94}Mo and ^{96}Mo, i.e. the isotopes which do not occur by fission, reveals the degree of sample contamination by natural Mo. To isolate Mo from an irradiated nuclear fuel sample, extraction with acetylacetone, α-benzoinoxime, etc. is used.

The IDA of uranium and plutonium in burnt fuel has been described in many papers, and the methods and results presented have been tested and verified in a series of international intercomparison experiments.

During these experiments, the influences of individual steps of the analytical procedures on the results were evaluated and the contributions of particular errors to the total error of analysis were discussed. The intercomparison exercises proved that the overall error was mainly due to the error in the measurement of the isotopic composition of the spiked samples. The errors in the determination of uranium (plutonium), expressed as relative standard deviations, were 0.01–0.1 %. The interlaboratory means differed by 0.1–1 % relative, the reason in most cases being the differences in calibration of the spike and the spectrometer, and in the discrimination factor.

Determination of 10–100 µg/g americium and curium in burnt fuel by using ^{243}Am and ^{248}Cm as spikes has been described. Ion-exchange chromatography with α-hydroxyisobutyric acid (AHIBA) as eluent was applied to the isolation of Am and Cm. The error of the determination of the main element isotopes was $\sim 0.5\%$.

In the laboratory of the authors of this section, methods have been developed for determining uranium, plutonium, americium and neodymium in the same sample aliquot, using ^{233}U, ^{242}Pu, ^{243}Am and ^{150}Nd as spikes. Evaporation with H_2O_2 and HNO_3, instead of the usual evaporation with $HClO_4$ or the hydroxylamine–nitrite redox step, was used to ensure isotope exchange. The separation procedure invólved (a) sorption of the sample solution on a column (Dowex-1 X8), (b) elution of the fission products, including Nd, Am and Cm, with $9M$ $HCl + 0.4M$ HNO_3, (c) the elution of plutonium with $9M$ $HCl + 0.1M$ HI, (d) the elution of uranium with $0.1M$ HCl. The decontamination factor U : Pu was $\sim 10^8$ and the coincidence of the ^{238}U–^{238}Pu isobars was eliminated. A group of rare-earth elements (REE) and the transplutonium elements (TPu) was isolated from the fission-product fraction on a column (Dowex-1 X8) with a mixture of $CH_3COOH + HNO_3$. The individual REE and TPu were isolated by chromatographic separation (Dowex-50 W X8) with AHIBA (pH 4.62). The elution of alpha and gamma emitters was examined continuously with flow-through detectors, with recording of the elution curves. The contamination of the Nd fraction with Sm or Ce is $< 0.01\%$; the contamination of the Am fraction with Cm is $< 0.1\%$. For the determination of U (mg), Pu (µg), ^{148}Nd (ng–µg), and Am (ng–µg), the errors were 0.1, 0.2, 0.2 and 0.6 % relative, respectively.

IDA has also been applied to the determination of some physical constants. In an international programme, IDA was used for determining the half-life of decay of ^{240}Pu and the value $t_{1/2} = 6574 \pm 12.8$ years was found. Similarly, the half-life for ^{239}Pu was determined.

IDA has been employed for estimation of the volumes of nuclear fuel solutions in storage tanks. To an unknown volume of the solution, a known amount of tracer (Li, Mg or other element absent in the solution) is added; the solution is thoroughly stirred, a known amount of it is taken and the content of the tracer is determined by IDA with the spikes ^{6}Li, ^{26}Mg, etc. The error in the determination of the volume depends on the degree of dilution of the tracer, the error of weighing, blank of the tracer, etc. and varies from several tenths to a few % relative (see also Section 3.8).

A number of papers dealing with the applications of IDA in nuclear chemistry have been published in monographs [189] and the IAEA proceedings.

3.9.4.4 Determination of geological age

In absolute geochronology, the duration (in hundreds of millions of years) of geological ages, periods and times, and the time that has elapsed since a given geological event, are established by determining the absolute age of minerals of known relative ages. This determination is possible because the minerals contain measurable amounts of radioactive elements and their decomposition products, and the type and half-life of the radioactive decay is known. The latter is presumed to have been constant during the geological ages.

Isotopic analysis of argon and potassium provides a means of determining geological age. Argon present in uranium ores contains ^{38}Ar in great excess over ^{36}Ar. ^{38}Ar is produced by the reactions

$$^{35}Cl\,(\alpha,p)^{38}Ar \text{ and } ^{35}Cl(\alpha,n)^{38}K \xrightarrow{\beta^+} {}^{38}Ar$$

Some potassium minerals also contain argon enriched with heavier isotopes.

Mass spectrometric measurements have indicated values of the ratio ^{40}Ar/^{36}Ar greater than 400 or even 500 in sylvinites (KCl) and in other potassium minerals of various origins, as compared to 239 for atmospheric argon. The excess of ^{40}Ar indicated by these ratios originates from the conversion

$$^{40}K \xrightarrow{K\text{-capture}} {}^{40}Ar$$

which takes place simultaneously with the β^+-decay of ^{40}K to ^{40}Ca. Consequently, the age can be calculated by means of the equation

$$N_{40} = N_K \frac{\lambda_e}{\lambda}(e^{\lambda t - 1})$$

where N_{40} is the number of ^{40}Ar atoms per gram, created from ^{40}K by electron capture, N_K the number of ^{40}K atoms at time t, λ_e the decay constant of the capture process, and λ the overall decay constant for ^{40}K.

The accuracy of the method has been markedly increased by the introduction of isotope dilution into the mass spectrometric determination of argon. Accurately known amounts of spike and air are mixed, and the isotopic composition of the mixture is determined by mass spectrometry.

Geological age can be calculated on the basis of the determination of the total number of moles of radiogenic argon and potassium present in the mineral, by using the equation

$$ t = \frac{1}{\lambda} \ln \left[1 + \frac{^{40}Ar}{^{40}K} \left(\frac{1 + R}{R} \right) \right] $$

where t is the age in millions of years, λ the total decay constant of ^{40}K ($\lambda = \lambda_\beta + \lambda_K$), ^{40}Ar the number of moles of ^{40}Ar in the sample, ^{40}K the number of moles of ^{40}K in the sample (the atomic fraction of ^{40}K in potassium is 0.000119), and R is the branching ratio, λ_K/λ_β (0.584 × 10^{-10} year^{-1}/4.72 × 10^{-10} year^{-1} = 0.1237).

The U–Th–Pb dating method is based on the radioactive decay of the isotopes ^{238}U → ^{206}Pb, ^{235}U → ^{207}Pb and ^{232}Th → ^{208}Pb. The isotope ^{204}Pb which is present in the natural mixture does not occur by radioactive decay and serves for checking. Lead is separated by distillation at high temperature, with subsequent extraction or, after the decomposition of the sample by mineral acids, by a multistep extraction with various reagents, e.g. diethyldithiocarbamate, dithizone, etc. The authors of this chapter separated Pb, after mixing the sample with graphite powder, by distillation and subsequent electrolytic isolation from the 'condensate' (electrolysis at a glassy-carbon electrode coated with a mercury film, mixed electrolyte of hydrochloric, tartaric and ethylenediaminetetra-acetic acids). The blank was < 10^{-11} g of Pb. For measuring the isotopic composition of Pb (sample ≤ 0.1 μg), the emitter technique is mostly used; the Pb solution is loaded onto a filament covered with an emitter (suspension of Al_2O_3, SiO_2, H_3BO_3, H_3PO_4) with which Pb, after thermal treatment, forms a vitrified material. The results of determination of the isotopic composition of Pb are usually related to the N.B.S. standard reference materials 981, 982 and 983. To evaluate the concentration, ^{210}Pb is used as spike.

A further dating method, very often used, is based on the radioactive decay ^{87}Rb → ^{87}Sr. The total concentration of ^{87}Sr at time t is given as

$$^{87}Sr = {}^{87}Sr_0 + {}^{87}Rb_0(1 - e^{\lambda t})$$

where ^{87}Sr is the number of ^{87}Sr atoms at time t (time of analysis), $^{87}Sr_0$ and $^{87}Rb_0$ are the numbers of ^{87}Sr and ^{87}Rb atoms, respectively, at time $t = 0$ (i.e. of the origin and homogenization of the rock), and λ is the decay constant (1.42×10^{-11} year^{-1}). The Rb–Sr evolution diagram is given in Fig. 3.16. A mixture of HF + $HClO_4$ or HF + HNO_3 is usually employed for dissolution of the samples. After separation of fluorides, Rb and Sr are generally isolated by chromatography on columns packed with cation-exchange resin (Dowex-50 W X8), with elution by HCl of increasing concentration; for the isolation of Sr, chromatographic separation with a mixture of HNO_3 + CH_3OH, AHIBA solu-

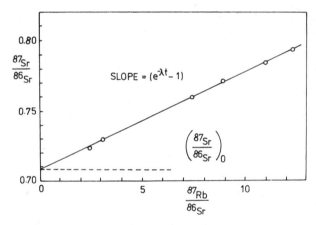

Fig. 3.16 — Rb–Sr evaluation diagram.

tion, etc. is sometimes used. The isotopes ^{87}Rb as well as ^{84}Sr or ^{86}Sr are used as spikes. For calibration of the spectrometer and the spike, the N.B.S. standard materials 998 (RbCl) and 987 ($SrCO_3$) are mostly used; in other cases the values for the isotopic composition of Sr are normalized with respect to $^{86}Sr/^{88}Sr$ ($= 0.1194$). Considerable attention should be paid to the purity of the chemicals used, and the amounts of Sr and Rb contaminating the sample from the chemicals should be $< 10^{-10}$ g. Further details on the application of the method are given in [190].

For most dating methods described, the mother and daughter elements have different chemical and physical properties; this can result in migration of both elements during natural geological changes, and the results of the age determination may then be incorrect. This disadvantage is largely eliminated in the Sm–Nd dating method based on the radioactive decay ^{147}Sm → ^{143}Nd. For isolation of the REE group, ion-exchange resins in HCl or in $CH_3COOH + HNO_3$ mixture are used; the individual REE are separated chromatographically with AHIBA or di(2-ethylhexyl)phosphoric acid. To check the elution of the elements, the radioactive tracers ^{244}Cm, ^{243}Am or ^{154}Eu are used, with corresponding flow-through alpha and gamma detectors at the outlet of the column; thus calibration of the chromatographic column is not needed. The results of the isotopic composition are normalized, as a rule, to the values of the isotope ratios ^{145}Nd/^{146}Nd = 0.7219, ^{146}Nd/ ^{142}Nd = 0.63615, or ^{148}Sm/^{154}Sm. To evaluate the concentration of the isotopes, ^{143}Nd (^{150}Nd) and ^{147}Sm (^{149}Sm) are used as the spikes. Because of the high demand on accuracy, great attention should be paid to calibration of the spike. The calibration method has been discussed by Wasserburg et al. [191]. The values of the ^{143}Nd/^{144}Nd isotope ratios and ^{147}Sm/^{144}Nd concentration ratios which form the evaluation diagram axes are given with relative errors (2 σ) of < 0.01 % and < 0.1 %, respectively.

REFERENCES

[1] O. Hahn, Z. Phys. Chem. (Leipzig), 1923, **103**, 461.

[2] R. Ehrenberg, Biochem. Z., 1925, **164**, 183.

[3] G. Hevesy and R. Hobbie, Z. Anal. Chem., 1932, **88**, 1.

[4] I. E. Starik, Problemy Sov. Geol., 1933, **3**, 70.

[5] G. Hevesy and E. Hofer, Nature, 1934, **134**, 879.

[6] D. Rittenberg and G. L. Foster, J. Biol. Chem., 1940, **133**, 737.

[7] A. S. Keston, S. Udenfriend and R. K. Cannan, J. Am. Chem. Soc., 1946, **68**, 1390.

[8] F. C. Henriques and C. Margnetti, Ind. Eng. Chem., Anal. Ed., 1946, **18**, 477.

[9] K. Bloch and H. S. Anker, Science, 1948, **107**, 228.

[10] A. S. Keston, S. Udenfriend and M. H. Levy, J. Am. Chem. Soc., 1950, **72**, 748.

[11] E. Bojesen, Scand. J. Clin. Lab. Invest., 1956, **8**, 55.

[12] J. Růžička and P. Beneš, Collection Czech. Chem. Commun., 1961, **26**, 1784.

[13] N. Suzuki, Nippon Kagaku Zasshi, 1959, **80**, 370.

[14] I. E. Zimakov and G. S. Rozhavskii, Zavodsk. Lab., 1958, **24**, 922.

[15] J. Růžička and J. Starý, Talanta, 1961, **8**, 228.

[16] I. Obrusník and A. Adámek, Report ÚJV 1970, 1967, Inst. Nuclear Research, Řež, Czechoslovakia.

[17] T. Braun and L. Ladányi, Acta Chim. Acad. Sci. Hung., 1968, **55**, 361.

[18] J. Tölgyessy, T. Braun and M. Kyrš, Isotope Dilution Analysis, Pergamon Press, Oxford, 1972.

[19] J. Klas, Collection Czech. Chem. Commun., 1966, **31**, 3119.

[20] D. A. Lambie, Analyst, 1959, **84**, 173.

[21] V. I. Shamaev, Zh. Analit. Khim., 1967, **22**, 988.

[22] E. Broda, Radioactive Isotope in der Biochemie, F. Deutsche, Vienna, 1958.

[23] M. D. Kamen, Radioactive Tracers in Biology, Academic Press, New York, 1947.

[24] D. M. Grodzinski: *Metodika primeneniya radioaktivnykh izotopov v biologii*, Izd. UASKHN, Kiev, 1962.

[25] J. Růžička and J. Starý, *Substoichiometry in Radiochemical Analysis*, Pergamon Press, Oxford, 1968.

[26] V. Kriváň, *Angew. Chem.*, 1979, **91**, 132.

[27] T. T. Gorsuch, *Radioactive Isotope Dilution Analysis*, RCC Review No. 2, Revised Ed., Feb. 1967, UKAEA Radiochemical Centre, Amersham.

[28] J. Brandštetr, I. Zvára, T. S. Zvárová, V. Knobloch, M. Křivánek, J. Malý and Su Chun-Guj, *Radiokhimiya*, 1964, **6**, 47.

[29] J. Krtil and M. Bezděk, *Talanta*, 1969, **16**, 1423.

[30] F. M. Bathie and B. A. Burden, *Analyst*, 1968, **93**, 1.

[31] K. Kudo and K. B. Kobayashi, *J. Radioanal. Chem.*, 1979, **53**, 163.

[32] R. S. Large and L. P. O'Connor, *Anal. Chim. Acta*, 1968, **40**, 123.

[33] L. A. Currie, G. M. France and H. L. Steinberg, *Int. J. Appl. Radiat. Isotopes*, 1963, **16**, 1.

[34] E. L. Geiger, *Health Physics*, 1967, **13**, 607.

[35] N. Yamagata, *Nature*, 1963, **200**, 157.

[36] J. W. Morgan and J. F. Lovering, *Anal. Chim. Acta*, 1963, **28**, 405.

[37] D. Monnier, W. Haerdi and A. Rouèche, *Anal. Chim. Acta*, 1964, **31**, 413.

[38] M. Curie, *J. Chim. Phys.*, 1930, **27**, 1.

[39] P. W. Urnezis and R. B. Holtzman, in R. E. Rowland and A. F. Stehney, *Argonne Nat. Lab. Annual Rept.* ANL-80-115 (1981), Part 2, p. 97.

[40] H. Renault, *Bull. Soc. Chim. France*, 1963, 657.

[41] R. B. Hahn and R. O. Backer, *Nucleonics*, 1956, **14**, 90.

[42] J. Brandštetr, V. V. Volkov, V. A. Ermakov, T. S. Zvárová, M. Křivánek, J. Malý and Su Chun-Guj, *Radiokhimiya*, 1963, **5**, 706.

[43] M. Kay, A. S. Jones and R. Smart, *Br. J. Nutr.*, 1966, **20**, 439.

[44] E. Broda and T. Schönfeld, *Radiochemische Methoden der Mikrochemie*, Springer-Verlag, Vienna, 1955.

[45] W. H. Clingman, Jr. and H. H. Hammen, *Anal. Chem.*, 1960, **32**, 323.

[46] C. Rosenblum, *Anal. Chem.*, 1957, **29**, 1740.

[47] H. Maier-Hueser, *Bull. Inform. Sci. Tech.* (*Paris*), 1970, No. 147, 37.

[48] J. C. Hardouin, *Bull. Inform. Sci. Tech.* (*Paris*), 1970, No. 170, 45.

[49] M. Kyrš, *Metod osnovannyi na izotopnom razbavlenii proizvodnogo soedineniya*, in M. Kyrš and G. N. Bilimovich (eds.), *Novye metody radioanaliticheskoi khimii*, Energoizdat, Moscow, 1982.

[50] A. S. Keston, S. Udenfriend and M. Levy, *J. Am. Chem. Soc.*, 1950, **72**, 748.

[51] F. Bayard, A. Kowarski and V. V. Weldon, *J. Lab. Clin. Med.*, 1970, **75**, 347.

[52] H. G. Löppert, *Radiochem. Radioanal. Lett.*, 1970, **5**, 209.

[53] H. Veen, *Int. J. Appl. Radiat. Isotopes*, 1974, **25**, 355.

[54] B. Hoffmann, R. Claus and H. Karg, *Acta Endocrinol.*, 1970, **64**, 377.

[55] E. Gard, *Mikrochim. Acta*, 1970, 983.

[56] W. Riess, *Anal. Chim. Acta*, 1974, **68**, 363.

[57] R. Svendsen, *Acta Endocrinol.*, 1960, **35**, 161.

[58] E. I. Gruenstein and A. L. Pollard, *Anal. Biochem.*, 1976, **76**, 452.

[59] R. H. Gray and D. M. Steffensen, *Anal. Biochem.*, 1968, **24**, 44.

[60] P. Spear and B. Roizman, *Anal. Biochem.*, 1968, **26**, 197.

[61] P. T. Davies and J. H. Deterding, *Int. J. Appl. Radiat. Isotopes*, 1972, **23**, 293.

[62] H. G. Burger, J. R. Kent and A. E. Kellie, *J. Clin. Endocrinol.*, 1964, **24**, 432.

[63] A. M. Shefner, R. Ehrlich and H. C. Ehrmantraut, *Int. J. Appl. Radiat. Isotopes*, 1957, **2**, 91.

[64] R. L. Boeckx, J. Protti and K. Dakshinamurti, *Anal. Biochem.*, 1973, **53**, 491.

[65] B. Kliman, *Advan. Tracer Methodology*, 1968, **4**, 227.

[66] G. Ayrey, *Chem. Rev.*, 1963, **63**, 645.

[67] K. Kudo, H. Araki, M. Fujimoto and Y. Sato, *Radioisotopes*, 1967, **16**, 57.

[68] G. N. Bilimovich and N. N. Churkina, *J. Radioanal. Chem.*, 1971, **8**, 53.

[69] G. B. Briscoe and S. Humphries, *Talanta*, 1971, **18**, 39.

[70] K. Kudo and N. Suzuki, *Trends Anal. Chem.*, 1984, **3**, 20.

[71] S. M. Grashchenko and E. V. Sobotovich, *Radiokhimicheskie metody opredeleniya mikroelementov*, Nauka, Moscow, 1965, p. 75.

[72] N. Suzuki, S. Nakamura and H. Imura, *J. Radioanal. Chem.*, 1980, **57**, 37.

[73] I. E. Zimakov, *Trudy Vsesoyuz. Nauch.-Issl. Inst. Veterin. Sanit.*, 1973, **45**, 204.

[74] A. V. Pomerantseva, I. E. Zimakov and V. I. Spitsyn, *Zh. Analit. Khim.*, 1973, **28**, 1984.

[75] T. Kambara, J. Suzuki, H. Yoshioka and Y. Ugai, *J. Radioanal. Chem.*, 1980, **59**, 315.

[76] G. Röbisch, W. Schober and P. Bauer, *J. Radioanal. Chem.*, 1982, **75**, 63.

[77] G. Röbisch, W. Schober and R. Dietel, *J. Radioanal. Chem.*, 1983, **79**, 77.

[78] E. I. Grosheva, Yu. A. Bankovskii, O. E. Veveris and A. P. Sturis, *Izv. Akad. Nauk Latv. SSR*, 1980, No. 7, 117.

[79] N. Suzuki and K. Kudo, *Anal. Chim. Acta*, 1965, **32**, 456.

[80] K. Kudo, *Radioisotopes (Tokyo)*, 1966, **15**, 209.

[81] K. Kudo, *Radioisotopes (Tokyo)*, 1967, **16**, 199.

[82] K. Kudo, K. Yamanoto and S. Kuriyama, *Radioisotopes (Tokyo)*, 1967, **16**, 514.

[83] A. V. Stepanov and A. D. Gedeonov, *Zh. Analit. Khim.*, 1976, **31**, 2133.

[84] V. V. Atrashkevich, I. E. Sidoruk and G. N. Bilimovich, *Zavodsk. Lab.*, 1978, **44**, 977.

[85] J. M. Lo, J. C. Wei and S. J. Yeh, *Anal. Chim. Acta*, 1977, **93**, 301.

[86] K. Kudo and N. Suzuki, *J. Radioanal. Chem.*, 1980, **59**, 605.

[87] K. Shikano, K. Kudo and K. Kobayashi, *J. Radioanal. Chem.*, 1982, **74**, 73.

[88] R. A. Nadkarni, *J. Radioanal. Chem.*, 1974, **20**, 139.

[89] P. Aruscavage, *Anal. Chim. Acta*, 1976, **82**, 343.

[90] E. Gundersen and E. Steinnes, *Talanta*, 1976, **23**, 44.

[91] J. Starý, in *Nové směry v analytické chemii*, Vol. I, J. Zýka (ed.), SNTL, Prague, 1983.

[92] V. V. Atrashkevich, G. N. Bilimovich and I. P. Alimarin, *J. Anal. Chem. USSR*, 1975, **30**, 1765.

[93] L. D. Akol'zina, G. N. Bilimovich, I. P. Alimarin and N. N. Churkina, *Zh. Analit. Khim.*, 1975, **30**, 1743.

[94] G. N. Bilimovich and J. Starý, *Zh. Analit. Khim.*, 1979, **34**, 996.

[95] L. D. Akol'zina, I. P. Alimarin, G. N. Bilimovich and N. N. Churkina, *J. Radioanal. Chem.*, 1980, **57**, 279.

[96] C. B. Dissanayake, *Geochem. J.*, 1976, **10**, 71.

[97] A. V. Stepanov and A. D. Gedeonov, *Zh. Analit. Khim.*, 1973, **28**, 2302.

[98] A. Adámek and M. N. Chiriatev, *J. Radional. Chem.*, 1975, **26**, 51.

[99] D. Gorenc and L. Kosta, *J. Radioanal. Chem.*, 1978, **44**, 279.

[100] K. Kobayashi and K. Kudo, *J. Radioanal. Chem.*, 1978, **46**, 265.

[101] V. Jedináková and J. Sladkovská, *Isotopenpraxis*, 1980, **16**, 200.

[102] D. Cornell, E. H. Klehr and J. Tölgyessy, *Radiochem. Radioanal. Lett.*, 1978, **38**, 391.

[103] H. Chermette, J. F. Colonat, H. Montes and J. Tousset, *Anal. Chim. Acta*, 1977, **88**, 339.

[104] Yu Yu-fu and Tang Jing-guan, *Radiochem. Radioanal. Lett.*, 1982, **51**, 103.

[105] N. V. Bogdanov and V. I. Shamaev, *Soviet Radiochem.*, 1977, **19**, 226.

[106] M. L. Reddy, A. S. Reddy, P. C. Reddy and B. Rangamannar, *Radiochem. Radioanal. Lett.*, 1983, **57**, 227.

[107] K. Ashok Rao and B. Rangamannar, *Radiochem. Radioanal. Lett.*, 1982, **54**, 325.

[108] K. Kudo, private communication.

[109] M. Yuzawa and N. Suzuki, *J. Radioanal. Chem.*, 1981, **62**, 115.

[110] Y. Kanda and N. Suzuki, *J. Radioanal. Chem.*, 1979, **54**, 7.

[111] H. Imura and N. Suzuki, *Talanta*, 1981, **28**, 73.

[112] E. Hoentsch and J. Flachowsky, *Isotopenpraxis*, 1977, **13**, 423.

[113] V. I. Shamaev, Yu. P. Korchagin and T. V. Chudinovskikh, *J. Anal. Chem. USSR*, 1978, **33**, 223.

[114] T. Shigematsu and K. Kudo, *J. Radioanal. Chem.*, 1981, **67**, 25.

[115] M. Jimenez Reyes, A. G. Maddock, G. Duplatre and J. J. Schleiffer, *J. Inorg. Nucl. Chem.*, 1979, **41**, 1365.

[116] K. Přikrylová, E. Makrlík and M. Kyrš, *J. Radioanal. Nuclear Chem., Articles*, 1986, **97**, 13.

[117] K. A. Rao and B. Rangamannar, *Radiochem. Radioanal. Lett.*, 1983, **59**, 293.

[118] K. Shikano and K. Kudo, *J. Radioanal. Chem.*, 1983, **78**, 71.

[119] I. V. Golubtsov and V. P. Ignatov, *Zh. Analit. Khim.*, 1975, **30**, 1170.

[120] V. I. Golubtsov and V. P. Ignatov, *J. Anal. Chem. USSR*, 1978, **33**, 405.

[121] T. Kambara, J. Suzuki, H. Yoshioka and Y. Shibata, *Bunseki Kagaku*, 1979, **28**, 231.

[122] T. Kambara and H. Yoshioka, *Chem. Lett.*, 1978, 1225.

[123] T. Kambara, J. Suzuki, H. Yoshioka and T. Nakamura, *Chem. Lett.*, 1975, 927.

[124] T. Kambara, J. Suzuki, H. Yoshioka and T. Nakamura, *Radioisotopes (Tokyo)*, 1975, **24**, 756.

[125] T. Kambara, J. Suzuki, H. Yoshioka and V. Nakajima, *J. Radioanal. Chem.*, 1978, **43**, 139.

[126] K. A. Rao and B. Rangamannar, *Radiochem. Radioanal. Lett.*, 1983, **59**, 285.

[127] H. Akaiwa, H. Kawamoto and K. Ogura, *Talanta*, 1977, **24**, 394.

[128] J. Starý, M. Kyrš and M. Marhol, *Separation Methods in Radiochemistry*, Academia, Prague, 1974.

[129] S. Pošta and F. Kukula, *Radioisotopy*, 1976, **17**, 559.

[130] G. A. Perezhogin and L. P. Gavrilova, *J. Anal. Chem. USSR*, 1975, **20**, 608.

[131] T. Shigematsu and K. Kudo, *Radioisotopes (Tokyo)*, 1976, **25**, 641.

[132] Yu. V. Yakovlev, O. V. Stepanets, V. P. Kolotov and B. Ya. Spivakov, *J. Radioanal. Chem.*, 1975, **28**, 85.

[133] A. Carmichael and J. E. Whitley, *Analyst*, 1970, **95**, 393.

[134] J. W. Mitchell and R. Ganges, *Anal. Chem.*, 1974, **46**, 503.

[135] G. N. Bilimovich, L. D. Akol'zina and I. P. Alimarin, *Zh. Analit. Khim.*, 1973, **28**, 2417.

[136] M. Katoh and K. Kudo, *J. Radioanal. Chem.*, 1983, **79**, 23.

[137] M. Kyrš and J. Hálová, *J. Radioanal. Chem.*, 1983, **78**, 29.

[138] B. C. Haldar and B. M. Tejam, *J. Radioanal. Chem.*, 1976, **33**, 23.

[139] H. Jaskólska, *J. Radioanal. Chem.*, 1975, **25**, 89.

[140] A. D. Sawant and B. C. Haldar, *Radiochem. Radioanal. Lett.*, 1978, **33**, 323.

[141] A. D. Sawant and B. C. Haldar, *Radiochem. Radioanal. Lett.*, 1979, **39**, 373.

[142] G. Pfrepper and N. A. Malyshev, *Proc. 3. Tagung, Nucleare Analysenverfahren*, 11

April 1983, Dresden, p. 246.

[143] H. Wertenbach and A. von Baeckmann, *Symposium Nuclear, Radio- and Radiation Chemistry*, 3 October 1966, Jülich, *Angew. Chem.*, 1967, 399.

[144] H. Lefèvre, G. Metzger and M. Le Nagard, *Radiochem. Radioanal. Lett.*, 1975, **23**, 155.

[145] G. N. Bilimovich, in *Novye metody radioanaliticheskoi khimii*, M. Kyrš and G. N. Bilimovich (eds.), Energoizdat, Moscow, 1982, p. 159.

[146] I. Obrusník and A. Adámek, *Talanta*, 1968, **15**, 433.

[147] T. Braun, L. Ladányi and M. Marothy, *J. Radioanal. Chem.*, 1971, **8**, 263.

[148] H. L. Polak, *J. Radioanal. Chem.*, 1971, **9**, 241.

[149] H. L. Polak, H. Groot and E. E. Challa, *J. Radioanal. Chem.*, 1973, **13**, 319.

[150] G. N. Bilimovich, in *Novye metody radioanalitcheskoi khimii*, M. Kyrš and G. N. Bilimovich (eds.), Energoizdat, Moscow, 1982, p. 150.

[151] N. Muralidhar, B. Rangamannar and V. R. Krishnan, *Radiochem. Radioanal. Lett.*, 1982, **55**, 221.

[152] J. H. Fremlin, J. L. Hardwick and J. Suthers, *Nature*, 1957, **180**, 1179.

[153] V. I. Shamaev, *Anal. Chim. Acta*, 1979, **104**, 327.

[154] V. I. Shamaev, *Anal. Chim. Acta*, 1979, **106**, 333.

[155] V. I. Shamaev, *Zh. Analit. Khim.*, 1980, **35**, 885.

[156] V. I. Shamaev, *Zh. Analit. Khim.*, 1985, **40**, 2145.

[157] I. E. Zimakov and L. L. Vinokurova, *Trudy Vsesoyuz. Nauch.-Issled. Inst. Veterin. Sanit.*, 1974, **49**, 206.

[158] Z. Maksimovič and M. Kyrš, in *Solvent Extraction Research*, A. S. Kertes and Y. Marcus (eds.), Wiley, New York, 1969, p. 131.

[159] M. Kyrš and R. Kandiyotti, *J. Radioanal. Chem.*, 1973, **13**, 181.

[160] N. V. Bogdanov, T. V. Chudinovskikh and V. I. Shamaev, *Zh. Analit. Khim.*, 1976, **31**, 1163.

[161] M. Kyrš and K. Přikrylová, *Chem. Listy*, 1981, **75**, 586.

[162] I. A. Perezhogin, *Zh. Analit. Khim.*, 1980, **35**, 1012.

[163] P. C. A. Ooms, U. A. Th. Brinkman and H. A. Das, *J. Radioanal. Chem.*, 1978, **46**, 255.

[164] J. Klas, J. Tölgyessy and E. H. Klehr, *Radiochem. Radioanal. Lett.*, 1974, **18**, 83.

[165] J. Lesný, J. Tölgyessy and E. H. Klehr, *Radiochem. Radioanal. Lett.*, 1977, **28**, 77.

[166] M. Kyrš and K. Přikrylová, *J. Radioanal. Nucl. Chem.*, 1984, **81**, 227.

[167] J. Lesný, J. Tölgyessy, O. Rohoň, J. Štefanec, J. Klas and M. P. Chacharkar, *Radiochem. Radioanal. Lett.*, 1980, **42**, 9.

[168] J. Lesný, J. Tölgyessy, O. Rohoň and Z. Zacharová, *Radiochem. Radioanal. Lett.*, 1981, **47**, 293.

[169] J. Tölgyessy, J. Lesný and J. Klas, *Radiochem. Radioanal. Lett.*, 1974, **20**, 159.

[170] V. R. S. Rao, Ch. Pulla Rao and G. Tataiah, *Radiochem. Radioanal. Lett.*, 1977, **28**, 43, 77.

[171] J. Klas, J. Tölgyessy and J. Lesný, *Sub-superekvivalentová izotopová zrieďovacia analýza*, Veda, Bratislava, 1985.

[172] Z. Šmejkal and Z. Teplá, *J. Radioanal. Chem.*, 1983, **77**, 49.

[173] V. P. Guinn and H. R. Lukens, Jr., in *Trace Analysis, Physical Methods*, G. H. Morrison (ed.), Interscience, New York, 1965.

[174] I. P. Alimarin and G. N. Bilimovich, *Khim. Nauka i Prom.*, 1956, No. 1, 74.

[175] S. J. Bresler, *Radioaktivnye elementy (Radioactive Elements)*, Gos. Izd. Tekh. Teor. Lit., Moscow, 1957.

[176] W. D. Ralph, T. R. Sweet and I. Mencis, *Anal. Chem.*, 1962, **34**, 92.

[177] W. S. Lyon, *Anal. Chem.*, 1982, **54**, 227R.
[178] F. Kulcsár, D. Teherani and H. Altmann, *J. Radioanal. Chem.*, 1981, **63**, 217.
[179] L. M. Mikheeva and N. B. Mikheev, *Radioaktivnye izotopy v analiticheskoi khimii*, Gosatomizdat, Moscow, 1961.
[180] J. A. Marinsky and J. Krasner, in *Radioisotopes in the Physical Sciences and Industry*, IAEA, Vienna, 1962, p. 503.
[181] E. J. DeWitt, C. T. Lester and G. A. Ropp, *J. Am. Chem. Soc.*, 1956, **78**, 2101.
[182] R. D. H. Heard, L. Thompson and C. H. Yates, *Rev. Can. Biol.*, 1951, **11**, 66.
[183] C. G. Clayton, R. Spackman and A. M. Ball, *Symposium Radioisotope Tracers in Industry and Geophysics*, Prague, 21 Nov. 1966, SM-84/39.
[184] D. J. Bornhop and S. E. Manahan, *Anal. Lett.*, 1980, **13**, 1041.
[185] A. Unger and S. Poler, *Holztechnologie*, 1983, **24**, 2.
[186] L. A. Lenskii and V. V. Rachinskii, *Zh. Fiz. Khim.*, 1982, **56**, 2096.
[187] J. Cabicar, *Stabilní izotopy*, Academia, Prague, 1983.
[188] I. Björkhem, O. Lantto, L. Svensson, E. Äkerlöf and R. Blomstrand, *Clin. Chem.*, 1981, **27**, 733.
[189] H. Sorantin, *Determination of Uranium and Plutonium in Nuclear Fuels*, Verlag Chemie, Weinheim, 1975.
[190] G. Faure, *Principles of Isotope Geology*, Wiley, New York, 1977.
[191] G. J. Wasserburg, S. B. Jacobsen, D. J. De Paolo, M. T. McCulloch and T. Wen, *Geochim. Cosmochim. Acta*, 1981, **45**, 2311.

SELECTED BIBLIOGRAPHY

A. J. Ahearn (ed.), *Trace Analysis by Mass Spectrometry*, Academic Press, New York, 1972.
E. W. Blautt, *Dynamische Massenspektrometer*, F. Vieweg & Sohn, Braunschweig, 1965.
B. R. Doe, *Lead Isotopes*, Springer-Verlag, New York, 1970.
G. Faure and J. L. Powell, *Strontium Isotope Geology*, Springer-Verlag, Berlin, 1972.
E. V. Sobotovich, *Isotopy svintsa v geokhimii i kosmokhimii*, Atomizdat, Moscow, 1970.
I. E. Starik, *Yadernaya geokhronologiya*, Academy of Science of the USSR, Moscow, 1961.
H. Kienitz, *Massenspektrometrie*, Verlag Chemie, Weinheim, 1968.
R. M. Abermathey, G. M. Matlack and J. E. Rein, in *Analytical Methods in the Nuclear Fuel Cycle*, IAEA, Vienna, 1972, p. 513.
I. L. Barnes, K. M. Sappenfield and W. R. Shields, in *Proc. Recent Development in Mass Spectrometry*, Koreichi Ogata and Terno Hayakawa (eds.), University Park Press, Baltimore, 1970, p. 862.
J. R. Beckman, S. F. Marsh, R. M. Abermathey and J. E. Rein, *Int. J. Appl. Radiat. Isotopes*, 1984, **35**, 163.
A. Fudge, in *Nuclear Mass Spectrometry 1978*, J. G. Van Raaphorst (ed.), Netherlands Energy Research Foundation, Petten, 1978, p. 56.
A. J. Fudge, A. J. Wood and M. F. Banham, in *Proc. Analytical Chemistry in Nuclear Reactor Technology*, 10–12 Oct. 1981, Gatlinburg, p. 152, Report TID 7629, 1962.
H. Hinterberger, *Ann. Rev. Nucl. Sci.*, 1962, **12**, 435.
P. J. Hooker, R. K. O'Nions and R. K. Pankhurst, *Chem. Geol.*, 1975, **16**, 189.
H.-J. Knab and H. Hinterberger, *Anal. Chem.*, 1980, **52**, 390; *Geochim. Cosmochim. Acta*, 1981, **45**, 1563.

F. L. Lisman, R. M. Abermathey, W. J. Maeck and J. E. Rein, *Nucl. Sci. Eng.*, 1970, **42,** 191.

S. F. Marsh, R. M. Abermathey, R. J. Beckman, R. K. Zeigler and J. E. Rein, *Int. J. Appl. Radiat. Isotopes,* 1978, **29,** 509.

L. J. Moore, E. F. Heald and J. J. Filliben, in *Advances in Mass Spectrometry,* N. R. Daly (ed.), Vol. 7A, Heyden, London, 1978, p. 448.

D. A. Papanastassiou, D. J. De Paolo and G. J. Wasserburg, *Proc. Lunar Sci. Conf. 8th,* 1977, 1639.

D. A. Papanastassiou and G. J. Wasserburg, *Earth Planet. Sci. Lett.,* 1972, **17,** 52; 1973, **19,** 324.

W. E. Pereira, R. E. Summons, T. C. Rindfleisch and A. M. Duffield, *Clin. Chim. Acta,* 1974, **51,** 109.

P. Rickard, N. Shimizu and C. J. Allegre, *Earth Planet. Sci. Lett.,* 1976, **31,** 269.

D. H. Smith, H. S. McKnown, W. H. Christie, R. L. Walker and J. A. Carter, *Report ORNL/TM-5485,* 1976.

M. F. Thirlwall, *Chem. Geol.,* 1982, **35,** 155.

M. Wantschik, L. Koch and B. Gasner, *Int. J. Mass Spectrom. Ion Phys.* 1983, **48,** 405.

4

Radio-reagent methods

4.1 INTRODUCTION, PRINCIPLES, CLASSIFICATION AND USES

4.1.1 History of the development of the radioactive reagents method

The paper by Hevesy and Paneth published in 1913 [1] is considered to be the basis of all methods using labelled compounds. It was based on an unsuccessful attempt by Hevesy a year earlier, in Manchester, to separate RaD from inactive lead. The fact that this work contributed to the award of the Nobel prize 30 years later shows how much it was ahead of its time [2].

The papers by Ehrenberg [3] in the 1920s represent a starting point for the present method of radioactive reagents, and reflect all the variety and flexibility of the method. However, the method was not highly developed at that time. The accuracy of the analysis was lower than that of classical chemical determinations, and the existence of only one available radionuclide, ^{212}Pb, led to a rather complicated and cumbersome variant of analysis in many cases.

The discovery of artificial radioactivity (1934), leading eventually to increasing number and availability of various radionuclides (1946), resulted in an upsurge of interest in, and the development of, the method of radioactive reagents [4].

Methods based on the release of radioactive components were developed mainly in the sixties (see Section 4.5). They are characterized by a strong orientation towards solving practical problems and by a tendency to automation.

The radioactive reagent method received another stimulus from the papers by Starý *et al.* [5] on homogeneous isotope exchange in the organic phase. These procedures exhibit high selectivity, sensitivity and speed.

Concentration-dependent distribution and similar approaches have been developed, e.g. in the papers by Kyrš, DeVoe, Klas, Das, Zimakov, Alimarin, Grashchenko, Sobotovich, Shamaev, Tölgyessy, Klehr, Bilimovich and others [6]. This direction is probably most useful in the determination of anions and cations exhibiting little tendency for complex formation.

The first application of radioactive reagents in the separation of volatile chelates by a gas chromatographic method appeared in 1975 [7].

4.1.2 Definition of the radioactive reagents method

A meaningful definition should reflect the differences between the radio-reagent method and indicator analysis in addition to setting the boundary with isotope dilution analysis.

As stated in Section 2.3, in indicator analysis (sometimes called the method of labelled atoms) an appropriate isotope of the analyte is added before the chemical operations are started. After mixing, it is practically impossible to separate radioactive and inactive ions, molecules and atoms of the same chemical composition, so the amount of substance in every part of the system can be determined at any time by measurement of the radioactivity. The conditions for the application of this method are that

(a) all the analyte is in reactive form at the start of the analysis;
(b) the proportionality constant does not change during the chemical operations.

The main feature of the *isotope dilution method* is that complete isolation of the analyte is not necessary (see p. 44). Isotope dilution can be called the method of corrections. Another characteristic of isotope dilution analysis is its use of a change in specific activity.

The *radio-reagent method* is based on use of a radioactive species in a quantitative reaction, and measurement of the change in activity of that species in the course of the reaction. The radioactive species may be a labelled reagent, the analyte, or a substance able to give an exchange reaction with some compound of the analyte.

Consequently, characteristic features of this method are as follows:

(a) the determination is based on a well-known chemical reaction;
(b) the fraction of the labelled substance consumed in the reaction passes into a second phase;
(c) this part is determined by the decrease, increase or appearance of radioactivity in the selected phase;
(d) calculation of the analyte concentration is based on a knowledge of the corresponding factor relating radioactivity to concentration or mass (g, g/l., mole, mole/l., etc.).

For methods based on this principle, the term *radiometric analysis* or *radiometric microanalysis* has been used [8]. This term now seems too general, and therefore we use (in agreement with earlier authors [9, 10]) the term *radio-reagent methods* or *analysis with the use of radioactive reagents*. However, for reasons given above, this term cannot be considered as ideal, either. A more exact term would be *analysis based on a chemical reaction, the degree of which is determined by means of radioactive substances*, which is somewhat cumbersome.

In some cases distinguishing between the isotope dilution analysis and radio-reagent methods appears problematic. These intermediate procedures involve the addition of a radioactive isotope of the element to be determined, and the absence of independent analytical operations for determination (weighing, spectrophotometry, etc.). Thus, for example, Mazière [11] stated in 1972, "the boundary between radiometric analysis and isotope dilution is rather vague in certain types of analyses of organic substances". The similarity of the methods is mentioned in the review by McMillan [12]. Tousset considers the relationship of the radioactive substance to the analyte as the most important criterion [13].

A similar point of view has been stated by Crespi [14]. He has divided the applications of labelled substances into two large groups, namely,

(a) isotope methods (isotope dilution and its variants, as well as methods of isotope exchange), and
(b) non-isotopic methods, or methods where radioactive reagents are used (methods using two radionuclides, methods of radioactive component release, radiometric titration).

A similar opinion has been presented in a short paper by Lyon [15] dealing with radioimmunological analysis: "as a matter of fact, this is only the isotope dilution applied to biological systems ... the procedure differs slightly from that of the standard technique of isotope dilution. The difference is caused only by a biological character of the system".

However, in our opinion, the differentiation between the methods of isotope dilution and the methods of radioactive reagents made here is justified, even when overlap may arise (see Section 4.1.3.3).

4.1.3 Classifications of radio-reagent methods

4.1.3.1 The basic principles of classification
and a tentative approach

The classification of individual types of radio-reagent methods is relatively difficult, as shown by Soaman's book, where the section on miscellaneous methods is about as long as the part on classified methods [16]. We consider that determinations using radio-reagents are based on numerous principles, and therefore it is difficult to find some general rules.

In our opinion, the abundance of attempts to classify radio-reagent methods arises from the use of non-commensurable criteria in the classifications. We think that a useful contemporary classification would be a compromise between a historical classification and a systematic choice and application of classification criteria.

Our classification is as follows:
(1) Radiometric titrations.
(2) Methods based on release of a radioactive component.
(3) Methods using radioactive reagents.
    (3.1) Methods where the reagent is labelled with a radioisotope.
        (3.1.1) The variant where the reaction proceeds quantitatively (classical variant).
        (3.1.2) The variant where the reaction does not go to completion.
        (3.1.3) The variant where either the reagent or the analyte is present in large excess (analysis without saturation).
    (3.2) Methods where the radioactive substance is able to exchange with the substance to be determined.
        (3.2.1) The variant where the reaction proceeds quantitatively (classical variant).
        (3.2.2) The variant where the exchange reaction does not go to completion.
        (3.2.3) Analysis with the use of excess reagent.
    (3.3) Methods where the labelled substance is the same as the substance to be determined.
        (3.3.1) The variant where the reaction proceeds quantitatively.

(3.3.2) The classical variant of the method of concentration-dependent distribution.

(3.3.3) Methods based on isotope exchange.

(3.4) Methods using two different radionuclides.

(3.4.1) Methods where the ratio of the radioactivities of the two radionuclides in the product is measured (see p. 57).

(3.4.2) Shamaev's radioisotope stoichiometric method of analysis [17] (see Section 4.5).

(3.5) Methods is which two elements (or their ratio) are determined by use of one radionuclide.

Radiometric titration usually involves use of radioactive reagents, to find the equivalence point, but there are variants that do not use a radio-reagent method; instead they are based on a change in the intensity of radiation caused by its absorption or scattering in a medium containing the analyte (see Chapter 5).

The group of methods based on *release of a radioactive component* is dealt with in Section 4.5. The analyte reacts with a labelled solid or liquid substance to release the radioactive component into a phase in which it has a higher degree of freedom of particle movement, i.e. from the solid to the liquid or gaseous phase, or from a liquid to a gaseous phase. In our opinion, such methods can be characterized by the fact that the release of the radioactive component is not accompanied by an equivalent movement of the inactive substance to be determined into the initial radioactive phase. This fact is the basis for the difference between the methods of radioactive component release and some methods based on exchange reactions (3.2.1).

Let us give examples. The determination of cyanide based on the reaction

$$2\,CN^- + {}^{203}Hg(IO_3)_2 \downarrow \quad \rightarrow \quad {}^{203}Hg(CN)_2 + 2\,IO_3^-$$

can be considered as a radio-release method (^{203}Hg passes from the precipitate into the liquid phase). On the other hand, the determination of fluorine according to the reaction

$$F_2 \uparrow + 2\,KI^* \quad \rightarrow \quad I_2^* \uparrow + 2\,KF$$

should be considered as an exchange method (I_2 is released but F_2 goes from the gaseous to the solid phase).

A third large group is formed by all the other methods which cannot be included in groups 1 or 2. No special explanations are necessary, as the subdivisions are self-explanatory. However, in order to subdivide it, the criteria applied take account of the type of chemical interaction, and relationship between the substance to be determined and radioactive reagent.

Several types of interaction can occur. (i) The radioactive substance is a typical reagent which is able to react with the compound to be determined [methods (3.1)]; (ii) interaction is based on a 'non-equivalent' competition [methods (3.2)]; (iii) interaction is based on an 'equivalent' competition, i.e., the labelled substance is chemically identical with that to be determined [methods (3.3)]. The difference depends on which of the substances involved is radioactive.

For example, if the precipitation of barium or radium sulphate is used for the determination, the following variants may occur: in methods (3.1), the inactive barium is determined by precipitation with a surplus of radioactive sulphate (or inactive sulphate is determined with surplus of labelled barium); in methods (3.2), the amount of barium is determined by precipitation with an insufficient amount of sulphate in the presence of radium acting as an indicator. The higher the barium content the lower the fraction of barium precipitated and, correspondingly, of co-precipitated radium; in methods (3.3), a constant, but insufficient, amount of sulphate can be used for the determination of barium. The indicator of completeness of precipitation is radioactive barium. The similarity of the methods (3.2) and (3.3), as shown in our example, demonstrates that methods (3.3) are an integral part of the methods of radioactive reagents.

The group of methods (3.4) can be considered as a synthesis of methods (3.1) and (3.3). The subgroup (3.4.1) can also be envisaged as a combination of the methods of isotope dilution analysis and radioactive reagents (see p. 168).

Category (3.5) covers a rather small number of methods where two substances are determined by means of one radionuclide. In the majority of cases, one of the substances is identical with the radioactive substance, and the second is able to exchange with it. It is again some kind of combination of the two simple methods mentioned above, (3.2) and (3.3). Two substances can be determined by means of one radionuclide provided that the weighed amount of the sample is constant, or an additional analytical operation is introduced. The analysis of zinc–cadmium alloy with a radioactive zinc isotope, described by Shamaev and Martynov [18] will be used as an example. In the presence of iodide, zinc is extracted by dithizone much more efficiently than cadmium. If the extraction is carried out from solutions where the sum of the zinc and cadmium concentrations is constant and if the amount of added dithizone is lower than that necessary for complete reaction with zinc, then the distribution ratio of zinc between the phases (determined from the distribution of radioactive zinc) decreases with increasing ratio of zinc to cadmium. The dependence of the zinc distribution ratio on the ratio mentioned above (calibration curve) can be used for the determination of zinc in cadmium (1–25 %), provided that the sample weight and the concentrations of iodide and extraction agent are kept constant. Naturally, there also exist methods for a simultaneous determination of two substances by means of two corresponding isotopes.

We shall now consider the criteria which split the groups (3.1), (3.2) and (3.3) into special subgroups. They relate to the quantitative or non-quantitative nature of the basic chemical reaction, i.e. the question of quantitative consumption of one of the reacting components. This way of viewing the problem was suggested by Kyrš [19] in connection with the proposed idea of 'concentration-dependent distribution'. As any method of radioactive reagents depends upon the use of a certain chemical reaction, it can easily be understood that the question of a quantitative or partial completion of this reaction is of great importance. When the reaction proceeds to full consumption of one of the components, the determination obviously does not depend, within certain limits, on changes in temperature, pH, reaction time, pressure, etc. Conversely, all the factors mentioned above should be kept strictly constant in order not to diminish the reproducibility and accuracy of the results in the event of a partial reaction.

This criterion can be used (in typical cases) for differentiation between the substoichiometric variant of analysis by isotope dilution, and the method of radioactive reagents. If the given reaction can be carried out quantitatively at the given concentration level, it is usually possible to work out the determination of the given element by the substoichiometric variant of the method of isotope dilution.

Applying this criterion to the groups of methods (3.1), (3.2) and (3.3), we obtain the subgroups (3.1.1), (3.2.1) and (3.3.1), which correspond to quantitative reaction.

The subgroups (3.1.2), (3.2.2) and (3.3.2) include the cases where the reaction proceeds only partially, mostly because the corresponding equilibrium constants are rather low. Such methods of determination usually require the use of calibration curves. If the reaction does not proceed quantitatively and the concentrations of both components differ (or can differ) by several orders of magnitude, such methods of determination are ascribed to subgroup (3.1.3).

The determination of chloride by using a radioactive AgCl precipitate [20] can be given as an example. A solution containing an unknown concentration of chloride is shaken with an excess of radioactive AgCl and the radioactivity of silver is measured in the equilibrium solution. The higher the chloride concentration the lower the concentration of silver in the solution (owing to the constancy of the solubility product of AgCl). The Cl⁻ ions do not react with the AgCl precipitate, they only suppress its dissolution. The amount of chloride ions in the precipitate is not directly related to the concentration of the Cl⁻ ions in the solution.

As a second example of the method (3.1.3), we mention the use of a calibration curve (expressing the dependence of the distribution coefficient of traces of caesium between the aqueous and organic phase) for the determination of tetraphenylborate ions in an aqueous (or organic) solution under standardized conditions [21]. The portion of the tetraphenylborate reacting with the caesium is negligible and the caesium concentration is, within limits, irrelevant.

Criteria which relate to the degree of completion of the reaction have also been applied in methods (3.2.3) and (3.3.3). A paper by Spitzy [22] can be mentioned as an example of method (3.2.3). Copper is determined by extraction with an excess of dithizone in CCl_4, followed by addition of a certain excess of a radioactive zinc salt (^{65}Zn). As the Cu–dithizone complex is more stable than the zinc complex, the latter is extracted only in an amount which corresponds to the amount of dithizone not bound with copper. In the 'classical' variant (3.2.1), the analysis would be performed by shaking the organic phase (containing only labelled zinc dithizonate) with an aqueous solution containing the copper salt. The copper concentration could be measured by the amount of radioactive zinc transferred into the aqueous phase. It is evident that the difference between these variants is that in the former the analysis involves two successive operations while in the latter a stoichiometric equivalence exists between the amount of dithizone and zinc.

The group of methods (3.3.3) (isotope exchange) can be considered analogously. The determination of silver (in lead), based on shaking the solution to be analysed with an organic phase containing silver diethyldithiocarbamate labelled with ^{110m}Ag, can be mentioned as an example [23]. The larger the amount of silver present in the initial aqueous phase, the greater the radioactivity in this phase after equilibration.

At present, there is, in our opinion, no doubt that isotope exchange analysis is an integral part of the method of radioactive reagents, and differs substantially from isotope dilution analysis. However, different isotopes of the substance to be determined undergo mixing in both methods. The proposal by Langer [24], founder of the isotope-exchange method, that these methods differ only in the exchange mechanism (of ionic or molecular character) seems to be outdated.

4.1.3.2 Other categories of the method of radioactive reagents

The classification of variants within the method of radioactive reagents does not include explicitly all variants described in the literature. Some of these are mentioned below and their connection with the proposed classification is considered.

Radiometric microanalysis. This term was used by Ehrenberg as early as the 1920s [3]. He used the natural radionuclide ^{210}Pb (RaD) for the determination of lead and other substances. His procedures were based on precipitation reactions in various combinations, sometimes rather complicated. Thus, potassium was precipitated with sodium cobaltinitrite, the excess of cobalt reagent was oxidized by permanganate, and the excess of the latter reduced with oxalic acid. The excess of oxalic acid was determined by precipitation with labelled lead. This and similar procedures seem, at present, to be rather cumbersome, but Ehrenberg's method exhibits an appreciable resemblance to the method of isotope exchange. In this, to an unknown amount of lead (x), an amount, m, of labelled lead (activity A_1) is added. Then lead is precipitated with a calculated amount of carbonate, equivalent to the amount m. The activity, A_2, of the liquid phase (after centrifugation) is measured. The unknown amount of lead is calculated by using the equation:

$$x/m = A_2(A_1 - A_2)$$

The methods developed by Ehrenberg belong, according to our classification, to several categories such as (3.1.1), (3.1.2), (3.2.1), (3.3.3). They reflect an unusual degree of inventiveness in overcoming problems connected with the limited choice of radionuclides available at the time.

For reasons mentioned above, the term 'radiometric microanalysis' cannot be recommended for use in the sense of 'radio-reagent methods'. However, the reader can still find this term with that meaning, or in a somewhat modified sense, in the literature [11, 25].

Method of concentration-dependent distribution. The method of concentration-dependent distribution (CDD) mainly includes the variants (3.1.2), (3.1.3), (3.2.2) and (3.3.2). However, some CDD methods can also be found in other categories. The method will be described fully in Section 4.4; hence only a brief characterization is given here.

To define the CDD method, the criterion of the degree of completion of the reaction should be applied. In the event of total consumption of one of the components, the method does not belong to the group of CDD methods. If the values of the corresponding equilibrium or reaction rate constants play an important role in the actual distribution, then the procedure is a CDD method.

Characteristic features of all the CDD methods are set out below.

(1) The analysis is based on the distribution of a radioactive substance between two phases or between several parts of the system used (paper chromatography, electrophoresis).

(2) The ratio of the activities in both phases or parts of the system depends strongly on the concentration of the substance to be determined.

(3) The distribution is governed by the appropriate equilibrium or reaction rate constants.

In addition to the examples mentioned above (chloride determination, p. 161, tetraphenylborate determination, p. 162) we shall give an example of barium determination based on its adsorption on iron(III) hydroxide [26]. It has been found that the distribution coefficient of barium between $Fe(OH)_3$ and the solution (under controlled conditions of phase ratio, pH, concentration of accompanying substances and temperature) depends on the total concentration of barium in the system. If this dependence is experimentally determined, it can be used as a calibration curve for the determination of barium. In this case, a known amount of a suitable barium radionuclide is added to the unknown amount of a barium salt, and the adsorption is carried out under standard conditions. From the measured activity of each phase and by use of the calibration curve, the unknown concentration of barium corresponding to the measured ratio $Ba_{precipitate}/Ba_{solution}$ can be found.

The CDD method can be divided into two subclasses, namely, 'saturation analysis' and 'analysis without saturation'. In the former, the change in the distribution of the radioactive substance results from an increase in the saturation of the reagent by the given substance, i.e. the distribution ratio gradually decreases with decreasing concentration of the free reagent. In the example above, the $Fe(OH)_3$ adsorbent is saturated with barium ions.

In the case of 'analysis without saturation', the cause of the change in the distribution ratio is the shift of the chemical equilibrium as a function of the concentration of the substance to be determined without participation of the phenomenon of saturation. The concentration of the unknown substance can differ from that of the reagent by several orders of magnitude. The determinations of chloride and tetraphenyl-borate mentioned above can be considered as examples of 'analysis without saturation'.

The term 'saturation analysis' appeared in the literature in connection with the determination of biochemically important substances (steroids, hormones, vitamins) and is equivalent, to some extent, to the term 'radioimmunoassay'. Starting from the period 1957–1960, these methods have been developing rapidly, and the number of published papers probably exceeds ten thousand. This number is higher than that of papers dealing with all other variants of the radio-reagent method combined. This field is described in Chapter 6.

Method of radiometric correction. This term, used in a rather broad sense, was introduced by Shamaev [27]. Its characteristic feature is the fact that the amount of added radionuclide can be neglected compared to the amount of the inactive substance. In other words the criterion of the relative amounts of the inactive element and its radioisotope is used for classification purposes. As we do not use this criterion in our classification, it is evident that methods of 'radiometric correction' belong to various categories of our classification [(3.3.1) and (3.3.2)]. Shamaev has suggested that, from the historical standpoint, the method of radiometric correction can be considered to be a variant of the isotope dilution method. However, in our opinion, it is methodologically more correct to consider this method a special radiochemical method of analysis.

We consider that some methods worked out by Shamaev and called methods of 'radiometric correction' indeed belong to the category of isotope dilution methods. However, other methods bearing the same name (where the activity of the portion of the substance not isolated is measured) could, in our opinion, be included in the category of methods of radioactive reagents.

The list of various terms suggested for special variants of radio-chemical methods belonging to categories (3.3.2) or (3.5) of our classification can be extended. Thus Landgrebe *et al.* [26a] introduced the term *substoichiometric radiometric analysis* and latter started to use the term *method of concentration-dependent distribution.* The term *method of calibration curves* has been used by Shamaev *et al.* for similar determinations [27]. Klas *et al.* [28] have introduced the term *method of sub-superequivalent isotope dilution,* which belongs to category (3.3.2). The principle of the method, and its mathematical description, are given in Section 3.6.

Das *et al.* [29] introduced the term *spiking isotope dilution* (isotope dilution with the addition of a trace amount of a radioisotope of the element to be determined to the standard and to the sample). The method assumes linearity of the calibration curve (see also p. 209).

In our opinion, all the methods mentioned here are radioactive reagent methods. However, in the opinion of their originators they are seen mostly as a variety of the isotope dilution method. As a matter of fact, during the modification of some typical procedures of the isotope dilution method, variants have arisen which lie outside the limits of the isotope dilution method, but within the scope of the method of radio-active reagents. The essential features of these developments will be explained in the next section.

4.1.3.3 The boundary between the isotope dilution method and the method of radioactive reagents

The basic differences between the two methods have been explained above and are sufficiently clear. The confusion arises only with regard to subdivisions (3.3.1) and (3.3.2) of our classification and sub-stoichiometric isotope dilution. This question has been fully discussed in the literature [10, 30], and we present here only the principal theses and some examples.

A typical feature of the isotope dilution method, unlike the radio-reagent method, is that the correct analytical result can be obtained without complete isolation of the substance to be determined, before the final measurement. Thus, in order to draw the boundary between these two groups of methods, we propose to use the following criterion. A given method is an *isotope dilution method,* if it is possible to insert (after the operation of mixing the active and inactive isotopes) an additional operation leading to a non-reproducible loss of part of the substance to be determined and, in order to obtain correct results, it is not necessary to change the number and course of all other operations.

A thorough analysis of the literature shows that this simple criterion can easily be applied and that it provides a sufficiently accurate and logical distinction between the two methods.

The determination of calcium in calcium–strontium mixtures by the 'method of radiometric correction with substoichiometric isolation' can be taken as an example. The analytical procedure consists of the following operations:

(a) dissolution of the sample;
(b) labelling with ^{45}Ca of known activity, A_N;
(c) addition of an insufficient amount (g) of oxalate for complete precipitation, followed by filtration;
(d) measurement of the activity of the filtrate, A_F;

(e) calculation of the unknown amount of calcium, m_{Ca}, according to
the formula

$$m_{Ca}/kg = A_N/(A_N - A_F)$$

where k is the conversion factor.

Does this procedure belong to the isotope dilution method or to the
method of radioactive reagents?

If, for example, part of the calcium is 'lost' between operations (b)
and (c), it is evident that the value of A_F will be lower and the result will
be lower too. Thus, this procedure belongs to the method of radioactive
reagents and not to isotope dilution.

The second example is concerned with the same analytical problem.
However, operations (d) and (e) are replaced by (d'), measurement of
the activity of the precipitate, $A_{prec.}$ and (e'), calculation of the unknown
amount of calcium, m_{Ca}, according to the formula,

$$m_{Ca}/kg = A_N/A_{prec.}$$

If part of the calcium is 'lost' between operations (b) and (c), the
activity of the precipitate does not change and the result is correct
provided that all the oxalate added has reacted quantitatively. Thus, it
is possible to insert [after operation (b)] a step for removal of strontium,
in which a certain loss of calcium is allowable. This procedure is
therefore an isotope dilution method.

Thus, the characteristic feature of the isotope dilution method in
such procedures is the measurement of the activity of that part of the
substance to be determined which is bound with a substoichiometric
amount of reagent. If we measure the unbound part of the analyte, we
lose the essential advantage of the isotope dilution method and are
using the method of radioactive reagents.

We can quote Shamaev et al. [31] in a paper dealing with radiometric
determination of small amounts of cadmium in copper by extraction
with a substoichiometric amount of dithizone: "As the stage of the
copper isolation (where some losses of cadmium are possible) precedes
the substoichiometric determination of cadmium, it is necessary to base
the correction of the results on the activity of the isolated part (organic
phase)."

The second feature which, as a rule, distinguishes the isotope dilution method from the method of radioactive reagents is the principle of
substoichiometry, i.e. the isolation of practically equal quantities of
substances from solutions of different concentrations. If this principle
is used and the isolated part is measured, the method can be ascribed
(in most cases) to the isotope dilution class. If the principle of substoichiometry is not strictly met, the procedure becomes a method of

radioactive reagents. This is quite clear if it is taken into account that the irreproducible 'loss' of analyte leads to a decrease of its total concentration in the solution. The effect of this decrease on the result of the analysis will not be eliminated (when the principle of substoichiometry is not used) by using the dependence of the amount of the isolated substance on its concentration in the system at the moment of isolation.

It follows that the isotope dilution method can be combined with any of the variants of the method of radioactive reagents, especially (3.3.2). A relatively small number of methods of this type have been published. Such an analysis includes the following operations:

(a) dilution of the analyte with a known amount of a radioisotope of known activity;
(b) separation of a pure compound of the analyte (losses are permitted);
(c) radiometric determination of the yield following the purification operations;
(d) measurement of the absolute amount of the analyte, after purification, by one of the methods from group (3.3.2) without the addition of a further quantity of the radioisotope.

In other words, this is an application of the method of concentration-dependent distribution to the purified preparation (at lower concentration, owing to losses). To calculate the initial concentration, a correction is made by use of the same radionuclide as in determination by the method of concentration-dependent distribution.

4.1.3.4 Alternative approaches to the classification

Menon [32] introduced the term *radiometric analysis* for procedures in which the assay of a non-radioactive substance in a given system is based on the use of radioactive tracers. Radiometric analysis is classified by him as follows:

(a) radioisotope dilution;
(b) radiometric titration;
(c) radio-reagent methodology.

It is obvious that Menon's scheme is very close to ours. The slight difference arises from our treatment of radiometric titration as a subdivision of radio-reagent methods and not independent of such methods, and our omission of the general concept of radiometric analysis as a synonym for both radioisotope dilution and radio-reagent methods.

An interesting independent approach to the classification of radio-reagent methods was proposed by Shamaev [33] who reconciles the term *radio-reagent method* with the group of methods in which the radioactive substance is, in some of its variants, not the reagent, but a radioactive form of the substance to be determined. This is done by introduction of the term *radioindicator method*. His classification of this category could be summarized as follows:

(1) methods of addition of a known quantity of the test element;
(2) double isotopic label method;
(3) method of isolation with different quantities of reagent;
(4) method of isolation from multiple volumes of a solution to be analysed;
(5) method of isolation with two different reagents;
(6) calculation method;
   (6.1) substoichiometric variant of radiometric correction;
      (6.1.1) Suzuki and Kudo method [34];
(7) calibration curve method;
   (7.1) interpolation method;
   (7.2) comparative method;
(8) method of Pomerantseva *et al.* [35];
   (8.1) method of Grashchenko and Sobotovich [36];
      (8.1.1) successive-approximation method;
(9) concentration-dependent distribution;
(10) graphical calculation method;
(11) analytical calculation method;
(12) method of radioactive reagents;
   (12.1) radio-release methods;
(13) method of isotope exchange;
(14) radiometric titration;
(15) radiometric correction.

It can be seen that there are many common features between our approach and that of Shamaev. The main difference is that Shamaev classifies many of the categories as parallel (independent) variants whereas we prefer to define some of the categories in a broader sense so that others form only a subdivision of them, e.g. according to our concept, categories (7), (10), (11) and others, above, would belong to (9). The other differences result from our regard for a consistent application of selected criteria, paying less attention to the names given by various authors to their own procedures. On the other hand, Shamaev is concerned with a graphical representation of the mutual relationships between various modifications. His original classification is basically a pictorial one.

In the interest of brevity, Shamaev's classification cannot be discussed here in detail, but it serves to give the reader an idea of the richness

of the modifications and ramifications proposed by various authors in
this field of radioanalytical chemistry. Moreover, any serious future
attempt to propose a new classification cannot disregard this scheme.

4.1.4 General principles and characteristic features
of methods using radioactive reagents

4.1.4.1 Requirements for the radioactive reagent
and general recommendations

The reagent

The radioactive reagent should have an appropriate half-life and energy
of radiation. In the majority of cases, it is difficult to work with
radionuclides with half-lives shorter than 1 hr (see Tables 4.1 and 4.2).
The isotope ^{18}F ($t_{1/2} = 1.87$ hr) can be mentioned as an example. Only
laboratories which are close to a reactor or cyclotron can work with this
isotope. In addition, the reactions involved might proceed too slowly
compared to the decay rate. For example, special conditions should be
found in order to ensure a sufficiently rapid reaction for the determination of boric acid

$$H_3BO_3 + 4\,HF \quad \rightarrow \quad HBF_4 + 3\,H_2O$$

On the other hand, Tables 4.1 and 4.2 indicate that a rather long
half-life (e.g. 1000 yr) leads to a low specific activity and, consequently,
to lower sensitivity of the method. Carbon-14 belongs to this kind of
nuclide; its specific activity is insufficient because of the long half-life
(5370 yr). Thus, for example, for the determination of testosterone by
double labelling, the usual variant where testosterone is mixed with
labelled testosterone (^{14}C) and then acetylated with acetic anhydride
containing ^{3}H, has been replaced by a reverse variant (^{3}H-testosterone,
^{14}C-anhydride). The specific activity of labelled testosterone can be
much higher in the second case. However, the new variant has not
found wide application, owing to the high cost of acetic anhydride
labelled with ^{14}C.

Table 4.1 — Detection limits and specific activities of commonly used radionuclides (data
from [37]).

Nuclide	Specific activity (Bq/mg)	Detection limits	
		Bq	mg
^{131}I	4.80×10^{10}	4	8×10^{-12}
^{32}P	1.11×10^{13}	4	10^{-12}
^{35}S	1.48×10^{12}	40	10^{-11}
^{14}C	1.85×10^{8}	40	2×10^{-7}

Table 4.2 — Approximate detection limits of radionuclides as a function of their half-lives (data from [37]).

Half-life	Detection limit* (mole)
1 hr	2×10^{-20}
1 d	8×10^{-19}
1 month	2×10^{-17}
1 yr	2×10^{-16}
10^3 yr	3×10^{-13}

* Radioactivity measurement with a counter.

Low radiation energy leads to the necessity for using special methods for radioactivity measurement. (^{3}H, ^{14}C and ^{35}S have maximum beta-particle energies of 0.018, 0.158 and 0.167 MeV, in that order.) However, it can decrease exposure of personnel. Thus, for example, in determining fatty acids by the method based on their conversion into soaps, the traditionally used ^{60}Co (1.17 and 1.33 MeV, gamma emitter) has been replaced by ^{63}Ni (0.067 MeV, beta emitter).

In the case of some elements such as aluminium, boron, zirconium, silicon and titanium, no suitable radioisotopes exist.

When the method of radioactive reagents with two radionuclides is used, an easily detectable difference in the radiation from the two isotopes is desirable. For example, the measurement of a mixture of ^{3}H and ^{115m}Cd or of ^{14}C and ^{65}Zn radionuclides is a serious problem because of the complex decay schemes of the cadmium and zinc isotopes.

The radioactive reagent should be firmly bound to the substance to be determined. Furthermore, its excess should be easy to remove. Thus, for example, a relatively high solubility product or a low extraction constant can seriously hinder the use of the corresponding method.

Troubles connected with the removal of reagent excess very often occur, and vigorous efforts are sometimes needed to overcome them. If, for example, the radioactive reagent is used for location of a substance on a paper chromatogram, it should not react with the active groups of the paper. In determination of chloride and bromide with ^{203}Hg(NO$_3$)$_2$ by the method based on extraction of HgCl$_2$ or HgBr$_2$, difficulty is caused by an insufficiently effective separation of the mercury excess. (The distribution coefficients are $D_{Hg^{2+}} = 0.07$; $D_{HgCl_2} = 0.562$; $D_{HgBr_2} = 6.8$.)

One way of overcoming such difficulties is to use only a limited excess of the radioactive reagent. In such cases, it is useful to know, for the blank, the dependence of the activity in the corresponding phase on the reagent concentration.

Difficulties connected with the removal of excess of reagent generally appear to be an important factor, in many cases limiting the sensitivity of the method. In such situations increasing specific activity of the radioactive reagent cannot improve the sensitivity.

In some cases the concentration of the radioactive reagent in its active form has to be decreased by the addition of an auxiliary substance which forms a complex with the reagent. The reason for this rather unusual approach is that some other substances can react with the reagent if its concentration is relatively high, and thus may interfere with the determination.

An example is the determination of iodide [38] by the reaction $I^- + C_6H_5Hg^+ \rightleftharpoons C_6H_5HgI_{org}$. The equilibrium constant of this reaction is so high that the distribution ratio of iodide between benzene and water reaches a value of 300 when the equilibrium concentration of the phenylmercury cation is only $5 \times 10^{-11}M$. The separation factor for the iodide–bromide pair is 10^3. The concentration of the reagent has to be somewhat higher than that of the iodide. It is evident that in order to determine, say, $10^{-8}M\,I^-$, the required concentration of phenylmercury would greatly enhance the distribution ratios of both iodide and bromide (e.g. $D_I = 6 \times 10^4$ and $D_{Br} = 60$), and the determination would not be selective for iodide. In the presence of $3 \times 10^{-4}M$ thiourea, which reacts with phenylmercury, the degree of extraction of iodide is $\sim 97\%$, and that of bromide $\sim 3\%$.

An additional basic requirement for a radioactive reagent is that it should not undergo side-reactions. It should react with the substance to be determined in only the expected manner. For example, it is impossible to determine the number of —COOH groups in some materials by using radioactive silver ions, because the silver salt is reduced by its interaction with the material.

The radioactive reagent, naturally, should also fulfil other obvious requirements such as stability during storage, availability and, if possible, low cost.

The primary requirement of high specific activity is connected with the characteristic feature of the method of radioactive reagents, the wide calibration range. In gravimetry, polarography, spectrophotometry and other methods, the calibration slope is generally a constant, controlled by such factors as the molecular weight of the precipitating agent, diffusion current of the corresponding oxidation-reduction reaction at a given concentration, molar absorptivity of the complex in question, etc. These factors control the sensitivity of the method. In the method of radioactive reagents, it is possible, for example, to precipitate barium with a solution of radioactive sulphate of various specific activities (Bq/g) so that unit mass of barium can correspond to different counting rates of the labelled precipitate.

A radiochemical analogue of a constant of the molar absorptivity type is the *radiostoichiometric factor* or the *radioactivity equivalent* given by

$$F = (nM_r/M_x)\,S$$

where M_r is the molecular weight of the radioactive reagent, M_x the molecular weight of the substance to be determined, n the number of moles of the radioactive reagent interacting with one mole of the substance to be measured, and S the specific activity (cpm per mole).

The factor F relates the initial amount of the substance (x) to be determined and the radioactivity of the product (A) through the relation $xF = A$.

A favourable situation, with n as high as 12, can be illustrated by the reaction

$$AsO_4^{3-} + 12\,MoO_4^{2-} + 27\,H^+ \quad \rightarrow \quad H_3AsMo_{12}O_{40(org)} + 12\,H_2O$$

used as the basis of an extraction determination of arsenic with radio-molybdenum [39].

One of the tests used to ensure that the radio-reagent methods work is to investigate the influence of the value of S for the reagent on the net activity of the product (A). A fully satisfactory test should yield a constant ratio A/S ($\pm$ 10 %) for a range of S values over at least one order of magnitude. The generally valid rule that high specific activities of the reagent ensure high sensitivities of determinations can be violated under special circumstances, such as fluctuation of the blank, or considerable solubility of the precipitate. Thus, the working solution of the radioactive reagent is sometimes prepared by a dilution of the initial reagent with an inactive isotope, or the unlabelled reagent.

The determination of sulphur in oil by precipitation as $BaSO_4$ labelled with ^{133}Ba is an example [40]. As the sensitivity of this method is limited by fluctuations in the blank, it is reasonable to diminish the specific activity of the commercial barium reagent by a 350-fold dilution with inactive $BaCl_2$.

In some cases, high specific activity may lead to radiation decomposition of the substance to be determined or of the reagent during storage, as well as to undesirable side-reactions. The determination of the number of olefin bonds in organic substances by bromination with ^{82}Br can be mentioned as an example. At a specific activity of 8 Ci/g, the expected addition of bromine occurs as desired; at a specific activity of about 950 Ci/g, the substitution reaction with bromine also occurs, initiated by free radicals formed by radiolysis. Thus, there exists an upper limit to the specific activity of bromine which can be used.

As far as autoradiolysis of radio-reagents is concerned, most authors believe that the presence of radioactive atoms in the molecule of the radio-reagent usually does not diminish the stability of the reagent, unless its specific activity is excessively high. For example, a ^{36}Cl specific activity of $0.4\,\mu$Ci per g of chloramine-T does not influence the stability of the reagent.

In the case of organic compounds containing several atoms of the label element, both homogeneous labelling (the same specific activity in all positions) and labelling only in one selected position can be envisaged. Radioactive reagents should be labelled in positions which do not undergo undesirable isotope exchange. However, it is sometimes difficult to check commercial reagents in that respect. For example, homogeneous distribution of the labelled carbon atoms, even in such a simple molecule as acetic acid, can be ensured only by a rather complicated series of chemical reactions.

To obtain information on the position of the ^{2}H, ^{3}H and ^{13}C nuclides, nuclear magnetic resonance spectroscopy has been used in recent years. This method gives results in a much shorter time than chemical methods do. For example, it has been proved that in the molecule of DL-(4-^{3}H) propranolol (21 H atoms)

$$\text{O-CH}_2\text{-}\overset{\overset{\displaystyle \text{OH}}{|}}{\text{CH}}\text{-CH}_2\text{-NH-CH(CH}_3)_2$$

only position 4 is occupied by tritium (one peak in the NMR spectrum at $\delta = 7.54$).

The radionuclide of the radioactive reagent should either decay to stable (non-radioactive) products or its daughter product should be of sufficiently short half-time for radioactive equilibrium to be established in the decay chain. Otherwise, measures should be taken to prevent an undesirable effect, caused by the shift of the radioactive equilibrium between the mother and daughter (^{90}Sr $\rightarrow$ ^{90}Y; ^{140}Ba $\rightarrow$ ^{140}La), on the results of the radioactivity measurements. The relatively rare application of uranium as a radioactive reagent can be explained by this fact. Actually, the method of radioactive reagents is rarely used for determinations with radioactive substance giving rise to daughter radionuclides.

An important requirement is the radiochemical and chemical purity of the radioactive reagent. The analytical chemist should have a full description of the reagent as well as certificate data for a commercial reagent. Knowledge of these data facilitates dealing with possible radiochemical contamination of the material.

The methods of preparation of most radioactive reagents were worked out and tested long ago, so their use has few complications, since the sources of contamination have been discovered already and removed. Problems usually arise in the case of labelled compounds prepared by new (unusual) methods. The considerable attention paid to methods of control of radiochemical and chemical purity of labelled compounds has led to a situation in which radioactive reagents are often chemically purer than ordinary inactive reagents. Special attention should be paid to the chemical purity of radioactive reagents if a possible inhibition of enzymatic reactions is suspected or when undesirable physicochemical processes are likely to occur. Certain inactive admixtures can also be the cause of an incorrect determination of the specific activity of the corresponding compound.

General recommendations

If it is necessary to decrease the specific activity of radioactive preparations, they should be diluted immediately after delivery. This lowers the risks resulting from radiation decomposition and of losses caused by adsorption on the walls of vessels. Favourable conditions for storing the radioactive reagent should also be secured.

When purifying radioactive reagents, it is desirable to use the same reactions as in the corresponding methods of chemical analysis. The procedure is, in most instances, simplified in this way and, generally, the removal of the most seriously interfering species is accomplished.

When working with extremely low concentrations of radioactive substances, it is necessary to take into account any special behaviour of the substances under these conditions (see Beneš and Majer [41]).

In highly diluted solutions, surface effects can seriously influence the chemical behaviour of substances. The fraction of substance adsorbed by the walls of the vessel can increase greatly with decreasing concentration. Thus, the choice of vessel materials and their pretreatment (purification, deactivation) are of great importance. Glass vessels are usually suitable for storage of diluted solutions of organic substances and inorganic ions at low pH values.

In the case of microquantities of $^{32}PO_4^{3-}$ its adsorption may strongly increase when the pH of the solution is changed. To prevent such adsorption, a known amount of inactive carrier of the active element or substance should be added in advance, provided that it is compatible with subsequent operations.

When using the method of radioactive reagents, it is necessary to have complete mixing of the radioactive and corresponding inactive forms of the substance in order to prevent undesirable errors. A suitable method should be chosen for individual cases (see also Section 2.3). For example, the equilibrium between inactive niobium and ^{95}Nb in aqueous solution is attained (under certain conditions) only after several days. Thus, the ^{95}Nb radionuclide should be added to the crucible before fusing niobium pentoxide and other sample constituents with $K_2S_2O_7$, or to the hydrofluoric acid (when used for dissolution), where isotope exchange is rapid [42]. To obtain complete mixing of the ^{207}Bi isotope with inactive bismuth, it is necessary to boil the solution in the presence of H_2SO_4. In dealing with organic substances, complete mixing can be hindered by inert bonding of the substance with a component of the mixture. Weak bonding, such as the hydrogen bond between thyroxine and various components of plasma, does not interfere with the exchange.

In the method of radioactive reagents the blank plays an extraordinarily important role, owing to partial passage of the excess of radioactive reagent into the reaction product phase, partial reaction of the radioactive reagent with other substances, and other processes. It is always necessary to take into account the blank correction. The relation between the activity of the product phase (A) and the amount of the element (x) to be determined is expressed by the formula $A = Fx + a$, where a is the correction for the blank and F the radiostoichiometric factor. However, the formula $x = (A - A_0)/F$ can also be used, where A_0 is the activity obtained in the blank experiment. The correction formula cannot be applied without a thorough investigation, however, as it is necessary to know which relationship is the more suitable for the problem in hand. The necessary correction should not depend on those quantitative characteristics of the experiment which do not remain constant (the degree of isolation of the substance, volume of the initial sample). The blank experiment should, as precisely as possible, correspond to the real experiment. Sometimes even small deviations from the methodology of the experiment can change the true value of the blank.

If the analyte undergoes a sufficiently complicated purification scheme, it is desirable to use for the blank experiment the final product of the *blank purification,* where any contamination from the reagents can be concentrated. When determining small quantities of nickel by paper chromatography, the zone with the invisible nickel spot is immersed in an ammonium sulphide solution labelled with radioactive sulphur. For the blank experiment, a zone without nickel is immersed in the same solution to determine the adsorption of sulphide.

The value corresponding to the blank, and its variations, often determine the overall sensitivity of the method of radioactive reagents.

In addition to the blank, it is sometimes recommended to perform a parallel experiment with the aim of verifying the value of F. Some authors recommend using the method of standard additions. In this case, the effect of temperature is eliminated, correction for radioactive decay is not necessary, and the effect of some species present in the initial solution (for example, a kinetically inactive form of the reagent) can be eliminated.

It is usually useful to take the mean value of several replicate determinations for calculation of the amount of the substance. This is also necessitated by the statistical nature of radioactive decay, by variations in the preparation of samples for radioactivity measurement, and by possible losses of the substance or an accidental interference in the case of the analysis of substances of low concentration.

It should always be borne in mind that application of radioisotopes is meaningful only when the given analysis cannot be done more simply and cheaply by another method. Thus, for example, chromate can be determined by measurement of its absorbance at 364 nm instead of by radiometric measurement as chromate labelled with $^{31}CrO_4^{2-}$ [43].

In conclusion, let us note that for a useful radio-reagent procedure it is not absolutely necessary to know exactly the composition of the product, or to ensure quantitative consumption of the substance to be determined.

As an example note the determination of submicrogram amounts of mercury by the displacement of ^{65}Zn from an organic solution of $\sim 2 \times 10^{-5}M$ zinc dithizonate $[Zn(HDz)_2]$ in CCl_4 [44]. It is believed that most of the mercury dithizonate formed is the primary chelate $Hg(HDz)_2$, but the secondary chelate $HgDz$ (in formation of which, two mercury ions displace only one zinc ion) can also be present in significant amounts. This is in accord with the fact that the net activity released is in the range 68–80% of the activity expected for 100% chemical yield. Although the cause of this phenomenon has not been unambiguously explained, the method gave good precision and accuracy for the determination of mercury in air or water samples.

4.1.4.2 Characteristic features and possibilities
of the method

The reaction of $Et_2NS(S^*)S^*-S^*(S^*)CNEt_2$, tetraethylthiuram disulphide, abbreviated to $(ET^*)_2$, with substances containing thiol groups is an interesting example in which the number of components determined can be increased or decreased, depending on the choice of reaction product. A symmetrical disulphide is formed according to reaction (I) with low molecular weight thiols; a mixed disulphide is formed according to reaction (II) with proteins containing thiol groups.

$$2\,RSH + (ET^*)_2 = RSSR + 2\,(ET^*) + 2\,H^* \tag{I}$$

$$RSH + (ET^*)_2 = RS(ET^*) + (ET^*)^- + H^* \tag{II}$$

where $(ET^*)^-$ is radioactively labelled diethyldithiocarbamate. If the activity of the $(ET^*)^-$ is measured, the total concentration of compounds containing the thiol group is determined. If the disulphides are isolated and measured, only proteins are determined.

In some cases, it is reasonable to prepare the radioactive reagent directly in the solution containing the substance to be determined. Thus, for example, radioactive iodine is prepared by the oxidation of iodide with H_2SO_5. A smaller excess of the reagent can be used, removal of the excess of reagent is simplified and the radiation hazard is decreased [45].

In some cases the radioactive reagent can be prepared, even in a relatively complicated way, directly in the laboratory where it is used. For example, radioactive carbon used for the determination of water by the reaction

$$H_2O + {}^{14}C \quad \rightarrow \quad {}^{14}CO + H_2$$

has been obtained by the following sequence of operations [46]:

$$Ba^{14}CO_3 \quad \rightarrow \quad {}^{14}CO_2 \xrightarrow{H_2} {}^{14}CH_4 \xrightarrow{Cl_2} {}^{14}C + (4\,HCl)$$

One of the features of the method of radioactive reagents is the possibility of using various carriers to improve the separation or purification of the reaction product. An example of an isotope carrier is the compound of phenobarbitone with mercury added to a precipitate resulting from the reaction of phenobarbitone with ${}^{203}Hg(ClO_4)_2$ [47].

A non-isotopic carrier is used, for example, in the determination of traces of yttrium with labelled phosphate where an AgCNS precipitate is formed and acts as a collector [48]. Another example is the selective adsorption of PdI_2 (for determining palladium with labelled iodide) on a $Zr(HPO_4)_2$ precipitate. Such procedures are impossible in classical gravimetric methods.

When determining some biochemically active substances, the reaction should be done *in vivo*. Thus, for example, a method for the determination of deoxycorticosterone has been proposed, based on its ability to accelerate the secretion of potassium from the organs of rats. Labelling with ^{42}K was used. Analogously, erythropoietin favours the passage of ^{59}Fe to red blood corpuscles of mice, and this is made the basis for its determination [49].

An interesting application of radioactive reagents to the quantitative determination of a radioactive substance, is the reaction of excess methanol-^3H with fatty acids containing ^{14}C. Milligram quantities of oestrone-6,7-^3H have been determined by the reaction with pipsyl chloride labelled with ^{35}S [50]. Such a method requires the measurement of the radiation from both radionuclides in the same sample.

The method of radioactive reagents is used, sometimes, for the determination of relative quantities of substances. The absolute quantities may remain unknown. The determination of the relative amount of alkyl radicals of the $C_3H_7^{\cdot}$ type can be mentioned as an example. Reaction with $^{131}I_2$ leads to the formation of labelled alkyl iodides, which can be separated by fractional distillation [51]. The relative amounts of various sterols can be determined by conversion into ethers by reaction with iodobenzyl chloride and chromatographic separation. The peak area on the chromatogram is proportional to the molar amounts of individual sterols [52].

A logical step in this direction is the application of the method of radioactive reagents to the identification of substances. The substance to be investigated, labelled with a radionuclide, is mixed with an authentic specimen of the substance it is thought to be, labelled with another radionuclide. Then both substances undergo repeated separation and purification (paper chromatography, counter-current extraction, etc.). If the locations of both radionuclides coincide throughout, then it can be assumed that the substances are identical. This method (using two different radionuclides) should be considered more reliable than traditional radiochromatography where the identity of substances is judged according to the location of peaks on chromatograms of standard and unknown substances.

We remind the reader (see Section 2.3) of the important methodological problem arising from the effect of replacement of one nuclide by another on the chemical reactions of the corresponding substances. Obviously this effect can be, as stated earlier, neglected to a first approximation.

At present, it is not necessary to work as carefully as investigators in 1951 carrying out experiments aimed at finding differences between inactive cobalt and ^{60}Co on elution from paper (inactive cobalt should be eluted more easily) [53]. According to the early papers by Hahn, radionuclides are adsorbed on glass surfaces or on various colloid particles more strongly than are the corresponding inactive species of the same element. This effect later proved to be much smaller than expected. To characterize it quantitatively, for example, the value of the $^{59}Co/^{60}Co$ separation factor on the cation exchanger Dowex-50 X12 is 1.00009 (^{59}Co is concentrated in the resin). For the anion exchanger Dowex-1 X10 (in $4M$ HCl), it is 1.00004. In the case of light isotopes (^1H–^3H), the isotope effect can be larger. For example, if the method of determination includes breaking the C–^3H bond, and the rate of this reaction controls the kinetics of the whole process, the isotope effect can be significant, especially in the case of methods where the reaction proceeds only for a predetermined time.

4.1.4.3 Radioactivity measurement

Generally, all methods of ionizing radiation detection described in handbooks on the application of radioactivity in chemical investigations can also be used in the method of radioactive reagents. We shall present here only some remarks, examples, and data relating to the radio-reagent method.

One possibility is combination of a chemical reaction with the radioactivity measurement. Gorbushina et al. [54] have described an extraction–scintillation method for the determination of radionuclides. Uranium is extracted from aqueous solution with tributyl phosphate in toluene, and the extract is used directly for the liquid-scintillation measurement. A similar idea was published by McDowell and co-workers [55, 56]. The gamma emitters are extracted to remove the interfering components (and components causing a quenching effect) and to transfer the emitter to a medium suitable for scintillation measurement. It can be appreciated that the application of this principle in the method of radioactive reagents could lead, in some cases, to a shortening of the time of the analysis.

Melikhov et al. [57] have proposed a radiometric method for the investigation of adsorption, without a phase separation. One phase (a suspension) is closer to the counter than the second phase (a solution). It is evident that this original idea can be applied to the method of radioactive reagents.

By electrophoresis or chromatography, it is possible to determine traces of substances by densitometry of autoradiograms. At an exposure of 2–24 hr, it is possible to detect 10^{-6}–$10^{-1}\,\mu Ci$. Autoradiography can also be used for the location of spots on a chromatogram before their removal for the radioactivity measurement by another method.

In use of radioactive reagents labelled with ^{14}C, the question arises as to the most efficient method for radioactivity measurement. The substance is very often converted into CO_2 (wet oxidation, oxidation in a Parr bomb with Na_2O_2 and $KClO_4$, or combustion in oxygen). The carbon dioxide can be absorbed in NaOH and converted into $CaCO_3$, which is measured, for example, by a scintillation counter after the addition of hydroxypropylmethyl cellulose to obtain homogeneous dispersion in the sample holder. $BaCO_3$ is less suitable for activity measurement, as its carbon content is only about half that of $CaCO_3$. Carbon dioxide can also be absorbed by organic solvents such as phenethylamine or a mixture of monoethanolamine and the monoethyl ether ethylene glycol, all of which can be used in liquid scintillation counting.

It is also possible to disperse the initial substance labelled with ^{14}C (without oxidation) in a liquid phase in the form of a stable emulsion, or to dissolve it in an appropriate solvent.

In the determination of substances labelled with ^{14}C and ^{3}H, it is possible, after separation by thin-layer chromatography on silica gel, to dissolve the sorbent in hydrofluoric acid.

Cherenkov radiation is used for the measurement of ^{32}P. This provides the possibility of using large volumes of the solution for the radioactivity measurement, preserving high efficiency and non-dependence of the counting rate on pH or the presence of salts.

In the method of radioactive reagents there is usually no need to measure the absolute activity of a substance. To calculate the amount of an unknown substance, only counting rates of two or more samples are necessary. When using two radionuclides, it is almost always necessary to know only the ratio of their activities.

4.1.5 Uses for the analysis of samples
The aim in this section is to illustrate the scope and usefulness of this method. Many examples of practical applications of radio-release methods are given in Section 4.5. Here analyses by other procedures will be illustrated.

A portable apparatus has been developed [58] for the determination of chloride in water (0.001 %) based on the principle of precipitation of chlorides with silver labelled with ^{110m}Ag. Traces of anionic surfactants in river and drinking waters are determined by extraction as ion pairs with ferroin (labelled with ^{59}Fe) into chloroform.

A rapid routine method for the determination of sulphate in soils is based on precipitation with a barium salt labelled with ^{133}Ba [59]. The ^{131}Ba isotope has been used for the determination of sulphur in small samples of animal tissue [60]. The phosphate content of fertilizers and minerals can be determined by precipitation with ammonium molybdate (^{99}Mo). This procedure is rapid, increases the sensitivity, and extends the range of determinable concentrations [61].

Traces of metals (Cu, Pb, Ag) on the surface of stainless steel can be determined by autoradiography after their reaction with radioactive iodine (^{131}I) in the form of amyl iodide [62].

A number of earlier papers [63–65] reported the determination or detection of alkyl radicals, which react selectively with metallic mirrors of radioactive lead and bismuth (^{210}Bi) to form stable compounds of the Pb(CH$_3$)$_4$ type. The extraction of silver (^{110}Ag) dithizonate with chloroform has been used for the determination of the solubility of dithizone in $0.1M$ HClO$_4$ [66].

The method of radioactive reagents is widely used in the field of applied organic chemistry.

The iodine number of fats can be determined by reaction with the excess of iodine bromide labelled with ^{131}I [67]. Fatty acids (for example, oleic acid) can be precipitated with cobalt (^{60}Co) acetate. The excess of cobalt reagent is removed by paper chromatography [53]. Higher fatty acids (C$_{16}$) present in serum can be determined by extraction of their nickel (^{63}Ni) salts into a mixture of chloroform and heptane (4 : 1) [68]. The reduction of sugars, oligosaccharides and, after hydrolysis, glycoproteins with sodium borohydride (NaBH$_4$ labelled with ^{3}H) is a routine, universal method for their determination [69]. The application of compounds labelled with ^{14}C and ^{3}H to determination of fatty acids has been described by Koch and van Tilborg [70].

The carbonyl groups formed at the surface of thin films of polyethylene during oxidation, and which therefore affect its quality, can be determined by the reaction

$$\text{>C=O} + \text{H}_2\text{N-NH}\!-\!\!\left\langle\!\!\bigcirc\!\!\right\rangle\!\!-\!\text{NO}_2 \longrightarrow \text{>C=N-NH}\!-\!\!\left\langle\!\!\bigcirc\!\!\right\rangle\!\!-\!\text{NO}_2 + \text{H}_2\text{O}$$
$$\qquad\qquad\qquad\quad\text{NO}_2 \qquad\qquad\qquad\qquad\quad\text{NO}_2$$

The dinitrophenylhydrazine is labelled with ^{14}C [71].

Hydroxyl groups which are sometimes formed (in low concentrations $\simeq 0.1\%$) in polyethylene, can be determined by reaction with acetic anhydride labelled with ^{14}C [72].

Very low concentrations of carboxylic groups in cellulose can be determined by their reaction with excess of ^{144}Ce(III) [73]. The reaction of thiol groups in proteins with chloromercurybenzoate (^{14}C) is used for the selective and sensitive determination of these groups in microquantities of proteins [74].

The determination of the number of groups of a given type by the method of radioactive reagents is also applied to the study of important characteristics of polymers and elastomers, such as molecular weight and degree of cross-linking. etc.

For example, the number of the C–Li bonds in polybutadiene lithium can be determined by a fast and irreversible reaction

$$—CH_2^-Li^+ + RO^3H \quad \rightarrow \quad —CH_2{}^3H + ROLi$$

where ROH is an aliphatic alcohol [75].

The number of double bonds in the small quantities of polyisobutene present in butyl rubber can be determined by chlorination (^{36}Cl). By use of the method of radioactive reagents, the molecular weight of synthetic elastomers and polymers can be found, provided that the stoichiometry of the reaction is known and the labelled group can be introduced into every molecule of the polymer. For example, in the case of polystyrene, exactly two fragments of the labelled initiator are present in each molecule. The molecular weight can be calculated from the specific activity of polystyrene and that of the initiator [76]. The molecular weight of polyethers is found according to the number of OH-groups at the ends of the chains formed by the reaction with p-toluenesulphonyl chloride labelled with ^{35}S [77]. The molecular weight of butadiene and the degree of cross-linking are determined by means of a tertiary alkylmercaptan (^{35}S). One molecule of alkylmercaptan enters each unit of the elastomer [78].

Radioactive reagents play an important role in the determination of hormones, steroids and other biochemically important substances. For example, the application of radioactive reagents to the determination of steroid hormones has led to solution of a number of problems arising from the insufficient sensitivity of standard chemical methods [79]. Furthermore, the sensitivity of the gas chromatographic method has been increased [80].

4.1.6 Advantages and shortcomings

One of the most important advantages of the method of radioactive reagents is its high sensitivity, which is due, in principle, to the large amount of energy released during radioactive decay of the nucleus. In some instances, even the decay of individual atoms can be detected.

In the case of precipitation reactions, methods have been worked out where practically invisible amounts of a precipitate are determined by measuring the radioactivity. For the determination of gold precipitated as $AuCr(SCN)_4(NH_3)_2.H_2O$ by a traditional gravimetric method the lowest amount of product that is convenient is about 2.5 mg. However, the variant using labelled chromium can determine amounts as low as 25 µg. In determining fatty acids by the reaction with copper or cobalt salts, a detection limit of 50 nmoles has been obtained. By extraction of the radioactive soap containing ^{60}Co, 0.08 nmole of fatty acid can be determined provided that the specific activity is high enough.

Among the factors limiting the sensitivity of the method, besides the specific activity of the radioactive reagent, are variations of the blank, chemical decomposition of the radioactive reagent or of the inactive substance at low concentration (10^{-6}–$10^{-8}M$), low equilibrium constants (solubility of the radioactive reagent in the method of radioactive component release, low extraction constant in the method of concentration-dependent distribution), or adsorption of one of the reactants, on the walls of the vessels.

The method of radioactive reagents is relatively simple and direct. For example, the methods or radioimmunological determination are simpler than the corresponding biochemical methods without radioactive reagents.

Owing to the high energy of the ionizing radiation, the methods can easily be automated (see Vol. II).

When comparing the method of radioactive reagents with activation analysis, it should be noted that the apparatus is much cheaper and simpler: a nuclear reactor, cyclotron or other radiation source is not necessary. The method of radioactive reagents is sometimes more sensitive than activation analysis. This is because it can use radionuclides which are formed in an insufficient amount during the activation, or radio-reagents can be used with radionuclides of higher specific activity than those obtained in activation.

A typical example in which the method of radioactive reagents uses radionuclides other than those formed by activation, and is more sensitive, is the determination of nickel by an exchange reaction, in which nickel replaces radioactive zinc in its complex with EDTA. The zinc released is extracted with dithizone [81].

The advantage of the method of radioactive reagents (especially the automated variant) is particularly evident when traces of an element are determined in materials of complex composition by a routine method. In such circumstances, it is more reasonable to add to the material to be analysed a small quantity of the radionuclide of the element to be determined. It is easier to observe the behaviour of the added radionuclide than that of the same nuclide produced by activation of the sample. In the latter case, radionuclides interfering with the determination can be formed as a result of the nuclear reactions of matrix and other elements. Moreover, the radiation hazard is lower in the first case. An important advantage of the method of radioactive reagents is the possibility of determining an element by a variety of procedures using different radioactive reagents.

A serious disadvantage, compared with activation analysis, is the fact that the method of radioactive reagents sometimes requires a special purification of all reagents used, in order to decrease the blank. The importance of the blank as well as difficulties connected with its control can be demonstrated by the determination of sulphur in the light fraction of oil by burning the sample, oxidizing SO_2 to H_2SO_4, and determining the latter by the method of radioactive reagents. With a 2-ml sample, the blank was $1.9 \pm 0.3\,\mu g$. Of this, 40 % came from the furnace and sealing, 20 % from the hydrogen peroxide used as oxidizing agent, 20 % from the water used, and 20 % from the air. Variations in the blank were caused by accidental contamination of the air in the laboratory [40].

In activation analysis, of course, contamination with the inactive form of the element to be determined, after the activation, does not affect the result of the analysis.

The advantage of the simpler apparatus necessary for the method of radioactive reagents is offset in some cases by the high complexity of the chemical processes used in the analysis.

In comparing the method of radioactive reagents with the corresponding methods not using radionuclides, it should be noted that radionuclides are usually used in order to make the determination more sensitive and rapid.

Compared to emission spectroscopy, the method of radioactive reagents is not a multielement technique. However, there are some radioactive reagents which are able to react with numerous elements. For example, the precipitation of a number of metals with sulphide (^{35}S) possesses certain features of a multielement method. Moreover, the various metals can be determined with a single universal standard. An analogous group reagent for a number of biochemically important substances is acetic anhydride labelled with ^3H or ^{14}C.

An example in which a radio-reagent method suffers less interference than the analogous colorimetric determination (because coloured ions can be tolerated in the former case) is the determination of As(V) with ^{99}Mo and solvent extraction of arsenomolybdate. Moreover, the reduction step needed for the colorimetric method is not required.

A comparison with substoichiometric isotope dilution is given on pp. 166 and 167. Again the main advantage of the method of radioactive reagents is the higher sensitivity and wider variety of available reagents, but the reproducibility and accuracy are usually lower than for substoichiometric IDA. Rudenko [82] notes that "the isotope dilution method is appreciably more valuable" than the method of radioactive reagents. In our opinion, this evaluation is not universally true.

An important advantage common to the methods of radioactive reagents, isotope dilution and destructive activation analysis is the ability to discover and observe the behaviour of accidental losses.

The method of radioactive reagents, especially in some of its variants, is similar (in principle) to several methods of physicochemical analysis where other properties of the substance, for example the colour or fluorescence, are used instead of the radioactivity to generate an analytical signal [83]. A comparison leads to the conclusion that the radioanalytical methods are usually much more sensitive. It is possible to work with the majority of radionuclides (in the quantities necessary for the method of radioactive reagents) in well equipped ordinary chemical laboratories. Removal of the radioactive wastes is not a serious problem; they can generally be washed down the sink without violating the safety regulations, if enough water is used.

The method of radioactive reagents can also provide a concentration effect. For example, in the determination of mercury in litre volumes of water by exchange for silver in silver dibutyldithiocarbamate (in chloroform), the mercury can be concentrated by proper choice of phase–volume ratio.

This eliminates the problem of transporting large samples. A 1-litre sample can be extracted with 20 ml of chloroform solution of the reagent, at the sampling site, and the extract taken back to the laboratory. Furthermore, there is no danger of the loss of mercury due to adsorption or evaporation during prolonged storage.

The reaction

$$Hg(IO_3)_2 + 2\,Cl^- \quad \rightarrow \quad HgCl_2 + 2\,IO_3^-$$

can be used for the determination of Cl^- either by titrating the IO_3^- transferred into the solution, or by measuring the radioactivity of the $HgCl_2$ in the solution if the labelled form of $Hg(IO_3)_2$ is employed. The sensitivity of the former method ($0.003M$ Cl^-) is naturally limited by the blank value due to the solubility of mercuric iodate. If radioactive mercury is used, the solubility of the $Hg(IO_3)_2$ can be decreased by adding excess of iodate. The sensitivity can be increased by two orders of magnitude [84].

An early evaluation of the method of radioactive reagents is contained in a quotation by Meinke, who said in his introductory lecture at the Symposium on Radioanalytical Chemistry (Salzburg) in 1964: "With the exception of activation analysis, insufficient attention has been paid to other methods using radioisotopes, especially in the United States. Until now, the analytical chemists have not understood that using radionuclides, they can work out a new procedure which is often cheaper and faster. At present, there is no enterprise producing special instruments or equipment for radioanalytical methods and which would be interested in the development of radioanalytical systems. These instruments are simple and give great possibilities in the development of new approaches and original principles." Others [11, 85] have stated that the potential of the method of radioactive reagents has not been fully explored. The reason is an unjustified fear of working with radionuclides as well as misinformation on the cost of radioactive reagents and instruments. Furthermore, conservatism and insufficient knowledge of the possibilities have contributed to a relatively low interest in these methods.

The radioanalytical indicator methods (including the method of radioactive reagents) went through a revival in the sixties and the early seventies. Illaszewicz *et al.* [86] considered in 1968 that "during the last years, the radiometric methods of determination (the analysis of inactive substances by the use of radioactive indicators), as an independent technique, have occupied a leading position in radioanalytical chemistry".

The number of papers proposing new variants and approaches is now relatively small, however, and most investigations in this field are directed to solution of practical problems. This is an indication of a coming period of relative maturity for this field of analytical chemistry. Its relatively wide application has also been supported by development of the necessary instruments. At present, we may say that in the field of radioactive reagents, practice has caught up with the previous development of theory and principles of the method.

In microanalysis and analysis of traces of inorganic substances there is much competition between activation analysis, atomic-absorption spectrometry and other methods, including the radio-reagent procedures. It is necessary to select the most appropriate method according to the individual problem.

The existence of such serious universal problems as environmental pollution, over-population and lack of food in a number of countries sets new challenges for the analytical chemist, and we conclude that the limits of useful applications of the radio-reagent methods have by no means been reached.

4.2 THE CLASSICAL VARIANT

This variant (3.1.1 on p. 158) is based primarily on quantitative reaction of inactive analyte with a radioactive reagent. The excess of the reagent is removed from the phase containing the reaction product, the radio-activity of which is then a measure of the initial concentration of the inactive substance.

4.2.1 Calculation of the result, suitable reactions and practical hints

In many cases the test concentration is not calculated from an analytical relationship, but found from a calibration curve such as that shown in Fig. 4.1 [87].

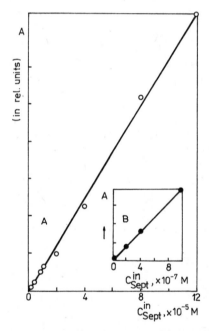

Fig. 4.1 — Calibration graph for the radiometric determination of Septonex by extraction with a chloroform solution of Rose Bengal labelled with [131]I: A — the activity of 1 ml of organic phase, C_{Sept}^{in} — initial molar concentration of Septonex in the aqueous phase, pH 9.91. A — $V_{org} = V_a = 5$ ml, $C_{RB}^{in} = 1.2 \times 10^{-4} M$; B — $V_{org} = 30$ ml, $V_a = 50$ ml, $C_{RB}^{in} = 2.44 \times 10^{-6} M$. Data for the plots obtained from [87].

If calculation is preferred, the following values must be known or determined:

(a) radioactivity of the product;

(b) specific activity of the radio-reagent;

(c) stoichiometric composition of the product;

(d) yield of the reaction.

Under exceptional conditions, the radioactivity of the excess of reagent is measured instead of that of the product. In that case the following values are used to calculate the result:

(a) the radioactivity corresponding to the initial concentration of the radio-reagent;

(b) the radioactivity of the unreacted portion;

(c) the specific activity of the radio-reagent;

(d) the stoichiometric composition of the product.

The formula for calculation of the amount of analyte, W_x mole, is

$$W_x = f(A_p/S_r)\,\Phi$$

where f is the yield of the reaction (for 70 % yield, $f = 0.7$), A_p the radioactivity of the product (cpm), S_r the specific activity of the reagent (cpm per mole, for the same conditions and time of measurement as for determination of A_p), and Φ the stoichiometric ratio of analyte to radio-reagent in the product.

In the method of standard addition the formula is

$$W_x = W_0 A_1/(A_2 - A_1)$$

where W_0 is the standard addition (known amount of substance added) in the same units as W_x, A_1 the activity of the product without the addition (cpm), and A_2 the activity of the product in the sample with the addition (measured under the same conditions as for A_1). (Plots of W_x vs. W_0, of W_x vs. A_2, should be essentially a horizontal line.)

The following properties are considered essential for the chemical reaction used:

(a) the reaction should be quantitative even if only a small excess of the radio-reagent is added;

(b) the kinetics should be favourable (only a few minutes needed to complete the reaction);

(c) the reagent should allow easy labelling (in organic substances the radioactive labelling atom should be covalently bound);

(d) the reaction should be sufficiently selective;

(e) the product should be easily separated from excess reagent;

(f) the blank should be low, with small variation;

(g) the course of the reaction, the composition of the product, and the conditions for a quantitative consumption of the analyte should be well known and

(h) the calibration curve should be linear.

Precipitation is the reaction most often used in this variant. Solvent extraction reactions are also used fairly frequently but other types of reaction only rarely.

The problems connected with the use of precipitation have been expounded by Krtil [88] and need not be repeated here. Typical examples of solvent extraction methods are given in Table 4.3, and details have been given in a recent review by Starý [98]. The reasons for the relatively infrequent use of extractants as radio-reagents are the high extractibility of many of these reagents even in the absence of the substance to be determined, the large excess of reagent necessary for full extraction, and problems associated with the labelled forms of many organic reagents (complicated synthesis of the labelled form, low energies of ^{14}C and ^{3}H, lack of a suitable N radionuclide).

Recent examples of other types of reactions used successfully in this type of analysis are as follows. A comparison has been made [99]

Table 4.3 — Examples of solvent extraction used for radio-reagent determination.

Substance to be determined	Radio-reagent	Solvent extraction	Remark; Reference
Cd	$K^{131}I$ in the presence of ascorbic acid	$Cd(pyridine)_2I_2$ into benzene	$> 10\,\mu g$ of Cd [89]
Bi	$K^{131}I$ (ascorbic acid $+ Na_2SO_3$)	$HBiI_4$ from $0.6M$ H_2SO_4 into butyl acetate	$> 0.04\,\mu g$ of Bi [90]
Na lauryl sulphate (NaLS)	$[Fe(II)(1,10\text{-phenan-}$ throline)$_3$)] labelled with ^{59}Fe	$Fe(phen)_3(LS)_2$ into chloroform	> 0.005 ppm in 20 ml of water [91]
H^+	$K^{131}I$; reaction $6H^+ + 5I^- + IO_3^- = 3I_2 + 3H_2O$	I_2 into pyridine + chloroform (1 : 10)	$> 0.0001M$ H^+ [92]
Ag and Cu in a mixture	Sodium diethyldithiophosphate, labelled with ^{32}P (NaA)	into CCl_4; CuA_2 from pH ~ 6, AgA from pH ~ 10.5	in rocks [93]
As, P	WO_4^{2-}(^{185}W) MoO_4^{2-}(^{99}Mo)	of heteropoly acid; e.g. into 1,2-dichloroethane	e.g. arsenomolybdate [94, 95]
Cl^-	$C_6H_5{}^{203}Hg^+$	$C_6H_5{}^{203}HgCl$ into benzene	~ 0.5 ng/ml [96]
H_3BO_3	$H^{18}F$	HBF_4 into 1,2-dichloroethane in the presence of Methylene Blue	$\sim 0.1\,\mu g$ of B [97]

recently of two methods for the determination of hydroxyl groups in coal, by acetylation with acetic anhydride. One of them used acetic anhydride labelled with ^{14}C; the acetylated coal was burnt at 925 °C and the $^{14}CO_2$ produced was measured with a scintillation counter. The classical method included saponification of the ester, separation of the acetic acid by distillation and titration with alkali. The radio-reagent method is less labour-consuming and more precise (relative standard deviation 3.9 % compared to 7.3 % for the classical method).

An example of an oxidant used as a radio-reagent is chloramine-T labelled with radiochlorine:

$$CH_3C_6H_4SO_2NCl^{*-} + 2\,H_2O + 2e^- \;\rightarrow$$
$$CH_3C_6H_4SO_2NH_2 + 2\,OH^- + Cl^{*-}$$

In this way many reducing substances (e.g. I^-, S^{2-}, sulphite, nitrite, xanthate) can be determined. The excess of radio-reagent is separated from radioactive chloride by first absorbing both substances at pH 8 on a column of Amberlite IRA 401 and subsequently eluting chloride with $0.3M$ $NaNO_3$ [100].

The classical method of radioactive reagents has also been used many times in the analysis of organic substances. Some practical examples were given in Section 4.1.4 and many more are given in the reviews by Krtil [88] and Schütte [101].

Electrons $(10^{-10}M)$ in semiconductors (CdS, ZnO, etc.) can be determined by the reaction $2e^- + 2\,(^3H^+) = {}^3H_2\uparrow$, which occurs after dissolution of the materials in tritiated water [102].

An additional recent example of the use of classical radio-reagent methods for routine analyses in industrial production and checking of chemical products is the proposal [103] to determine $\sim 5 \times 10^{-4}M$ meclophenoxate hydrochloride in aqueous solution by precipitation of AgCl with excess of $^{110}AgNO_3$ (3 mg/ml), with a precision of 0.1–0.5 %.

The following practical hints should prove useful in developing new classical-type radio-reagent procedures.

(1) Check the chemical and radiochemical purity of the stored radio-reagent from time to time.

(2) If possible, eliminate measurement of the radioactivity of precipitates.

(3) Pay special attention to optimizing separation of the reaction product from excess of reagent (lower the solubility of precipitates by adding suitable organic solvents, ensure full precipitation by the use of collectors, in solvent extraction use extractants with the highest possible difference in extractibility of product and reagent).

(4) If the composition of the product is not known or not constant, or

if the consumption of the analyte is not quantitative, pay increased
attention to preparation of the calibration graph.

(5) Use the method of standard addition if foreign substances are
    expected to affect the slope of the calibration graph.

(6) If using the method of standard addition, verify that the result of
    the analysis is independent of the amount of standard added.

4.2.2 Some applications of reactions with low yield

The high sensitivities of radiometric measurements are often in contrast
to the limited values of the equilibrium constants of the reactions
involved (incomplete precipitation or extraction). Many authors have
tried to overcome this problem (e.g. [104]), but, in our opinion, with
only mediocre success.

In recent years this problem has been tackled by Shamaev *et al.*
[105, 106], who concluded that the solubility product is not an adequate
yardstick for determinations based on precipitation, and prefer to use
the equilibrium constant of the hypothetical reaction in which all
species are regarded as in the aqueous phase:

$$S M^{n+} + n R^{S-} \;\rightleftharpoons\; SMR_{n/S} \qquad (K_{pr})$$

This approach can be characterized as a hypothetical dissolution of the
precipitate $M_S R_n$ and concomitant 'depolymerization' of the 'molecule'
$M_S R_n$ into S simpler molecules $MR_{n/S}$. This gives

$$K_{pr} = (C_M - [M^{n+}])^S / [M^{n+}]^S [R^{S-}]^n$$

which for binary $(1:1)$ precipitates leads to

$$C_M = C_R (1 - \alpha_R)(1 - \gamma_R / K_{pr} C_R \alpha_R) \qquad (4.1)$$

where α_R is the fraction of free radio-reagent (radioactivity of the phase
containing the free radio-reagent, divided by the initial total radioactivity), C_M and C_R are the initial molar concentrations of M and R, and
$\gamma_R = C_R / [R^{S-}]$.

Comparison between Eq. (4.1) and equations based on the assumption of complete precipitation or the validity of the solubility product
has been obtained [106] by determining chloride with radioactive silver
(solubility product 1.8×10^{-10}) with separation of the AgCl by dialysis
through a collodion membrane (see Table 4.4), and shows outstanding
superiority of the new approach, but final acceptance should be post-
poned until further corroboration is obtained.

Similar approaches (based on solvent extraction equilibria) have
been proposed by Shamaev [105] for solvent extraction of chelates and
ion-association systems, and experimental verification — as could be

expected in these cases — led to the conclusion that the relationships for incomplete extraction (in accordance with the extraction constants) yield acceptable results at concentration levels where the simple formulae based on full consumption of one component fail.

Table 4.4 — Determination of Cl^- with radioactive Ag (data from [106]).

Chloride taken $(10^{-5}M)$	C_{Cl}/C_{Ag}	$1 - \alpha_{Ag}$	Chloride found $(10^{-5}M)$		
			Assume complete pptn.	Assume validity of solubility product	Assume validity of Eq. (4.1)
7	1.0	0.530	4.240	4.72	6.200
5	1.0	0.370	1.850	2.42	3.900
1	0.5	0.310	0.155	5.40	0.950
0.7	0.5	0.202	0.070	6.47	0.518

$K_{pr} = 5.6 \times 10^4$ [105].

The significance of this basically sound approach should not, however, be overestimated. The use of calibration curves without elaborate calculations (which in any case neglect deviations from ideality of the phases) would probably yield analogous or even more reliable results.

Shamaev [107] has set out five possible approaches to determination of substances with radio-reagents at very low concentrations:
(a) successive approximations;
(b) interpolation;
(c) comparative procedures;
(d) calculation — analytical approach;
(e) calculation — graphical approach.

These procedures were again applied to the determination of chloride with radioactively labelled silver. Shamaev claims [108] that methods using a labelled reagent can generally be used for determining smaller amounts of substances than can methods using radioactive isotopes of the element to be determined. The reason for this is the higher sensitivity of the radio-reagent method. The determination limit for chloride with radioactive silver was found to be two orders of magnitude lower than that by isotope dilution.

4.2.3 Some advantages of the variant
If the product of the reaction is a precipitate the advantages of measuring the radioactivity over weighing are as follows:
(a) radioactivity can be measured for amounts of substance that could be weighed only with very low precision;

(b) it is not necessary to achieve a constant weight of precipitate;

(c) co-precipitation of inactive species is permissible provided they do not consume the reagent;

(d) other precipitates can be used as non-radioactive 'collectors' to ensure full precipitation of small amounts of radioactive precipitates [for AgI a suitable collector is $Fe(OH)_3$, $Zr(HPO_4)_2$ or $Be(OH)_2$];

(e) a non-radioactive form of the same precipitate can be used as an ideal collector if no undesirable isotope exchange occurs, e.g. in the determination of Mg with $^{32}PO_4^{3-}$ the precipitate $MgNH_4PO_4$ can be added;

(f) precipitation directly in the paper after previous chromatographic separation is possible [109, 110];

(g) instead of the radioactivity of the precipitate, that of the supernatant can be measured.

4.3 ANALYSIS BASED ON ISOTOPE EXCHANGE

One possible classification of the use of isotope-exchange reactions in radioanalytical chemistry is set out below.

(1) For separation (isolation):
    (1.1) separation with crystalline precipitates;
    (1.2) separation with amalgams;
    (1.3) separation on ion exchangers;
    (1.4) distribution between two liquid phases;

(2) For determination;
    (2.1) homogeneous isotope exchange;
        (2.1.1) in an aqueous phase;
        (2.1.2) in an organic phase;
    (2.2) heterogeneous isotope exchange;
        (2.2.1) between two immiscible liquid phases;
        (2.2.2) between other phases.

4.3.1 Isotope exchange as a method of separation

The use of isotope exchange for separation is illustrated in Table 4.5. This type of separation has no analogy in classical chemistry and is characterized by the fact that the addition of carrier followed by separation steps normally used in reverse IDA is replaced by an isotope exchange step, in which isotope dilution occurs. The advantage of this separation is its rapidity, simplicity, high selectivity in some cases (see Fig. 4.2) and easy automation. Increased selectivity is provided by the use of performed precipitates. Whereas in the course of precipitation,

Table 4.5 — Examples of the use of isotope exchange for separation.

Radioactive species separated	Isolating substance	Conditions	Remarks and uses; reference
$^{24}Na^+$	NaCl, NaClO$_4$ (solid)	Solution of HCl + acetone	Remove ^{24}Na after irradiation, to eliminate interference in activation analysis [111, 112]
$^{42}K^+$, Cu(I)	(a) KCl; (b) CuCl	(a) Solution of HCl + acetone; (b) Mineral acid	Remove ^{42}K after irradiation of biological samples [112]
I$^-$, Cl$^-$, Ag$^+$	AgI (preformed); AgCl	Other silver halides also studied	Separation from a mixture of fission products [113–116]
Co(II); Zr^{4+}	Co(NH$_4$)PO$_4$; ZrPO$_4$	Acid solution	Application in neutron activation analysis [117]
La(III); Sr(II); Ce(III)	La$_2$(C$_2$O$_4$)$_3$; SrC$_2$O$_4$; Ce$_2$(C$_2$O$_4$)$_3$	Mineral acid solution	Application in neutron activation analysis [117–121]
Sb(III); Mn(IV); Sn(IV)	Sb$_2$O$_3$(200 mg); (a) MnO$_2$; (b) SnO$_2$	Freshly prepared precipitates; exchange takes ~ 10 min (a) 0.1M HNO$_3$; (b) mineral acid	Radiochemical isolation; neutron activation separation [117, 122–125]
Cu(II); Cd(II); Hg(II); Sn(IV); Co(II)	CuS, CuCNS; CdS; HgS; SnS$_2$; CoS	Mineral acid	Application in neutron activation analysis [116, 125]
Ag$^+$	AgCl (electrolytically plated on a metallic net)	Exchange takes < 15 min	Separation of ^{110m}Ag in irradiated Cd, Pd, Si [114]
Sr^{2+}	SrSO$_4$ in a mixture with silica gel	Thin-layer chromatography, Sr does not travel	Separation of ^{90}Sr from ^{90}Y [126]
In, Sr, Pb, Ga, Sn, Zr	Amalgam of the element	Chloride media	Application in neutron-activation analysis [127]
Cd, Pb, Tl, Sr, Zn, Hg, In	Amalgam of the element	Concentration of element in amalgam ≫ concentration in solution	Radiochemical isolation of short-lived isotopes [128–130]

Table 4.5 (continued)

Radioactive species separated	Isolating substance	Conditions	Remarks and uses; reference
Rh	Amalgam	From 0.9M KCNS	Application in neutron-activation analysis [131]
I⁻, Br⁻	Solution of I₂ or Br₂ in CCl₄	Sample in aqueous phase contacted with the organic phase	Activation analysis of natural waters [132]
^{131}I⁻	Liquid scintillator contg. dissolved I₂	From aqueous phase	Direct radioactivity measurement [133]
Ag	Column supporting Ag dithizonate dissolved in tri-n-butyl phosphate	Organic phase fixed on polyurethane foam	Separation from aqueous solution [134]
^{128}I⁻	Resin saturated with solution of I₂ in KI	Solution of ^{128}I filtered through the resin	Isolate I from irradiated biological samples [135]

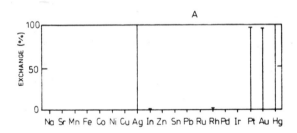

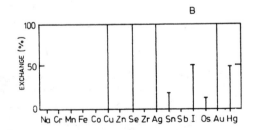

Fig. 4.2 — Selectivity of isotope exchange separation: A — between aqueous 0.1–2M H⁺ (mineral acid) and metallic Hg; B — between aqueous 1M H₂SO₄ and CuS precipitate. Based on data in [136].

foreign elements can be occluded or co-precipitated, during isotope exchange the chance of other elements entering the solid phase is lowered. In other cases the transfer of the radioactive tracer into the second phase is probably facilitated by the presence of weighable amounts of the inactive form in this phase. The prerequisites of a successful application of this type of separation are:

(a) a significantly higher ($\sim$ 100 times) concentration of the inactive form in the second phase than that of the radionuclide to be separated;
(b) sufficiently rapid isotope exchange (an average time required for separation by solvent extraction is 5–10 min, ion-exchange chromatography 10–15 min, precipitation 20–30 min);
(c) the absence of other chemical interaction;
(d) easy preparation for counting samples.

In contrast to the typical use of isotope exchange for determination purposes, full isotope-exchange equilibrium need not be attained. The method is used for the separation of

(a) isotopes in a complicated mixture in activation analysis;
(b) a radionuclide in a mixture with another of closely similar decay energy;
(c) elements of a matrix interfering with the measurement of the desired substance in activation analysis.

4.3.2 Isotope exchange as a method of determination

Recently this type of radioanalytical procedure has been reviewed [98, 136–139]. Consequently, our treatment can concentrate on the most outstanding features of the method. A tentative classification was given in Section 4.3.1.

This type of analysis is based on measurement of the fraction of the radionuclide in one form, the second form being either a different chemical compound or containing the radionuclide in a different oxidation state. If complete isotope exchange is reached, the distribution of masses and radioactivities is the same, i.e. $x : y = A_1 : A_2$ where x and y are the masses of the element in the different forms or phases and A_1 and A_2 are the respective radioactivities (count-rates). From the known value of y and the measured A_1 and A_2 values, the unknown mass x can be calculated. Very often, if $A_1 + A_2 = A$, the relationship can be modified $x = yA_1/(A - A_1)$ and if $y \gg x$ (then $A_2 \gg A_1$ and, of course $A \gg A_1$) we obtain $x = yA_1/A$. In a calibration procedure the values y and A are constant, so a calibration graph in the form $x = f(A_1)$ is linear.

The four requirements defined for isotope exchange as a method of separation (see p. 197) are also valid in this case. The absence of transfer mechanisms other than isotope exchange is imperative.

Table 4.6 — Heterogeneous isotope exchange for the determination of traces.

Species determined	Phase I	Phase II	Remarks (concentration range, time of exchange); reference
As	As(III) in $2M$ HCl	^{74}As(DDC)$_3$ in CCl$_4$	0.1–1.0 µg As [140]
Ag	Ag$^+$ in 0.5M H$_2$SO$_4$	^{110m}AgHDz in CCl$_4$ (dithizonate)	~ 2 ppm in Pb of high purity [141]
Hg	Hg^{2+} in 0.1M H$_2$SO$_4$	^{203}Hg(DDP)$_2$ in CCl$_4$ (dibutyldithiophosphate)	[142]
I	I$^-$ in H$_2$O	^{131}I$_2$ in benzene	~ 1 ng/ml I$^-$; combination of radio-reagent with IDA [143]
I	I$^-$ in 0.05M H$_2$SO$_4$	^{131}I$_2$ in CCl$_4$	[144]
I in 5-iodouracil or 3,5-diiodotyrosine	org. compound in H$_2$O phase	^{131}I$_2$ in CCl$_4$, 1,2-dichloroethane or n-hexane	At ~ 2 mg I ± 4%; non-equil. isotope exchange [145]
P	PO$_4^{3-}$ in 0.1M H$_2$SO$_4$	Tetraphenylarsonium molybdophosphate (^{32}P) in 1,2-dichloroethane	Excess of molybdate; exchange < 20 sec [146, 147]
Pb	Pb tartrate	^{212}Pb(DDC)$_2$ in CCl$_4$	pH 9–10; 1–2 min exchange 0.01–1 µg Pb in 5 ml [148]
Sb	Sb(III) in 0.5M H$_2$SO$_4$	^{124}Sb(DDC)$_3$ in CCl$_4$	0.01 µg [140]
Hg	Hg(II) in H$_2$O	^{203}Hg butylthiophosphate in CCl$_4$	0.1 µg; both direct and reverse exchange [142]
Cd	Cd in H$_2$O phase, tartrate present, pH 8–13	^{115m}Cd diethyldithiocarbamate in CHCl$_3$; Cd (DDC)$_2$	0.05–7 µg Cd [149]

The condition of separate measurement of the radioactivity of the two forms of the substance in question is automatically fulfilled in heterogeneous isotope exchange (see Table 4.6). The table illustrates the fact that exchange systems other than solvent extraction have rarely been used. Typical times necessary for equilibration, and the high sensitivity of the determination, are also shown.

Table 4.7 — Homogeneous isotope exchange for the determination of traces.

Species determined	Exchange between	Conditions of separation	Remarks; reference
C_6H_5HgCl	C_6H_5HgCl and $^{203}HgCl_4^{2-}$ in aqueous HCl solution	Extract C_6H_5HgCl into benzene from $3M$ HCl	Very selective; CH_3HgCl does not interfere; sensitivity 1 ng/ml in 5 ml [150]
CH_3Hg^+	$^{203}Hg^{2+}$ and CH_3Hg^+ in absence of Cl^- (excess Ag^+) in aqueous phase	Solvent extraction of CH_3Hg	Isotope exchange takes hours [151]
Hg(II)	$C_6H_5\,^{203}HgCl$ and Hg^{2+} (excess Ag^+ present) in aqueous phase	Solvent extraction of C_6H_5HgCl	Determination of traces of Hg in heavy metal salts [152]
Bi	Diethyldithiocarbamate of Bi and $^{210}BiI_3$ in CCl_4 [Bi(DDC)$_3$]	Back-extraction of BiI_3 in acid aqueous phase	Sensitivity 10^{-8} g of Bi in 7 ml [153]
Tl(I)	Diethyldithiocarbamates of Tl(I) and $^{204}Tl(III)$ in CCl_4 [Tl(DDC)$_3$]	Back-extraction of Tl(I)	0.02–10 µg of Tl; isotope exchange takes < 0.5 hr [154]
Ce	Di-(2-ethylhexyl)-phosphates of Ce(III) and $^{144}Ce(IV)$ in toluene	Back-extraction of Ce(III)	0.01–2 µg of Ce; exchange takes 5–10 min [155]
As	As(DDC)$_3$ and $^{76}AsI_3$ in CCl_4	Back-extraction of AsI_3	0.1–10 µg [156]
Sb	Sb(DDC)$_3$ and $^{124}SbI_3$	Back-extraction of SbI_3	0.5–20 µg [157]
Organic iodine	KI and org. I comp. in acetone or ethanol	Extraction of org. compound after addition of H_2O and benzene	> 1 mg of I $\pm 6\%$ [158]

The advantage of homogeneous isotope exchange is that it leads to better conditions for a rapid attainment of equilibrium but the two forms must be separated from each other by a special procedure. In

Table 4.7 (p. 199) three examples of exchange in an aqueous phase and five in an organic phase are given.

Some authors emphasize that in the event of slow attainment of equilibrium the analysis can be based on kinetic data. Ikeda's treatment is given in a short version in [10].

Gast and Das [159] have derived the equation for a simple case where X is exchanged between AX and BX,

$$\beta_t/\alpha_t = (1 - e^z)/[1 + (b/a)e^z]$$

where β_t is the specific activity of BX at time t, α_t the specific activity for AX, b the amount of BX, a the amount of AX, $z = -k(a+b)t/V^2$, where V is the volume of the system. The kinetic constant, k, of isotope exchange, is defined by the equation, rate of isotope exchange (mole/sec) $= k(a/V)(b/V)$. From the measured dependence of β_t/α_t on time, and the known values of k and V, the amounts a and b can be found by a computerized approximation program. The practical usefulness of this approach remains to be tested.

It is evident that the isotope-exchange procedures are selective for those elements which show a very high tendency to react with the reagent. An element forming complexes of higher stability will displace the test element and alter the results.

A theoretical basis for rational choice between procedures based on isotope exchange and those based on substoichiometric IDA with the same reagent has still to be developed.

4.3.3 Typical radio-reagent methods similar to isotope exchange
This section deals with the variant classified as 3.3.1 on p. 158.

A typical analysis could be described as follows. For determination of chloride by reaction with silver we spike the chloride solution (containing c_1 moles of Cl^-) with radioactive chloride (c_2 moles) of radioactivity A_0, so that the total amount of chloride is equal to $c_1 + c_2$. We add a moles (exactly known) of a solid silver salt, separate the precipitated AgCl, and measure the radioactivity of the solution. The value of c_1 can be calculated from the obvious relationship:

$$A_0/A_2 = (c_1 + c_2)/(c_2 + c_2 - a)$$

This idealized example illustrates the main features of the procedure:
(a) it is a radio-reagent method, not isotope dilution analysis (see pp. 77 and 168);
(b) the labelling substance is isotopic with the substance to be determined;

(c)   the labelled substance does not undergo isotope exchange (in the common sense of the word), because only one form of chloride is present before the precipitation step.

Next, we shall describe in detail a real case, from the recent literature, to illustrate the application of this type of analysis in a complicated procedure. It involves the determination of the total amount of caesium in waste solutions after the reprocessing of irradiated nuclear fuel, and was developed by Pfrepper and Malyshev [160]. The solution, containing $3M$ HNO_3 and $\sim 0.15\%$ caesium, is treated with a sub-stoichiometric amount of $2 \times 10^{-3}M$ molybdophosphoric acid in the presence of excess of sodium molybdate. The radioactivity of aliquots of the original and final solutions is measured by gamma spectrometry, and the extent of precipitation Φ is calculated from the data.

The amount of caesium (x) is calculated from the formula

$$x = fv/\Phi$$

where f is the amount of caesium (mg) precipitated by 1 ml of $2 \times 10^{-3}M$ molybdophosphate, and v is the volume of this precipitant used.

The main difficulty arises because the value of f depends on the initial caesium/molybdophosphate ratio, owing to the dependence of the composition of the precipitate on this ratio. A function $f = g(x)$ can be found for the conditions used, by an iteration procedure of the type: $x_1 = f_{stand}v/\Phi$; $f_1 = g(x_1)$; $x_2 = f_1v/\Phi$; $f_2 = g(x_2)$; $x_3 = f_2v/\Phi$, etc. It usually takes three iterative steps to obtain a constant x_i value.

Another interesting feature of this method (relative standard deviation 5%) is that in the presence of a tenfold excess of sodium molybdate (relative to phosphomolybdate), the degree of precipitation becomes independent of the concentration of nitric acid ($0-10M$ HNO_3). The determination is very selective and only Rb and Pu(IV) interfere.

The last typical feature of this procedure is that the element to be determined is radioactive because of its origin ($^{137}Cs + {}^{134}Cs$) and so special labelling is not involved.

The procedure is suitable for the determination of high levels of radioactivity in hot cells because of its simplicity. To the best of our knowledge it is the only procedure in the literature for the determination of caesium of high radioactivity.

Many interesting examples of this type of procedure from the early literature are summarized in a review by Kyrš and Krtil [139].

Nickel has been determined by use of ^{63}Ni and dimethylglyoxime and measurement of the radioactivity of the supernatant [104]. Other precipitates used in the same way include: $^7Be_3(PO_4)_2$, $Ba^{35}SO_4$, $Be_3(^{32}PO_4)_2$, $Ag^{131}I$ [co-precipitated with $Zr(HPO_4)_2$].

Similarly, vitamin B_{12} (labelled with ^{57}Co) can be partially sorbed on charcoal modified with albumin in the presence of an 'intrinsic factor' (a substance specifically binding the vitamin). Only the unbound form of the vitamin is subject to sorption. If the concentration of the vitamin is high, the substoichiometric concentration of the intrinsic factor is relatively low and the degree of sorption is high [161].

An unusual determination of rubidium (x) and potassium (y) by use of ^{86}Rb (x_0) and ^{42}K (y_0) in a mixture was described by Polevaya and Mirkina [162]. Three equal portions of sample are treated with different (substoichiometric) amounts of sodium cobaltinitrite, and from the weights of all three precipitates and the activities of potassium and rubidium in the products, the amounts of Rb and K can be calculated.

4.3.4 Pseudoisotope exchange methods

In this section procedures are described (category 3.2 on p. 158) which are based on competition between the analyte and a chemically similar labelled substance for an inactive reagent. The procedures are especially useful where no suitable radionuclide for labelling the analyte is available or an adequate radioactive form of the reagent cannot be obtained. It is evident that these procedures might also be named non-isotopic exchange procedures, non-isotopic displacement, or competitive radio-reagent procedures.

The basis of the method is that if element A can displace element B from a compound BX then A can be indirectly determined by determining B. It is assumed that the determination of B is significantly easier than that of A and that the amount of B replaced is strictly equivalent to the amount of A. Similar procedures were proposed long before the use of radionuclides in analytical chemistry, e.g. displacement reactions recommended for determination of chloride:

$$2\,Cl^- + Ag_2CrO_4 \quad \rightarrow \quad 2\,AgCl + CrO_4^{2-} \text{ or}$$

$$Cl^- + AgIO_3 \quad \rightarrow \quad AgCl + IO_3^-$$

The chromate is determined colorimetrically and the iodate by titration. The titrimetric determination of iodate is six times more sensitive than that of chloride. It is evident that if B is radioactive and is transferred to a phase different from that in which BX occurs, then this approach can be used for convenient determination of A.

It is naturally advantageous if the tendency of A to combine with X is much stronger than that of B. If not, complete substitution of A for B is made possible by removal of B as it is released.

This type of procedure could be divided into (i) procedures based on direct competition, (ii) procedures based on excess reagent determination, and (iii) procedures involving double decomposition. In case (i) there is a direct analogy with isotope exchange methods; in case (ii), the analyte binds one portion of the reagent and the rest is reacted with an excess of a radioactive substance. Double decomposition involves first the isolation of the substance in the form of a certain compound which is then treated with excess of labelled reagent. These three types of determination have been fully reviewed [88, 139]. Typical examples of all three variants are given in Tables 4.8–4.10.

Table 4.8 — Pseudoisotope exchange methods based on direct competition.*

Species determined	Competing radionuclide	Radioactive object of competition	Remarks; reference
Al	^{45}Ca	Ion-exchanger in Ca form	Detn. of Al in steel [163]
Pb	^{45}Ca	CaCO$_3$	^{45}Ca displaced into solution is measured [164]
La, Ce	^{45}Ca	Ca oxalate	The same as above [164]
Hg	^{110}Ag	Ag dithizonate or dibutyl thiophosphate	$10^{-6}M$ Ag compound in CCl$_4$, submicro amounts of Hg [142]
F$_2$	^{131}I	KI	Reaction $F_2\uparrow + 2KI \rightarrow I_2\uparrow + 2KF$; ng/ml F [165]
H$_2$O	^{36}Cl	AlCl$_3$	Reaction $3H_2O + AlCl_3 \rightarrow Al(OH)_3 + 3HCl\uparrow$ [166]

* The labelled substance should exhibit lower affinity than the analyte for the reagent.

The examples given can be generalized and shown schematically for comparison (Table 4.11).

Innovative contributions to this type of determination have been made by Shamaev and Malysheva [174]. An interpolation method has been proposed for the determination of bromide with carrier-free ^{131}I, a radionuclide exhibiting superior qualities to ^{82}Br ($t_{1/2} = 35$ hr). The method is based on the effect of the Br$^-$ concentration on the ^{131}I distribution between an AgBr precipitate and the supernatant solution. The essence of the method consists in adding equal amounts of AgNO$_3$ to three solutions labelled with ^{131}I and containing m_1, m_2 and m_x µg of Br. The distribution can be expressed in terms of $\alpha = A_{sol}/A_{initial}$ (non-separated fraction), or $D = (1 - \alpha)/\alpha$ or $K = 1/(1 - \alpha)$. If the difference

Table 4.9 — Procedures based on determination of excess of reagent.[+]

Species to be determined	Reagent added in excess	Labelled substance used for detn. of the excess	Remarks; reference
Ba	$NaSO_4$	$Pb^*(NO_3)_2$	$Na_2SO_4 + Pb(NO_3)_2$ added simultaneously [167]
NH_3	Na_2CO_3	$Pb^*(NO_3)_2$	[3]
Co	Zn diethyldi-thiocarbama-te in $CHCl_3$	$^{203}Hg(NO_3)_2$	> 0.8 µg Co [168]
Cu	Dithizone in CCl_4	^{65}Zn added after extrac-tion, Zn cannot displace Cu^+	[22]
Cu, Ni, Pb, Zn, Ca[a]	EDTA[b]	$^{110}AgIO_3$; EDTA partial-ly dissolves $AgIO_3$ at pH ~ 10	[169]

[+] Advantage: several species can be determination with one radionuclide.
[a] No need to prepare a radioactive precipitate or extract beforehand; relatively rapid reactions (determinations).
[b] If excess of reagent is not isolated, the competing radioactive substance should exhibit lower affinity than the test substance for the reagent.

between m_1, m_2 and m_x ($m_1 < m_x < m_2$) is small enough, m is a linear function of α, D or K. Therefore linear interpolation can be used:

$$m_x = [m_2(\delta_x - \delta_1) + m_1(\delta_2 - \delta_x)]/(\delta_2 - \delta_1)$$

where $\delta = \alpha$, D or K.

The choice between α, D and K depends on different factors, such as the absolute values of α, etc. Usually use of K gives the highest ac-curacy. Concentrations as small as 0.2 µg/ml of Br^- can be determined, almost an order of magnitude lower than that with ^{82}Br and the ra-diometric correction method. If bromide is determined in the presence of chloride an interpolation is possible by use of the linear dependence $\Phi = f(\alpha)$, where Φ is the Cl^-/Br^- ratio. The linearity applies only if the total amount of bromine and chlorine (m_0) is constant. The results follow from two equations:

$$m_{Br} = m_0(\varphi_x + 1)$$

and

$$\varphi_x = [\varphi_2(\alpha_x - \alpha_1) + \varphi_1(\alpha_2 - \alpha_x)]/(\alpha_2 - \alpha_1)$$

Table 4.10 — Double decomposition procedures.*

Species determined	Non-radioactive reagent[a]	Labelled substance used for the decomposition	Remarks; reference
Zn	$Hg(SCN)_4^{2-}$; $ZnHg(SCN)_4$ is pptd.	$^{59}FeCl_3$ in solution	The decrease in ^{59}Fe activity in solution is measured [170]
Co	Na ferrocyanide	$^{59}FeCl_3$	After paper chromatographic separation of Co; pptn. on paper [171]
Pb	Diethyldithiocarbamate; extn. into CCl_4	$^{204}Tl^+$ in aq. solution	The radioactivity of the extract after the second equilibration corresponds to Pb concentration [172]
In	3-Methyl-5-phenyl-pyrazoline-1-dithiocarbamate	$^{65}Zn^{2+}$	Similar to previous case; [173]

* The labelled substance should possess stronger affinity than the analyte for the reagent. Two-step procedures. One radionuclide can serve for the determination of several substances.
[a] The unreacted portion of the non-radioactive reagent must be removed before application of the labelled substance.

The value of m_0 can be determined by precipitating both Br^- and Cl^- with excess of ^{110m}Ag. It is possible to determine 3 µmoles of Br^- in the presence of $\sim$ 3 mmoles of Cl^- with an error of less than 10%. A theoretical approach to methods using radionuclides of competing elements has been given by Shamaev and Karetnikova [175].

Finally, let us describe in some detail a typical solvent extraction procedure for the determination of traces of mercury with ^{65}Zn dithizonate [Zn(HDz)_2] solution in chloroform [44]. The equilibrium constant for the two-phase reaction is approximately 3×10^{24}. The standard or an aliquot of sample is treated with 5 ml of pH 6 buffer, and diluted to 8 ml. After addition of 5 ml of $10^{-5}M$ $^{65}Zn(HDz)_2$ in CCl_4, the mixture is vigorously shaken for 1 min. Four ml of the clear organic phase are measured for radioactivity (gamma spectrometer). Similarly the activity is measured (under exactly the same conditions), for a blank prepared without added Hg^{2+}. The net activities of the sample and standard are compared. If the specific activity is 7.3×10^9 cpm/mM the blank activity is about 7.5×10^4 cpm and the net activity (total minus blank) released by 5 µg of Hg^{2+} is about 7.2×10^4 cpm. The calibration curve (net activity released vs. Hg^{2+} concentration) is fairly linear and

Table 4.11 — Schematic comparison of the three pseudoisotope exchange methods.

(i) Direct competition

$$A + \underline{\underline{B^*X}} \quad \rightarrow \quad AX + \underset{\sim}{B^*}$$

(ii) Excess reagent determination

$$A + \underline{\underline{X}} \quad \rightarrow \quad AX \qquad \text{(either remove AX or use B* with much lower affinity than A for X)}$$

$$X + \underline{\underline{B^*}} \quad \rightarrow \quad \underset{\sim}{XB^*}[B^*]$$

(iii) Double decomposition

$$A + \underline{\underline{X}} \quad \rightarrow \quad AX \qquad \text{(remove excess)}$$

$$AX + \underline{\underline{B^*}} \quad \rightarrow \quad \underset{\sim}{B^*X} \qquad \text{(remove excess B*)}$$

(The underlining $\sim$ indicates the radioactive substance which is finally measured, the underlining $=$ the substance added in excess.)

if the minimum acceptable is 3 times the standard deviation of the blank activity, the detection limit is estimated to be $\sim 0.2\,\mu g$ of Hg^{2+}, which compares favourably with $1\,\mu g$ for the spectrophotometric dithizonate method. It is believed that the method is suitable for the determination of mercury in air or polluted water samples. The precision and accuracy are fairly good.

In theory, any metal ion with an extraction constant larger than that of Zn^{2+} (see p. 86) can interfere. Such metal ions can be eliminated simply by a preliminary extraction of Hg^{2+} into $2 \times 10^{-5}M$ dithizone in CCl_4 from $0.1M$ EDTA at pH 4.5 followed by stripping into a mixture of $2\,ml$ of $0.25M$ H_2SO_4 and $1\,ml$ of 10% NaBr.

The authors claim that all other methods are insufficiently sensitive, or too costly and tedious, or require sophisticated or expensive equipment.

4.4. CONCENTRATION-DEPENDENT DISTRIBUTION

4.4.1 Definition, classification, relationship
to other categories of procedures

The usefulness [19] of this category of determination, its definition and sub-division into saturation analysis and non-saturation analysis together with several examples, have already been given in Sections 4.1.3.1 and 4.1.3.2.

Here some more detailed information will be given. The method of concentration-dependent distribution (CDD) has been reviewed [176–178].

The principle of the method can be shown by the graphs representing the theoretical dependence of the distribution ratios for solvent extraction of metals on the total (initial) concentration of the metal ion at constant extractant concentration, pH, temperature, volume ratio, etc. (Figs. 4.3 and 4.4). The area to the left of curve 5 corresponds to concentrations which are too low to be determined by the CDD method because the distribution ratio does not depend strongly enough on the metal ion concentration. The area to the right of curve 7 corresponds to full consumption of the extractant. This area corresponds to concentrations suitable for use of the extractant for substoichiometric IDA (see Section 3.5). The range of concentrations corresponding to the area (hatched) lying between curves 5 and 7 is suitable for determinations based on CDD (saturation variant). From the figures two additional conclusions can be drawn, namely, (i) the minimum concentrations allowing CDD determinations are about an order of magnitude lower than those treatable by substoichiometric analysis and (ii)

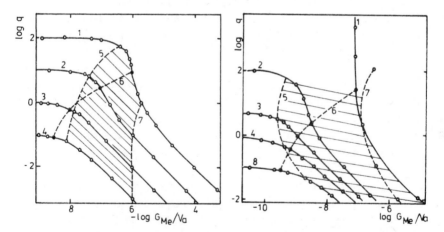

Fig. 4.3 — Theoretical calibration curves for extraction of Me^+ with a constant amount of extraction agent. Reaction $Me + R = \overline{MeR}_{org}$ (equilibrium constant $\beta_H = 10^8$); q — distribution ratio of labelled metal; G_{Me} — amount of metal; V_a — volume of the aqueous phase; q_0 — distribution ratio for tracer concentrations of metal. q_0: 1 — 100; 2 — 10; 3 — 1; 4 — 0.1; 5 — points with $d \log q / d \log G_{Me} = -0.5$; 6 — equivalence points; 7 — lower limit of substoichiometric determination. (Reproduced from [19] by permission of the copyright holders, Elsevier Science Publishers, Amsterdam.)

▶

Fig. 4.4 — Theoretical calibration curves for extraction of Me^{3+} with a constant amount of extracting agent. Reaction $Me + 3R = MR_{3org}$ ($\beta_H = 10^{26}$). q_0: 1 — 2×10^6; 2 — 100; 3 — 5; 4 — 1; 8 — 0.1. For curves 5, 6, 7 see Fig. 4.3. (Reproduced from [19] by permission of the copyright holders, Elsevier Science Publishers, Amsterdam).

the difference between these minimum concentrations (width of shaded areas) is larger for the species MR_3 than for MR (cf. Figs. 4.3 and 4.4).

The comparison of the theoretical sensitivities for these two types of determination, made by Alimarin *et al.* [179], is in general harmony with these conclusions drawn from Figs. 4.3 and 4.4. The sensitivity of the substoichiometric radioanalytical methods is generally lower than the sensitivity of the CDD methods, for a given error level. However, as the D_0 values (distribution ratio for solvent extraction of tracer amounts of the element) increase and the charge on the test cation decreases, this difference decreases quickly and for some systems practically disappears. Note that curves 5 and 7 have a tendency to approach each other with increasing D_0 values, as clearly illustrated, especially in Fig. 4.3. For the determination of cations with charge greater than two, the use of the CDD method with solvent extraction as the means of separation is generally justified. The price paid for this increased sensitivity, however, is that the pH must be strictly controlled [19].

It has already been mentioned that different authors have proposed methods very similar to CDD (in that incomplete consumption of the reagent or variable composition of the reaction product can be tolerated as in the sub-superequivalence method, spiking IDA, etc.). Consequently, comparison of the sensitivities of these related procedures is important. This has not been done in a general form but Shamaev [180] has compared the minimum amounts of caesium which can be determined by solvent extraction of its tetraphenylborate into nitrobenzene and various methods of determination, including the CDD method (Fig. 4.5). It can be seen that the sensitivity of the CDD method is surpassed only by use of the principle of successive approximations.

Besides the increased sensitivity, the possibility of achieving higher theoretical precision by CDD methods than by substoichiometric IDA or other radioanalytical procedures is suggested by Figs. 4.3–4.5. This follows from the relatively high values of the negative slopes (dD/dG_{Me}) in the region between curves 5 and 7 in Figs. 4.3 and 4.4. This indicates that a certain error in D leads to a relatively low error in G_{Me}. It should be emphasized that this increased precision relates mainly to errors due to the statistical character of the radioactive decay. Therefore, experimental comparison of the precision of the CDD method and other categories of radioanalytical determinations, carried out by the same laboratory with the same chemical reaction as the basis of the determination, are required before more general conclusions can be drawn.

The relationship of CDD methods to some other variants could be analysed theoretically. Instead, we prefer to enable the reader to judge the degree of resemblance of some radio-reagent determination types by briefly describing the essence of several other variants similar to CDD.

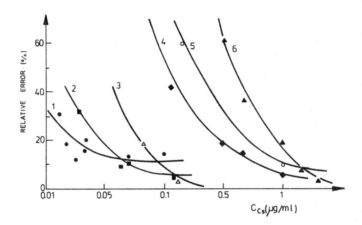

Fig. 4.5 — The dependence of error on caesium concentration for different radioreagent methods similar to CDD: 1 — method of successive approximation; 2 — concentration-dependent distribution; 3 — Grashchenko–Sobotovich method; 4 — analytical calculation method; 5 — DeVoe's method; 6 — radiometric correction. Prepared with data from [180].

Spiking IDA (see also p. 165) [181] has the following characteristics:
(a) the substoichiometric principle is not fully obeyed;
(b) the activity is added as (identical) spikes to both the standard and the sample solution;
(c) the count rates obtained for the fractions isolated from standard solutions are plotted against concentration in such a way as to yield a linear calibration curve;
(d) linearity of the calibration curve can also be achieved by using suitable empirical relationships.

As examples of equations for use when the liquid–liquid extraction is incomplete, the following are mentioned:

$$q = 1 - e^{-\alpha p}; \quad 1/q = \beta + \delta/p$$

where q is the proportion isolated from the total amount of the substance to be determined, α the ratio of substance to reagent in the compound formed by chemical reaction, p the ratio of amount of reagent to amount of substance to be determined, β and δ are adjustable parameters in the equations. Spiking IDA has been demonstrated by several examples.

Shamaev [182] has proposed, for determination of concentrations of substances low compared to the values of the corresponding equilibrium constants, a procedure termed *method of labelled test element isolation with different quantities of reagent aliquots of analysed solution* (IDQR). This method has some features in common with radiometric titration. They are:

(a) the aliquots are treated in parallel, with increasing amounts of reagent, and
(b) the element to be determined is labelled.

On the other hand, dissimilarities with radiometric titration are:

(a) the interphase distribution of the element is determined rather than the radioactivity of one phase, and
(b) the result is not found from the location of a break on the titration curve but by a numerical or graphical procedure.

The principle of the IDQR method can be illustrated with chelate extraction as a model reaction.

For this example and certain conditions, the equations for the mass balances and equilibrium constants yield the relationship

$$C_R/n(1 - \alpha_M) = C_M + D^{1/n}[H^+]/K_{ex}^{1/n}n(1 - \alpha_M)$$

where C_R and C_M are the concentrations of the reagent and labelled metal ion, respectively, K_{ex} is the extraction constant of the complex MR_n, D the distribution ratio of the metal, and $\alpha_M = 1/(D + 1)$.

The equation can be simplified to

$$y = C_M + kx$$

where $k = [H^+]/K^{1/n}n$, $y = C_R/n(1 - \alpha_M)$, and $x = D^{1/n}$. We know the values of C_R and n and measure D, so α_M is also known. Plotting y vs. x gives a straight line with slope k, and intercept C_M on the x-axis.

An illustration is given in Table 4.12, for the determination of $8 \times 10^{-7}M$ caesium by solvent extraction with tetraphenylborate into nitrobenzene at pH 9 ($n = 1$, $y = C_{TPB}/1 - \alpha_{Cs}$), $x = 1/K_{ex}\alpha_{Cs}$, $K_{ex} = 2.3 \times 10^5$). Shamaev [182] claims that IDQR "must have the highest sensitivity of determination in comparison with other radio-isotopic methods of analysis". The reason for this follows from the theory; briefly speaking, a very wide range of the ratio C_R/C_M can be used, so that large changes in α_M and D_M occur, and large C_R/C_M values enable small amounts of the metal to be determined.

On the other hand, in our opinion, possible drawbacks of the method include:

Table 4.12 — Determination of caesium by the IDQR method [182]; $C_{Cs} = 8 \times 10^{-7} M$.

C_{TPB} (mole/l)	$\dfrac{C_{TPB}}{C_{Cs}}$	a_{Cs}	$y \times 10^6$	$x \times 10^6$
3.92×10^{-7}	0.43	0.930	5.60	4.675
1.64×10^{-6}	2.03	0.745	6.43	5.840
4.01×10^{-6}	5.01	0.530	8.53	8.203
5.60×10^{-6}	7.00	0.462	10.41	9.410
8.02×10^{-6}	10.01	0.371	12.74	11.720
9.60×10^{-6}	12.00	0.318	14.08	13.670
1.20×10^{-5}	14.98	0.278	16.59	15.600
1.60×10^{-5}	20.02	0.218	20.48	19.940

Application of the method of least squares gives $C_{Cs} = 7.8 \times 10^{-7} M$ ($\pm 2.5 \%$).

(a) if for any reason the value of K_{ex} is not constant but gradually changes as C_R increases (higher ionic strength, etc.), the result may be erroneous although in classical CDD the result would not be affected;

(b) the same conclusion applies if with increasing C_R the value of n increases;

(c) for routine work, CDD is less laborious since a single calibration curve is used;

(d) in typical CDD, knowledge of the mechanism and stoichiometry is not obligatory, whereas in IDQR it is apparently required, at least to the extent indicated;

(e) in practice it may prove difficult to maintain the pH value constant over such a broad interval of C_R values and the use of a high buffer concentration for this purpose can lower the actual K_{ex} value.

Therefore, the choice between normal CDD and IDQR should be experimentally based in every case.

Shamaev also proposed [17] a radio-reagent method based on labelling both the analyte and the reagent. This new method is claimed to exhibit high sensitivity, and is recommended in cases where the degree of interaction between the reacting substances is very low and when the separation of a substance into the second phase is incomplete. The conditions for using this method include the following:

(a) the product of the reaction should pass into a second phase, but the test element and the free reagent should remain in the original phase;

(b) the stoichiometric ratio (n) by which the components react should be constant and known;

(c) the measurement of radioactivity should allow the independent

determination of the radioactivity of the unreacted test element and
of the unreacted reagent;

(d) the precision of the radioactive measurements must be high.
The last condition follows from the fact that this method exhibits an
increased (compared to substoichiometric IDA or the radio-reagent
method) influence of radiometric errors on the accuracy. The basic
equation for determinations by this method is:

$$m_M \Phi_M / m_R \Phi_R = 1/n$$

where m_M and m_R are the initial molar quantities of the test substance
and the reagent, respectively, Φ is the fraction of substance reacted and
n is the stoichiometric ratio of metal to reagent in the compound MR_n.

If we insert $\Phi_M = D_M/(D_M + 1)$, where D_M is the ratio of the concentration of the test substance in the product phase to that in the initial
phase (for solvent extraction with $V_{org} = V_{aq}$), and similarly for
$\Phi_R = D_R/(D_R + 1)$, we obtain:

$$m_M = m_R D_R (D_M + 1)/n (D_R + 1) D_M$$

where the D values are obtained from the radioactivity measurements.
The method [17] was tested on the reaction

$$^{110m}Ag^+ + {}^{131}I^- \rightleftharpoons {}^{110m}Ag^{131}I$$

where the product was isolated by electrodialysis. The 0.885–0.935
MeV photopeak of ^{110}Ag and the 0.384 MeV photopeak of ^{131}I were
used for measurement with a single-channel pulse-height analyser
and an NaI(Tl) crystal detector. The initial radioactivity of ^{131}I was
usually 5–10 times higher than that of ^{110m}Ag.

The results are given in Table 4.13. It is obvious that the determination is very sensitive and subnanogram amounts of substances can be
determined.

On the other hand, the ratios Φ_I/Φ_{Ag} which, theoretically, should be
equal to 1, become with decreasing concentration 1, 0.91, 0.92, 1.16,
1.81, 1.54, which indicates the presence of some unidentified error.
Probably for this reason, for silver the last three values found are
significantly higher than the amounts taken, and the opposite is true for
the iodine.

Shamaev uses the term *radioisotope-stoichiometric method of analysis*
for this type of procedure [17].

The method of *sub-super equivalence IDA* which exhibits certain
features in common with CDD is elucidated in Section 5.3.7. It has been

compared with CDD experimentally with the same extraction system
by Přikrylová *et al.* [183] whose work shows that the same set of
equilibrations can be the basis for calculation of the result by either
CDD or SSE.

The similarity between the CDD and RIA methods is illustrated on
p. 300.

Table 4.13 — The determination of silver and iodide by using both labelled iodide and
silver and precipitation of AgI [17].

Concentrations taken: $C_{Ag} = C_I$ (M)	Φ_{Ag}	Φ_I	Concentrations found (M)		Relative error of determination (%)
			C_{Ag}	C_I	
5×10^{-7}	0.730	0.73	5.000×10^{-7}	5.00×10^{-5}	0; 0
5×10^{-8}	0.550	0.50	4.500×10^{-8}	5.50×10^{-8}	$-10; +10$
1×10^{-8}	0.130	0.12	0.925×10^{-9}	1.08×10^{-8}	$-8; +8$
5×10^{-9}	0.104	0.121	5.820×10^{-9}	4.33×10^{-9}	$+15; -14$
1×10^{-9}	0.021	0.038	1.810×10^{-9}	0.55×10^{-9}	$+81; -45$
5×10^{-10}	0.022	0.034	7.730×10^{-10}	3.23×10^{-10}	$+55; -35$

4.4.2 Examples of CDD determination
Many examples can be found in the review by Kyrš and Rais [178].
Here we concentrate only on procedures developed recently and on
typical determinations.

4.4.2.1 Saturation analysis
Shamaev and Khudinovskikh [184] proposed the determination of
potassium in sea-water by solvent extraction (^{137}Cs or ^{86}Rb as tracer)
into nitrobenzene with substoichiometric amounts of tetraphenylborate
(note the relationship to Section 4.3.4). The results are affected by the
high concentrations of sodium present.

The calculation was done by three different procedures:
(a) the comparison method ($m_K = m_{stand}D_{stand}/D$);
(b) the method of radiometric correction ($m_K = m_{TPB}(\beta/D + 1)$);
(c) the method of calibration curves;
(here m denotes the amount of the substance, D the extraction coefficient, and β the solvent extraction exchange constant of the pair Rb/K
or Cs/K).

The results of all three procedures are in good agreement and the
errors for ~ 10 μmoles of K are in the region 0.5–1.5 %. The errors
associated with other methods for the determination of potassium in
sea-water are higher: spectrophotometry 2 %, flame photometry 4.8 %,

spectrography 6%. Only the gravimetric determination exhibits an error ($\pm$ 1.2%) comparable to that of the radiometric method, but is slow and labour-consuming. The results confirmed the theory of the constancy of the salt composition of sea-water.

The same extraction system with ^{134}Cs had been used earlier [185] to determine caesium in model solutions. The method of extraction with changing amounts of reagent makes it possible to determine caesium in the presence of 30 times as much rubidium and 70 times as much potassium; the calibration curve method was used in the presence of 100-fold ratio of Rb and 500-fold ratio of potassium to caesium.

Several papers have been devoted to the problem of determination by radioanalytical (radio-reagent) methods of an element in a mixture with a chemically very similar element (see review by Kyrš and Přikrylová [6]). The approach which attracted most attention is that using two distributions in two systems and obtaining the result either by calculations analogous to solving simultaneous equations or using sets of calibration curves equivalent to calibration 'surfaces'.

The theory of saturation analysis was developed by Kyrš and Hálová [186, 187] and is able to predict types of systems suitable for this method of determination, suitable labelling substances and expected errors. An attempt to find suitable reagents for the determination of pairs of neighbouring rare earths was published by Kyrš and Moravec [188].

Here, the principle of the determination using two calibration surfaces will be explained. The determination of two similar elements A and B is based on two calibration 'surfaces' which are shown in Figs. 4.6 and 4.7. The calibration surface for a given system consists of a family of curves, each showing the dependence of the distribution ratio of element A upon its initial concentration α, at a constant initial concentration of B (β). One system usually contains a certain constant amount of a reagent (R) able to react with both A and B, the stability constants of AR and BR being different. The second system contains another such reagent (Q). The ratios of the stability constants $r_1 = K_{AR}/K_{BR}$ and $r_2 = K_{AQ}/K_{BQ}$ should be as different from each other as possible. In Figs. 4.6 and 4.7 $r_1 = 1$ and $r_2 = 5$. The determination involves finding the distribution ratios $D^{(1)}$ and $D^{(2)}$ in the two systems, for the same sample. The lines I, II...VI comprise all possible combinations of α and β which can yield a given $D^{(1)}$ value. The lines VII, VIII...XII comprise all those which can yield the $D^{(2)}$ value shown as the horizontal line in Fig. 4.7. The points I...XII are plotted in a graph with α and β on the axes as shown in Fig. 4.8. The intersection of the two lines corresponds to the required values of α and β. The usefulness of this type of determination has to be verified experimentally.

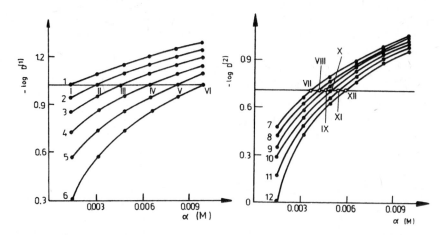

Fig. 4.6 — Calibration surface, system 1. $D^{(1)}$ — distribution ratios of element A in system 1; α — initial concentration of element A. Curve 1 — initial $[B] = \beta_1$; 2 — $[B] = \beta_2$, etc., $\beta_1 > \beta_2 > \ldots \beta_6$. The curves 1–6 are the projections of the intersections of the calibration surface for system 1 with horizontal planes corresponding to the values β_1, β_2, etc. (Reproduced from [186] by permission of the copyright holders, Akadémiai Kiadó, Budapest.)

▶

Fig. 4.7 — Calibration surface, system 2. $D^{(2)}$ — distribution ratios of element A in system 2; Curve 7 — $[B] = \beta_1$; curve 8 — $[B] = \beta_2$, etc. Otherwise analogous to Fig. 4.6. (Reproduced from [186] by permission of the copyright holders, Akadémiai Kiadó, Budapest.)

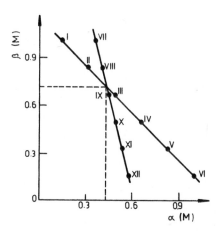

Fig. 4.8 — Determination of the two unknown concentrations. The figure represents the projection of the curves I, II ... VI and VII, VIII ... XII onto the plane α–β. (Reproduced from [186] by permission of the copyright holders, Akadémiai Kiadó, Budapest.)

4.4.2.2 Non-saturation analysis

This type of procedure has been used for the determination of the concentration of an extractant, by using the dependence of the distribution ratio for a suitable carrier-free radioactive metal on the extractant concentration under standard conditions. For the determination of di-isoamylphosphoric acid ^{95}Zr and ^{95}Nb have been used [189]; for dodecylbenzenesulphonic acid, ^{65}Zn [190]; for tetraphenylborate and molybdophosphate, ^{137}Cs [191]; for dibutylphosphoric acid formed from tributylphosphate by radiolysis, ^{95}Zr (a so-called Z-test for irradiated TBP) [192].

The distribution in question can also be one between different organs of a living organism. A tissue containing unknown amounts of thiouracil was fed to rats which afterwards received an injection of ^{131}I. After some time the rats were killed, the thyroid gland was cut out and its radioactivity measured. The higher the initial concentration of thiouracil, the higher the radioactivity found [193].

The dependence of the extent of co-precipitation (co-crystallization) of an element on its concentration has been used for its determination (^{210}Pb with NH_4I) [194]. The extent of zirconium (^{95}Zr) co-precipitation with a large excess of phenylarsonic acid increases with increasing zirconium concentration, enabling the determination of ~ 8 µg of zirconium [195]. A determination of trithionic acid $H_2S_3O_6$ (1 µg ± 4 %) labelled with ^{35}S can be based on the dependence of the ratio of peak height to peak area, in a thin-layer chromatographic procedure on silica gel [196].

In some cases, especially in sorption distribution, the distinction between saturation and non-saturation analysis becomes dubious. In a concrete case the number of adsorption sites with a certain activity can approach saturation, whereas the total number of all possible adsorption centres of the material may be far from saturated. This may be the case when 10^{-9} g of yttrium is determined by using the dependence of yttrium adsorption on a platinum surface on its concentration (labelling with ^{90}Y) [197].

The number of published procedures of non-saturation type is limited. This category is generally characterized by extremely high theoretical sensitivity but rather low precision. Further examples can be found in the review by Kyrš and Rais [178].

4.5 RADIO-RELEASE METHODS

Radio-release methods involve the application of radioactive emanation methods to analytical chemistry. Emanation methods have been known in radiochemistry for more than half a century and are also used in solid-state chemistry, physical chemistry, mineralogy, metallurgy, silicate chemistry, ceramics, and polymer chemistry. A recent review by Balek and Tölgyessy [198] covers the field from all basic aspects. Therefore, only the analytical applications will be discussed here.

Analytical applications of solids containing inert radioactive gases depend on the selective reaction of the inactive analyte or of an auxiliary reagent with the radioactively labelled solid agent. During the selective reaction the surface layers of the solid are disturbed, causing the release of the radioactive inert gas, e.g. ^{85}Kr having a half-life of 10.8 yr and beta energy of 0.7 MeV. Solids labelled with radioactive krypton are sometimes called "Kryptonates".

The determination can be made:
(a) with a calibration curve;
(b) by comparison with a standard (using an instrument calibrated by the standard beforehand);
(c) by titration.

The procedures based on radiometric titrations using Kryptonates or on the radio-release principle in general are described in the section on radiometric titrations (p. 273 ff.). In this section we shall describe the basic methods for preparing the necessary Kryptonates and using them by the application of calibration curves and/or standards, and also selected instruments and apparatus for the analysis.

4.5.1 Preparation of Kryptonates

4.5.1.1 Diffusion technique
The original method used for the preparation of ^{85}Kr-labelled solids was developed by Chleck et al. [199], who introduced krypton into various substances at elevated temperature and gas pressure. In this technique the essential element of the labelling apparatus is a pressure vessel in which the material to be labelled is placed. Detailed views of two pressure vessels, modified by Tölgyessy et al. [200], are shown in Fig. 4.9. The inert gas (krypton) is introduced at normal temperature and pressure, the vessel is closed off from the remainder of the system, and the temperature is raised to 300 °C. The pressure is controlled and may reach about 70 atm. The sample is kept under these conditions for several hours. At the completion of a run the material is quenched by immersing the pressure vessel in liquid nitrogen.

The amount, M, of the inert gas collected per cm^2 of sample surface is directly proportional to the pressure, p, and the square root of the reaction time, t, and exponentially proportional to the temperature, T.

The experimentally observed penetration depth is in the range
10–$1000\,\mu g/cm^2$, i.e. 10^2–$10^4\,nm$ and depends on the temperature exponentially and on the square root of exposure time (see p. 225 for
definitions).

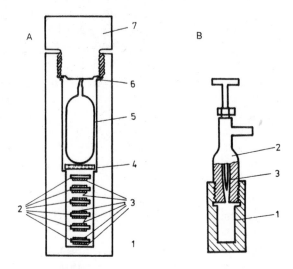

Fig. 4.9 — A: Steel vessel for kryptonation at high temperature and increased pressure.
1 — steel vessel body, 2 — sample to be labelled, 3 — metal (Pt) springs, 4 — perforated
steel plate, 5 — ampoule containing ^{85}Kr, 6 — rubber seal, 7 — screw plug. B: Pressure
vessel for diffusion kryptonation of metals. 1 — steel jacket, 2 — metal plug, 3 — needle
valve. (Reproduced from [200] by permission of the copyright holders, ÚVVVR, Prague).

The diffusion technique has been used to introduce ^{85}Kr into more
than 250 different solids, including metals, inorganic and organic compounds, in the form of foils, powders and single crystals [200].

Krypton atoms diffuse into the solids and remain, in general, in
substitutional positions in the metal lattice (the diameter of the Kr atom
being too large for ordinary interstitial accommodation). In some
solids, for example those with a layer structure, there are regular
interstitial voids of dimensions sufficient to accumulate inert gas atoms.
Materials of this type (e.g. boron nitride and molecular sieves) gave the
highest specific activities (6×10^5 and 7×10^6 cps/mg, i.e. 16 and
$200\,\mu Ci/mg$, respectively) of all materials prepared by Chleck et al.
[199]. The inert gas atoms can be fixed at various lattice imperfections,
including grain boundaries.

The diffusion technique is especially suited for the labelling of organic compounds. It has been pointed out in the literature that the

presence of hydroxyl groups which can be linked by hydrogen bonds is necessary for the formation of a crystal lattice permitting incorporation of the gas molecule. Krypton was found to be bound considerably more weakly in organic compounds than in metals labelled by the diffusion technique. It was proved that the presence of inert gas atoms in the crystal lattice has little or no effect on the chemical properties of the host solid.

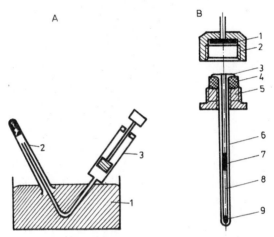

Fig. 4.10 — A: Filling of capillaries with ^{85}Kr. 1 — vessel with mercury, 2 — sample to be labelled, in a glass capillary, 3 — syringe. B: Diffusion kryptonation. 1 — matrix sealing, 2 — matrix, 3 — ground part of capillary, 4 — rubber seal, 5 — screw with an opening, 6 — glass capillary, 7 — movable mercury plug, 8 — gaseous ^{85}Kr, 9 — sample to be labelled. (Reproduced from [202] by permission of the copyright holders, Braun Verlag, Karlsruhe.)

In addition to that of Chleck *et al.* [199], further experimental arrangements for the diffusion technique have been proposed. Trofimov and Kazankin [201] prepared xenon-labelled *p*-cresol ([Xe][C$_7$H$_8$O]$_6$) by placing the cresol in a thick-walled glass vessel containing a xenon atmosphere, cooling it to the temperature of liquid nitrogen, and storing it at room temperature for a week.

Jesenák *et al.* [202] proposed a microdiffusion technique for labelling powders, permitting reduction of the losses of residual inert gas after labelling. The labelling was done in a thick-walled capillary, as shown schematically in Fig. 4.10A. After injection of krypton into the capillary containing the sample, the capillary was attached by a holder (see Fig. 4.10B) to a nitrogen bomb giving a pressure of 30–100 atm, and left for the necessary time.

If a higher quantity of the inert gas is to be incorporated into a solid, a technique can be used that is based on the labelling of powdered material by diffusion or ion bombardment, followed by pressing the labelled powder into a pellet of the desired size and shape. However, the solid will be locally inhomogeneous. This technique has been advantageously used for the preparation of labelled solids such as $AgIO_3$, KCl, $AgNO_3$, used for analytical purposes.

The specific activities of $AgIO_3$, KCl and NH_4NO_3 (grain size 0.1 mm) achieved by the diffusion technique are given in Table 4.14.

Table 4.14 — Kryptonation of powders; conditions and activity of the products [202].

Substance	Krypton pressure (Pa)	Temperature (°C)	Time of diffusion (hr)	Specific activity (cps/g)
$AgIO_3$	6.7×10^6	25	50	28 170
KCl	7.9×10^5	200	10	1 080
NH_4NO_3	8.8×10^5	20	5	1 630

4.5.1.2 Direct gas introduction during sample preparation or phase transition

In some cases, the inert gas can be advantageously incorporated during formation of the solid. Examples of such formation processes include
(a) crystallization from a melt or solution;
(b) sublimation of the solid;
(c) evaporation and condensation of a thin film of metal onto a cold surface;
(d) polymerization of monomers, and
(e) annealing of a solid in the temperature range where a phase transition or similar process occurs.

The choice of technique for inert gas introduction depends mainly on the character of the solid to be labelled and the reason for which the labelled solid is prepared.

Incorporation of inert gas atoms during the formation of the solid phase was first used for the preparation of radioactive Kryptonates of the clathrate type. Chleck [203] prepared [85]Kr-labelled hydroquinone (1,4-dihydroxybenzene) by slow crystallization from a melt under a positive pressure of krypton gas containing [85]Kr. The position of the krypton in relation to the hydroquinone units is shown in Fig. 4.11. An atom of [85]Kr is trapped in a trimer of hydroquinone molecules to give a typical clathrate $\{[C_6H_4(OH)_2]_3[^{85}Kr]\}_n$.

The condensation of a solid during sublimation offers very favourable conditions for the incorporation of an inert gas (usually [85]Kr) into the crystal lattice of the solid when a lattice capable of clathrate

formation is produced. The simple apparatus [204] used for this labelling technique is shown schematically in Fig. 4.12.

Tölgyessy *et al.* [205] demonstrated the formation of krypton-la-
belled polymers by polymerization in an atmosphere of inert gas.
Monomers such as methylmethacrylate, styrene and acrylonitrile were
polymerized in this way. During polymerization under static or
dynamic conditions, ^{85}Kr was captured by the polymer and was
found to be uniformly distributed in the resulting plastic.

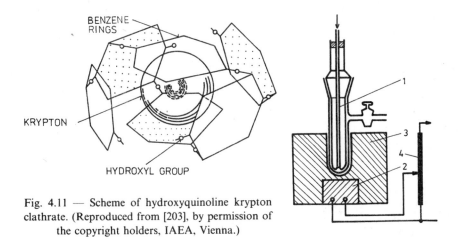

Fig. 4.11 — Scheme of hydroxyquinoline krypton
clathrate. (Reproduced from [203], by permission of
the copyright holders, IAEA, Vienna.)

Fig. 4.12 — Scheme of apparatus for sublimation kryptonation: 1 — glass apparatus,
2 — electrical heating, 3 — aluminium block, 4 — transformer. (Reproduced from [204]
by permission of the copyright holders, Slovak Technical University, Bratislava.)

Sizmann and Rupp [206] succeeded in incorporating ^{85}Kr into quartz
during its polymorphic transformation. If quartz is heated in an atmosphere of krypton, then during the phase conversion of beta quartz into
beta cristobalite, inert gas atoms enter the crystal during lattice re-
organization.

Hidalgo and Sizmann [207] labelled alumina powder (Al_2O_3) by
heating it at temperatures higher than 1323 K in an atmosphere containing ^{85}Kr. The gas inclusion in this temperature range is due to
recrystallization, closure of pores and sintering.

4.5.1.3 Ion bombardment
At present a whole range of techniques is available for ion bombard-

ment. Ionized atoms are accelerated with a high voltage, or ionized gas atoms are produced in an electrical discharge or a microwave plasma.

The ion bombardment equipment most frequently used for the implantation of inert gases into solids [198] can be characterized by the principles set out below.

(1) In a vacuum discharge tube, atoms of the inert gas are ionized by electrons emitted from a cathode or a high-frequency discharge. The ions formed are accelerated by the potential drop and strike the cathodic target with an energy which depends on the voltage drop. Some of the impinging ions are trapped in the material of the target. A diagram of this kind of apparatus is shown in Fig. 4.13.

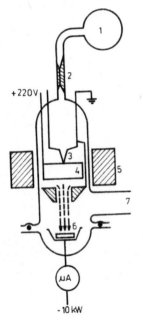

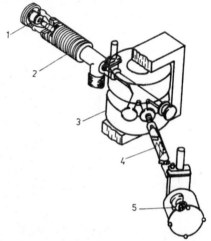

Fig. 4.14 — Schematic diagram of an electromagnetic separator: 1 — ion source, 2 — ion acceleration, 3 — mass separation, 4 — beam sweeping, 5 — multiple heated target changer for batch production. (Reproduced from [209] by permission of the copyright holders, Elsevier Science Publishers, Amsterdam.)

Fig. 4.13 — Diagram of apparatus for ion bombardment by an ion source with d.c. voltage: 1 — reservoir of the radioactive inert gas, 2 — capillary valve, 3 — filament, 4 — ion accelerator, 5 — permanent magnet, 6 — sample to be labelled, 7 — vacuum pump. (Reproduced from [198] by permission of the copyright holders, Elsevier Science Publishers, Amsterdam.)

The ion source described by Carter [208] operates in a rather high vacuum (10^{-4}–10^{-3} Pa). Equipment of this type makes it possible to produce a well concentrated beam of highly energetic ions (up to 100 keV) and is the most suitable for labelling metal targets or bulk material.

(2) Some authors [209, 210] have used ion sources with an electro-magnetic isotope separator and obtained very well defined beams of ions with well defined energy, mass, charge and entrance angles. A schematic diagram of an electromagnetic separator [209] is shown in Fig. 4.14. It consists of an ion source, mass separator and a target chamber. The gas atoms to be implanted are ionized to obtain the required type of ion. The ions are extracted from the source and accelerated in a strong electric field. The beam is then deflected by a magnet and slit configuration, so that the desired ions can be separated from unwanted ions of different masses. A second acceleration step, which is not shown in Fig. 4.14, is often applied after the mass separa-tion before the ions impinge on the target. The beam is scanned across the target to obtain homogeneous labelling.

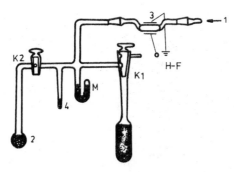

Fig. 4.15 — Scheme of apparatus for introducing a gas for ion bombardment in a high-frequency discharge: 1 — adsorption pump, 2 — reservoir of the inert radioactive gas, 3 — sample cell, 4 — metallic calcium for inert gas purification, K1 and K2 — valves in glass apparatus, M — manometer, H-F — high-frequency field generator.

(3) An apparatus for introducing traces of gas into solids in a high-frequency discharge was proposed by Jech [211]. A schematic diagram of the apparatus is shown in Fig. 4.15. The sample to be labelled is placed in a thick-walled glass cell connected to a vacuum apparatus. After the pressure in the cell has been reduced to about 1.3 Pa, a small amount of the radioactive gas is introduced so that the final pressure normally does not exceed 7–70 Pa. Ions of the inert gas are formed and accelerated by a pulsed high-frequency field with a maxi-mum voltage of 15 kV. The accelerated ions penetrate the surface layers of the solid to a depth of 1–10 nm.

Doubly and triply charged ions are produced in the gas discharge in addition to singly charged ions. It is impossible to focus the ions which are not mono-energetic and ions of residual gases in the rather poor vacuum also participate in the bombardment. However, apparatus in which the acceleration of ions is based on the discharge is simpler, and the labelling of samples can be done more rapidly since it does not require prolonged pumping to improve the vacuum. This method of introducing inert gas is convenient for labelling surface layers of solids, with minimal radiation damage. It is also more suitable for powders, since the unfocused beam of ions labels the whole surface of the powder. With ion bombardment, samples can usually be labelled by use of doses ranging from about 10^{10} ions/cm^2 (corresponding to a gas concentration of about 10^{-5} atom %) to 10^{16} ions/cm^2 (3 atom %).

In general, the ion bombardment technique has the advantages that the specimen can easily be handled and used without waiting time. On the other hand, damage to the surface layers of the crystal lattice can occur.

In using the method of ion bombardment, it must be remembered that not all ions reaching the surface are trapped in the substance. The degree of incorporation is characterized by the efficiency of trapping, η, sometimes called the trapping probability, which is defined as the ratio of the number of atoms which remain trapped within the solid to the total flux of ions reaching the sample surface during the bombardment.

We can suppose that entrapment of inert gas atoms by a solid consists of the following processes:
(a) penetration of the incident ion beyond the surface layer of atoms;
(b) migration within the lattice, giving rise to a certain possibility that the inert gas atom can escape through the surface;
(c) eventual trapping within the lattice, in interstitial or in regular sites or in other lattice defects produced by radiation.

Penetration through the surface layer necessitates the bombarding ion passing within a maximum distance of 0.5 lattice spacings from a surface atom.

The trapping efficiency, η, of ions in a substance, at constant temperature, depends on the character of the ion, its energy and the material of the target. Trapping efficiency normally increases with decreasing weight of the bombarding ion. For a given bombarding ion energy the trapping efficiency of inert gas ions is higher in aluminium than in tungsten, and considerably higher for glass ($\eta = 0.6$).

At energies below 5 keV, η is always less than 1 for many metals, but rapidly tends to unity above this energy.

The release of trapped atoms occurs at the target temperature and can be expected to increase as the temperature is increased, resulting in a reduction of the effective value of the trapping efficiency. On the other hand, an increase in η is observed as the target is cooled below room temperature. The η values for krypton ions of 45-keV energy are always close to unity for target atoms with atomic weight < 25.

The trapping efficiency for inert gas ions in a substance is also affected by the physical state of the material. The nature of the sites occupied by the gas atom introduced into the lattice depends on the structure and the state of the crystal lattice.

The energetic ions which have passed a solid surface are slowed down by elastic collisions with electrons and target atoms and by displacement cascades around the path of the ions penetrating the solid. The slowing down of energetic heavy ions, such as the inert gas ions, is strongly dependent on the energy, mass and charge of the ions, on the weight of the target atoms, and on temperature. For single-crystal targets it depends, moreover, on the crystallographic orientation of the lattice with respect to the direction of the incident ion beam. All these factors affect the resulting distribution of the inert gas atoms within the solid.

As the penetration of energetic inert gas ions is of general importance for the labelling of solids, its measurement will be briefly discussed.

The techniques for determining the penetration ranges of ions can be divided into three groups [198].

(1) Stripping techniques are based on the stripping away of a given thickness of material from a bombarded specimen and on the subsequent measurement of the amount of the remaining inert gas or the amount of the inert gas removed (anodic stripping, chemical and electrochemical stripping, mechanical vibratory polishing, sputtering to remove layers of known thickness from metals).

(2) Techniques based on the determination of the changes in the properties of the specimens caused by labelling bombardment, e.g. refractive index in transparent materials, electrical resistivity of semiconductors, etc.

(3) Techniques making use of the alpha or beta radioactivity of labelling atoms. The techniques are based on the fact that the energy spectra of alpha particles and low energy beta particles are degraded according to the depth of the radioactive atoms below the surface.

The penetration depth is usually expressed as the total weight of target per unit area (in $\mu g/cm^2$), as the number of atom layers, or as the distance (nm) from the surface. A depth of $1\,\mu m/cm^2$ corresponds approximately to 16 atom layers of aluminium, i.e. 4 nm.

The exact estimation of the penetration depth and the distribution profile of inert gas atoms introduced into the solids by ion bombardment is rather complex.

Experiments have shown that the dependence of the concentration of inert gas trapped within the solid on the flux of bombarding particles reaches a saturation value [198]. The saturation value, q, for a given inert gas and a given target material is usually expressed as the total weight of the sample per unit area in $\mu g/cm^2$ or in the number of atom layers. Target saturation occurs when the integrated flux of incident particles is sufficient to damage and to sputter away a significant amount of the target material, and some of the collected gas atoms escape.

The saturation value increases almost linearly with ion energy from 0.7 to 3.75 keV for inert gases in Ni, Mo and Pt. At the highest energies q for inert gas atoms differs for various materials. The variation of q with target atom weight is periodic, showing maxima and minima. The minima correspond to the minima of trapping efficiency, η, and to maxima in the sputtering.

It is interesting to note that q reaches a maximum for elements belonging to the same group of the periodic table, i.e. for group 4a (Ti, Zr, Hf). A minimum in q is found for Zn and Cd (group 2b). An approximate relationship between binding energies of elements and trapping efficiency can be noted. As might be expected, the less strongly held atoms are more easily removed by sputtering.

4.5.2 Analysis of non-radioactive gases

Most applications of radioactive Kryptonates have been determinations of gaseous components and traces of impurities in air. Methods have been developed for the determination of ozone [212], oxygen [212], sulphur dioxide [213], fluorine [214], chlorine [215], nitric oxide [215], amines [215], mixed hydrazine fuel [215], hydrogen fluoride [216], water vapour [217], carbon monoxide [218], hydrogen [203], etc. Some of these methods are intended for use in the determination of these components in the atmosphere of other planets (e.g. Mars and/or Venus). These methods are also of great importance in industrial hygiene for the determination of components (explosive and/or toxic) in the atmosphere of various hazardous workplaces.

4.5.2.1 Oxygen

Radioactive Kryptonates (kryptonated graphite, pyrographite, or copper) release ^{85}Kr during their oxidation because of the chemical destruction of the surface layers. The amount of ^{85}Kr released is proportional to the degree of oxidation and thus to the concentration of oxygen. The oxidation rate of the substance is generally a function of the temperature and oxygen concentration. It is an important limitation to the use of the most readily oxidized substances that they readily become coated with an oxide film. In the presence of oxygen this film is quickly formed and prevents further oxidation. In practice, the oxidation rate becomes independent of the oxygen concentration and approaches zero order. Kryptonated carbon (spectroscopically pure graphite rods) is a suitable material for the determination of oxygen.

Extensive tests have been made of oxygen determination by means of the radioactive Kryptonate of pyrolytic graphite at various temperatures, pressures and oxygen concentrations [212]. Figure 4.16 shows the relationships. It is obvious that changing the oxygen concentration will cause a change in the rate of decrease in activity of the Kryptonate. At high oxygen concentrations, the decrease of activity with time is not linear.

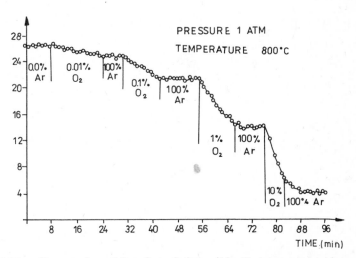

Fig. 4.16 — Decrease in activity of pyrolytic graphite Kryptonate at various oxygen concentrations. (Reproduced from [212] by permission of the copyright holders, Pergamon Press, Oxford.)

The rate of decrease in activity of the Kryptonate of pyrolytic graphite is a linear function of the logarithm of the oxygen concentration (at constant temperature) for various temperatures (Fig. 4.17) [212]. The rate at 1050 °C is sufficient for the determination of oxygen in the concentration range usually required (0.1 vol. % and total pressure of 13–130 mbars, which corresponds to 0.01–0.13 mbars partial pressure of oxygen). By using labelled pyrolytic graphite it has been possible to determine 2.2 ppm v/v oxygen in argon.

Use of copper Kryptonate (^{85}Kr) at different temperatures makes possible the determination of 10^{-5}–10^5 ppm of oxygen. This Kryptonate reacts faster with oxygen than the pyrolytic graphite Kryptonate does. Chleck and Cucchiara [212] proposed copper Kryptonate (which releases ^{85}Kr at a lower temperature than graphite or pyrographite) for the determination of the oxygen content in the atmosphere of Mars and other planets. Other components of the Earth's atmosphere (carbon dioxide and nitrogen), which may also occur in the atmosphere of Mars, do not interfere, nor does the density of the Mars atmosphere. The device monitors the activity of a Kryptonate by means of a sodium iodide crystal connected to a photomultiplier, power-supply, and readout.

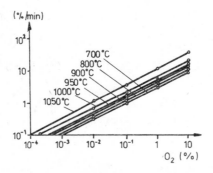

Fig. 4.17 — Response of pyrolytic graphite Kryptonate to oxygen at various temperatures. (Reproduced from [212] by permission of the copyright holders, Pergamon Press, Oxford.)

The practical problems involved in adapting the radioactive Kryptonate technique for use in a sensor on Mars include those set out below [219].

(1) Minimization of the electrical power to heat the radioactive Kryptonate to the temperature required to achieve adequate sensitivity for oxygen at the low levels expected in the Martian atmosphere. The optimum configuration for the kryptonated copper is a sandwich-type arrangement, in which the nichrome wire (which has a resistivity 100 times that of copper) is electrically insulated from the exterior kryptonated copper foil by two sheets of mica.

(2) Measurement of a wide range of oxygen concentrations by using two sources at different temperatures with separate Geiger–Müller tubes monitoring each. Another possible approach is the arrangement of two sources at different temperatures, but with the total activity monitored by one detector. At a high oxygen concentration the more

sensitive source would lose its ^{85}Kr rapidly until depleted, while the less sensitive source would lose its activity slowly, and the measured rate of loss of activity, or its average, would be the measure of the oxygen concentration.

(3) Improvement in sensitivity. A comparison of the reaction rates of copper samples indicated increase by a factor of 6–7 in the sensitivity when samples were prepared by bombardment instead of diffusion.

(4) Long-term exposure to high vacuum has no discernible effect on the sensitivity of copper Kryptonate as a detector for oxygen.

(5) Possibility of measuring ozone with the same kryptonated copper detector as used for oxygen. Although the radioactive copper Kryptonate prepared by ion-bombardment demonstrated a greater sensitivity to oxygen, increased sensitivity to ozone could not be achieved. The higher source temperature that appears necessary for adequate sensitivity at very low ozone levels would promote reaction with oxygen as well.

(6) Selection of a radiation detector that would withstand the contemplated pre-flight sterilization procedure (heating at 150°C for 24 hr).

4.5.2.2 Hydrogen

The increasing use of hydrogen in industry and for military purposes requires reliable instrumentation capable of measuring it, because it can react explosively with oxygen and halogens under certain conditions.

Chleck [203] developed a method for detecting hydrogen in concentrations $\leq 1\%$ aboard aerospace flight vehicles.

Among all Kryptonates of metal oxides investigated, the best hydrogen sensor seems to be that of platinum dioxide:

$$2H_2 + PtO_2[^{85}Kr] \quad \rightarrow \quad 2H_2O + Pt + {}^{85}Kr$$

During the measurement in the hydrogen detection system (Fig. 4.18) the reaction vessel containing the radioactive Kryptonate of platinum dioxide was filled with gases containing different hydrogen concentrations, at room temperature as well as at higher temperatures. For work at higher temperatures the Kryptonate was preheated in a furnace. The preheating makes the Kryptonate thermally stable at the furnace temperature and lower temperatures. Therefore, the activity of the Kryptonate will decrease only when reaction takes place. The work was carried out with milligrams of metal oxide Kryptonates with specific activities of 37–185 kBq/mg.

The logarithm of the rate of loss of activity of the Kryptonate (%/min) is directly proportional to the logarithm of the concentration of hydrogen above the $PtO_2[^{85}Kr]$, when the measurements are made at room temperature in an atmosphere of nitrogen. With air as the diluent, no decrease in activity was detected, because the hydrogen was oxidized by oxygen on the surface of the oxide:

$$2\,H_2 + O_2 \quad \xrightarrow{\;PtO_2[^{85}Kr]\;} \quad 2\,H_2O$$

During this oxidation, the surface of the oxide is not disturbed, and so krypton is not released. For the measurement of hydrogen concentration in its mixtures with oxygen, e.g. in air, this catalytic oxidation must be avoided. When the oxide is coated with an inert material such as aluminium oxide, the oxygen is excluded from the reaction, and once again reduction occurs with subsequent release of activity:

$$2\,H_2 + \text{coated } PtO_2[^{85}Kr] \quad \rightarrow \quad \text{coated } Pt + 2\,H_2O + {}^{85}Kr \uparrow$$

These coatings probably act by delaying the arrival of the relatively larger atoms at the surface of the $PtO_2[^{85}Kr]$. It is necessary to raise the temperature of the coated sample to 350 °C to obtain a sensitivity equal to that of an uncoated sample at room temperature.

Mixtures of kryptonated platinum dioxide and various inert materials were prepared to determine the optimum reagent material for hydrogen. The mixtures with HgI_2, AgI, PbS, MoS_2 and Al_2O_3, showed the greatest sensitivity, and were intensively investigated as to their ability to detect hydrogen under varying conditions of temperature and pressure. Although that with Al_2O_3 was the least sensitive of the five mixtures tested, it was the only reagent that did not lose sensitivity with increasing time, gas flow-rate or temperature.

The prototype of a hydrogen sensor [203] is shown in Fig. 4.19. The basic component consists of an end-window Geiger–Müller (GM) counter and a sensor element ($PtO_2[^{85}Kr]$–Al_2O_3 on a nichrome wire, fixed on a ceramic ring).

With a mixture of $PtO_2[^{85}Kr]$ and Al_2O_3 at the maximum working temperature of 435 °C (limited by the thermal stability of PtO_2), 1 % hydrogen can be determined in < 0.5 sec and 10 ppm hydrogen in < 10 min. The higher the specific activity of the Kryptonate used, the higher the sensitivity of the method.

Small amounts of hydrogen sulphide, sulphur dioxide, hydrogen chloride, hydrogen fluoride and water do not measurably interfere with the determination. Water vapour up to saturation, carbon dioxide up to 1 %, and sulphur dioxide up to 0.5 % have no influence. Hydrogen sulphide does not interfere up to a concentration of 100 ppm; at higher concentrations it does not react with the $PtO_2[^{85}Kr]$–Al_2O_3, but it prevents the reaction of hydrogen. Methane reacts only at high concentrations and temperatures.

This sensor demonstrates excellent specificity for hydrogen. A system for hydrogen detection by means of Kryptonates has been described; it can detect hydrogen in concentrations up to 3 % in air and up to 10 % in nitrogen, for use in the Saturn interstage units as well as in ground-base installations.

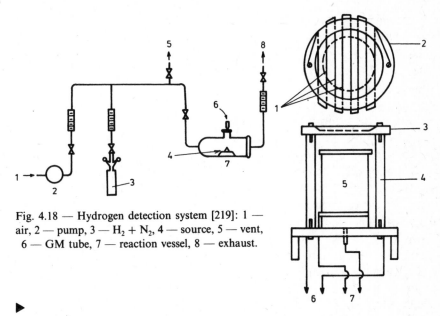

Fig. 4.18 — Hydrogen detection system [219]: 1 — air, 2 — pump, 3 — $H_2 + N_2$, 4 — source, 5 — vent, 6 — GM tube, 7 — reaction vessel, 8 — exhaust.

Fig. 4.19 — Prototype of hydrogen sensor. 1 — PtO_2 [^{85}Kr] – Al_2O_3 source on nichrome wire, 2, 3 — ceramic insulator, 4 — brass conducting and support rods, 5 — GM tube, 6 — source power supply, 7 — ratemeter and recorder. (Reproduced from [203] by permission of the copyright holders, IAEA, Vienna.)

4.5.2.3 Determination of air pollutants

Analysis for air pollutants is very difficult for a number of reasons; these include inhomogeneity of samples, the wide range of chemical and physical characteristics of pollutants, the low concentrations of many constituents, the instability of some pollutants, and interferences. No single method provides a general solution to all these problems. However, radiometric emanation methods show considerable promise in air pollution analysis [213].

Ozone. Ozone is an extremely reactive oxidant which has been found to produce an odour at 0.02–0.05 ppm, and irritation of the nose and throat at concentrations greater than 0.05 ppm. The World Health

Organization has suggested an upper safe exposure limit in industry of $100–200\,\mu g/m^3$ for an 8-hour day. The gas has also been implicited in the formation of smog, damage to flora and the premature cracking of rubber.

In ozone determinations with radioactive hydroquinone Kryptonate, the following reaction is used [212]:

$$[C_6H_4(OH)_2]_3[^{85}Kr] + O_3 \quad \rightarrow \quad 3\,C_6H_4O_2 + 3\,H_2O + {}^{85}Kr\uparrow$$

The logarithm of the activity of the ^{85}Kr released is proportional to the logarithm of the ozone concentration over a concentration range of $10^{-10}–10^{-6}\,g$ of ozone per litre of air (Kryptonate specific activity of about $3.7\,GBq/g$). A higher specific activity proportionally increases the sensitivity.

The general requirements for air analysers to be used in the upper layers of the atmosphere are high sensitivity, low weight, small dimensions and low price. These requirements are met in an ozone analyser which consists of the radioactive Kryptonate, a measuring cell with an end-window GM counter, and a small pump. The analyser, in a plastic case, is attached to the upper part of a normal weather radiosonde; the activity is recorded directly.

Another ozone analyser incorporates a highly efficient reagent produced by the improved process of adsorbing radioactive hydroquinone Kryptonate on calcium silicate [220].

For the determination of ozone in a gaseous medium also containing atomic oxygen, the apparatus consists of two reaction cells; one contains an untreated hydroquinone Kryptonate (without calcium silicate or platinum-black-silica catalyst), which responds to atomic oxygen in the gas stream, and the other a treated or catalysed hydroquinone Kryptonate, which responds to both ozone and atomic oxygen. The difference in ^{85}Kr readings in the measuring cells indicates the ozone concentration.

Chleck and Cucchiara [212] also investigated the feasibility of ozone determination by means of radioactive copper Kryptonate. In experiments at $200°$ and $100°C$, no reaction with oxygen was detected. However, it was shown that this Kryptonate can be used at temperatures below $100°C$.

Chromatographic papers impregnated with hydroquinone and kryptonated by diffusion were used for the determination of ozone by Pružinec *et al.* [221]. This emanometric determination is based on measurement of the residual activity after a suitable contact time.

Sheets of Whatman No. 2 paper ($20 \times 2\,cm$) were immersed in a concentrated solution of hydroquinone, allowed to dry at room temperature, and kryptonated. Air–ozone mixtures were allowed to impinge on 1.5-cm diameter circles of the paper. These circles were then counted to determine the decrease in radioactivity.

The dependence of relative residual activity on concentration was found to be linear in the ozone concentration range of 100–400 µg/l. The reproducibility of the method was approximately ± 4%.

Sulphur dioxide. Sulphur dioxide is an irritant gas that affects the eyes and the mucous membranes of the nose, throat and lungs. The generally accepted maximum allowable concentration that can be tolerated for several hours is 10 ppm [200].

Sulphur dioxide has been determined by a method based on the mechanism of double release [213, 222]. In the first stage, sulphur dioxide reacts with sodium chlorate to release chlorine dioxide, which is a strong oxidizing agent. The chlorine dioxide then oxidizes radioactive hydroquinone Kryptonate and gaseous ^{85}Kr is released. The following reactions are involved:

$$SO_2 + 2\,NaClO_3 \quad \rightarrow \quad 2\,ClO_2 + Na_2SO_4$$

and

$$ClO_2 + [C_6H_4(OH)_2]_3[^{85}Kr] \quad \rightarrow \quad ^{85}Kr\uparrow$$

(The first equation is an oversimplification, since one molecule of sulphur dioxide releases between 4 and 8 molecules of chlorine dioxide.)*

The activity of the ^{85}Kr released is linearly dependent on the concentration of sulphur dioxide, which can be determined down to 0.001 ppm. The diagram of the apparatus for the determination of sulphur dioxide in air is given in Fig. 4.20. The only gaseous substances which interfere are the oxides of nitrogen and ozone.

Chromatographic papers impregnated with hydroquinone and kryptonated by diffusion have been used for the determination of sulphur dioxide down to the same detection limit [223].

In another method, sulphur dioxide has been determined in air with the use of a radioactive Kryptonate of iodine [224]:

$$SO_2 + I_2[^{85}Kr] + 2\,H_2O \quad \rightarrow \quad H_2SO_4 + 2\,HI + ^{85}Kr\uparrow$$

---

* Much of the earlier work refers to use of sodium chlorite ($NaClO_2$) but as the chlorine in this is in oxidation state Cl(III), it is difficult to see how it can be oxidized to Cl(IV) by SO_2.

Fluorine, hydrogen fluoride. Fluorine, fluorides and hydrogen fluoride are produced from steel, aluminium and phosphate mills, pottery
and glass works, etc. They can concentrate in food chains, and also
cause damage to terrestrial plants and algae. Severely fluoride-polluted
forage can produce fatal fluorosis in cattle. Fluorides are used to
prevent tooth decay, but excessive amounts lead to severe discoloura-
tion of the teeth.

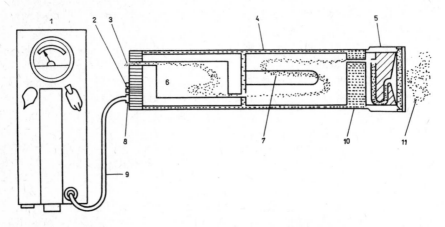

Fig. 4.20 — Scheme of instrument for determination of sulphur dioxide: 1 — indicating
meter, 2 — power supply, 3 — vent, 4 — detector, 5 — sensing head, 6 — pump,
7 — GM tube, 8 — power and signal connectors, 9 — signal cable, 10 — lead, 11 — air
intake. (Reproduced from [213] by permission of the copyright holders, IAEA, Vienna.)

Fluorine and fluorine-containing oxidants can be determined by
means of radioactive hydroquine Kryptonate. In the apparatus described by Hommel *et al*. [214], a radioactive Kryptonate with a specific
activity of 16.65 GBq/g was used.

An instrument that is capable of measuring both hydrogen fluoride
and fluorine by using a radioactive Kryptonate of silver iodide and
hydroquinone has been developed [225]. The following reactions are
involved:

$$AgI[^{85}Kr] + HF \quad \rightarrow \quad AgF + HI + {}^{85}Kr\uparrow$$

and decomposition of $[C_6H_4(OH)_2]_3[^{85}Kr]$ by F_2.

With this instrument it is possible to determine fluorine at concentrations of 0–10 ppm and hydrogen fluoride at concentrations of 0–50
ppm.

Jesenák *et al*. [216] suggested an emanometric method for hydrogen
fluoride, based on its reaction with kryptonated silica:

$$SiO_2[^{85}Kr] + HF \quad \rightarrow \quad SiF_4 + H_2O + {}^{85}Kr \uparrow$$

Discs of monocrystalline silica (13 mm in diameter and 1 mm thick) were kryptonated by ion bombardment. A series of HF solutions was prepared, covering the molar concentration range 0.0–0.15, which provided equilibrium vapour pressures of HF corresponding to the range 0.0–10.0 ppm. Ten-ml portions of the appropriate solutions were equilibrated in 50-ml plastic bottles at 25 °C for 2 hr. The kryptonated silica discs were suspended inside the bottle and allowed to react with the HF vapour. The change in activity of each disc with time was determined.

The dependence of the residual activity A_r on exposure time t and concentration of HF (C, ppm) is given approximately by the equation

$$\log A_r = -0.0072t + \log(1 - 0.0072C)$$

A_r is linearly related to the concentration of HF in the concentration range 0–2 ppm.

This method may also be suitable for the determination of the average effective HF concentration over longer exposure times. Hydrogen bromide will not interfere at concentrations up to 150 ppm and HCl at 115 ppm decreases A_r by less than 1 %.

Mercury. Air pollution by mercury vapour is a highly localized but serious problem. The material is moderately toxic, but it seems to have a short retention time in the human system. Methylmercury, on the other hand, is extremely toxic (long retention time). Certain micro-organisms are capable of converting elemental mercury into the methylated form.

Mercury vapour can be determined by an emanometric method using kryptonated selenium sulphide [226]. Freshly prepared selenium sulphide was kryptonated by a microdiffusion technique and applied as a coating to a filter paper. Upon exposure to air containing Hg vapour, the radioactivity decreased. The decrease, divided by the initial specific activity, was found to be directly proportional to the mercury vapour concentration and to the exposure time.

The apparatus used is shown in Fig. 4.21. Air was bubbled through a pool of mercury in thermostat I at 55 °C. The gas saturated with mercury vapour was then passed through thermostat II to reduce the concentration of mercury vapour to that corresponding to the temperature of thermostat II. After passage through a cotton filter (3) to remove any condensed mercury, the gas flowed into the reaction vessel (4) at 55 °C. The flow-rate of the gas was measured and controlled by a flowmeter (6). The concentration of mercury vapour could be varied by changing the temperature of thermostat II.

Papers impregnated with a known amount of kryptonated selenium sulphide were kept at 55 °C to achieve thermal stability for the Kryptonate. Their activity was counted, and the relative specific activity was calculated. In a given experiment, the activity of the sensor was monitored as a function of time for different mercury vapour concentrations.

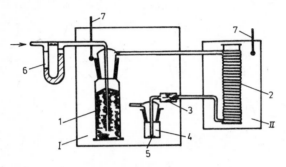

Fig. 4.21 — Diagram of the apparatus used for the determination of mercury vapour: 1 — mercury bubbler, 2 — rubber tubing wound around cylindrical flask, 3 — cotton filter, 4 — reaction vessel, 5 — sensor, 6 — flowmeter, 7 — thermometer; I — thermostat at 55 °C, II — adjustable thermostat. (Reproduced from [226] by permission of the copyright holders, Akadémiai Kiadó, Budapest.)

The results obeyed a simple linear relationship, showing that the decrease in the radioactivity divided by its initial specific activity, $\Delta A/S_0$, was directly proportional to the mercury vapour concentration and to the exposure time:

$$-\frac{\Delta A}{S_0} = 7.8 \times 10^{-4} C_{Hg}t$$

where t is in minutes, and C_{Hg} in ppm.

The excellent linearity of the calibration curves suggests good reproducibility. A test of the accuracy of the method gave the results listed in Table 4.15.

Carbon monoxide. Incomplete combustion of fuels in motor vehicles in one of the greatest sources of CO. Its toxicity to animals, including man, is due primarily to its affinity for haemoglobin. It forms carboxyhaemoglobin, which reduces the ability of blood to carry oxygen. Concentrations of 10–15 ppm for 8-hr periods are common in downtown streets of urban areas. The World Health Organization has suggested an 8-hr per day exposure limit of 50 ppm in industry [200].

Table 4.15 — Results of emanometric determination of mercury vapour. (Reproduced from [226] by permission of the copyright holders, Akadémiai Kiadó, Budapest.)

Mercury (ppm)	
Taken	Found
8.50	8.60
	8.40
6.00	6.15
	6.10
4.10	4.25
	3.98
2.80	2.60
	2.58
1.84	1.95
	1.85

Naoum *et al.* compared several oxidizing agents for the emanometric determination of CO; these included Hopcalite, I_2O_5, $PdCl_2$ and HgO [218].

In studies of the radioactive Kryptonate of $PdCl_2$, a fritted glass disc was impregnated with the Kryptonate, placed in a flow chamber, and exposed to a gas containing various concentrations of CO. The concentration dependence of the relative residual activity on the disc under conditions of constant flow-rate and exposure time (15 min) was linear in the range 25–100 ppm.

Filter paper was impregnated with HgO, kryptonated at 180 °C, and exposed to CO at 180 °C for 5 min. It was noted that the radioactivity on the paper decreased. This indicated that kryptonated HgO may be suitable for the determination of CO.

Results obtained with the first two agents and those with kryptonated nickel were not encouraging.

Water vapour. Naoum *et al.* have demonstrated [227, 228] that kryptonated discs of silica gel can be used for the emanometric determination of water vapour in the concentration range 1–3 % v/v. Silica gel powder was kryptonated by diffusion and pressed into pellets 1 cm in diameter and 1 mm thick, resulting in a pseudohomogeneous distribution of ^{85}Kr. A series of sulphuric acid solutions was prepared in the concentration range 1–14M; these had equilibrium water vapour concentrations in the range 1–3 % v/v.

In the study, 10 ml of the appropriate acid solution were placed in a 50-ml plastic bottle and allowed to stand overnight in a thermostat at 25 °C to achieve liquid–vapour equilibrium. The silica gel disc was placed inside the bottle in a plastic holder suspended by the bottle stopper and allowed to react with the water vapour inside the bottle for various periods of time. The time dependence of the activity of the disc was monitored and the relative residual activities were plotted against exposure time. A calibration curve for an exposure time of 8 hr is quite linear.

It was found that the experimental conditions for exposure times less than 4 hr could be approximated by a model of the initial period of diffusion into an infinite plane.

According to this model the concentration of water vapour is related to residual activity and exposure time as follows:

$$C_{H_2O} = k_1 \left(\frac{1 - A_r}{t^{1/2}} + k_2 \right)$$

where k_1 and k_2 depend on the characteristics of individual silica discs and the temperature.

The dependence of the relative residual activity on the vapour concentration and temperature for a constant exposure time of 8 hr is given by the equation

$$A_r = 1.00 - (0.254 - 0.00448t) \, C_{H_2O}$$

where t is in °C and C_{H_2O} has units of % v/v.

The adsorption isotherm of water vapour on silica gel is complex (e.g. it has a hysteresis loop). Therefore the equation developed is semi-empirical and the constants must be evaluated for a given set of experimental conditions.

4.5.2.4 Analysers
Two types of apparatus for the determination of oxidizing or reducing atmospheric impurities (F_2, Cl_2, $CClF_3$, SO_2, NO_2, NO, RNH_2, O_3), in concentrations of the order of ppm, have been developed in the research

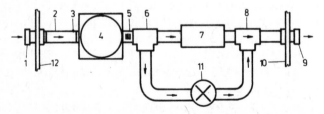

Fig. 4.22 — The flow diagram for the NO_2 monitor [215]: 1 — input, 2 — clathrate cell, 3 — red dot or arrow, 4 — counting chamber, 5 — flowmeter, 6 — T-junction, 7 — pump, 8 — T-junction, 9 — exhaust, 10 — panel, 11 — flow control, 12 — panel.

laboratories of Tracerlab. These are the NO_2 detector, which represents one of the simplest devices, and the UDMH (unsymmetrical dimethyl hydrazine) detector, which is much more complicated. With these instruments, gases which can react directly or indirectly by chemical, catalytic or thermal reactions with radioactive Kryptonates, are measurable [215, 229].

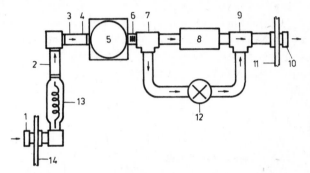

Fig. 4.23 — The flow diagram for the UDMH monitor [215]: 1 — input, 2 — $NaClO_3$ cell, 3 — clathrate cell, 4 — red dot or arrow, 5 — counting chamber, 6 — flowmeter, 7 — T-junction, 8 — pump, 9 — T-junction, 10 — exhaust, 11 — panel, 12 — flow control, 13 — cracking coil, 14 — panel.

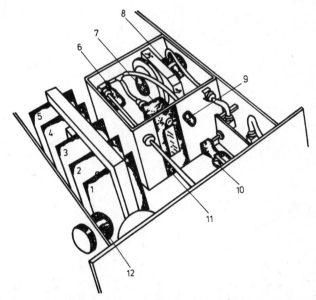

Fig. 4.24 — The internal view of the NO_2 monitor [215]: 1 — ratemeter, 2 — H.V. supply, 3 — charge monitor, 4 — discharge monitor, 5 — PRW and alarm, 6 — pump, 7 — GM tube, 8 — counter, 9 — flowmeter control valve, 10 — clathrate cell, 11 — flowmeter, 12 — sound alert.

The flow diagrams for the NO_2 and UDMH monitors are shown in Figs. 4.22 and 4.23. The basic system contains a sampling pump, flow-control valve, counting chamber and radioactive Kryptonates. Chemical and catalytic cells are included when necessary (Fig. 4.23). The sample gas is pulled through the system by the pump. As the gas of interest enters the instrument, it immediately comes into contact with the chemical and catalytic cells (if they are required), and then reacts with the radioactive Kryptonate. The ^{85}Kr released enters a counting chamber of known volume, where its concentration per unit volume is determined. This concentration can be related directly to the volume concentration of the gas of interest entering the system.

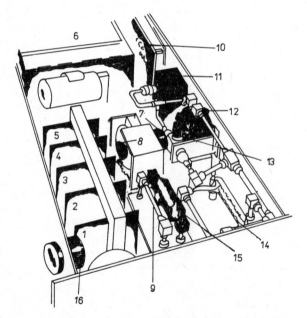

Fig. 4.25 — The internal view of the UDMH monitor [215]: 1 — ratemeter, 2 — H.V. supply, 3 — charge monitor, 4 — discharge monitor, 5 — PWR and alarm, 6 — cracking coil power supply, 7 — counting chamber, 8 — GM tube, 9 — NaClO$_3$ cell, 10 — power circuits, 11 — flowmeter, 12 — pump, 13 — flowmeter control valve, 14 — cracking coil, 15 — clathrate cell, 16 — sound alert.

An internal view of the NO_2 monitor is shown in Fig. 4.24, and of the UDMH monitor in Fig. 4.25. Multi gas systems have much poorer operational characteristics than the single gas units.

A lower detection limit of 0.5 ppm has been found. The upper limit is dependent on the required cell life and the saturation characteristics of the GM tube.

A schematic drawing of an instrument for the detection of three different gases (CO, NO$_x$ and hydrocarbon vapours) in automobile exhaust gases is shown in Fig. 4.26. Sensing substrates are [229]:

(a) PtO$_2$[^{85}Kr] for the determination of hydrocarbon vapours (detection limit for hexane is $\sim$ 16 ppm);

(b) [C$_6$H$_4$(OH)$_2$]$_3$[^{85}Kr] for the determination of NO$_x$ (minimum detectable concentration is $\sim$ 2 ppm);

(c) PdCl$_2$[^{85}Kr] for the determination of CO (minimum detectable concentration is $\sim$ 125 ppm).

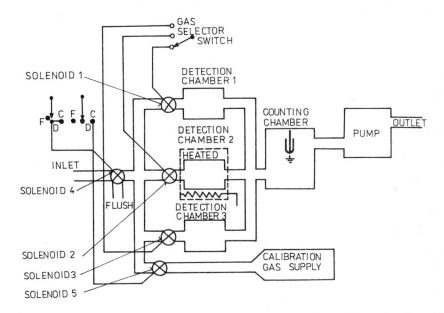

Fig. 4.26 — Schematic drawing of an instrument operating in the ^{85}Kr effluent mode for the detection of three different gases. (Reproduced from [229] by permission of the copyright holders, International Scientific Communications, Inc.)

4.5.3 Analysis of solutions

Radioactive Kryptonates can also be used for the analysis of solutions by direct reaction with the analyte or the reagent(s) used. Titration methods are of special importance, and are treated separately.

With the methods described in this part, analytical results were achieved by the use of calibration curves. Either the decrease in activity of the radioactive Kryptonate or the radioactivity of the ^{85}Kr released is measured.

4.5.3.1 Non-aqueous solutions

The effectiveness and useful life of jet fuels and fire-extinguishing fluids are affected by the amount of water present as an impurity.

A method for the determination of traces of water in organic liquids, particularly in methanol [203, 217], uses the apparatus shown in Fig. 4.27. The reaction vessel contains an excess of the water-sensitive radioactive Kryptonate of calcium carbide. The determination of water is based on the reaction

$$2\,H_2O + CaC_2[^{85}Kr] \quad \rightarrow \quad Ca(OH)_2 + C_2H_2 + {}^{85}Kr \uparrow$$

The total activity of the ^{85}Kr released is measured, and is directly proportional to the amount of water reacting with the radioactive Kryptonate added to the solution investigated. The method is applicable for the determination of 0.25–2.0 % H_2O.

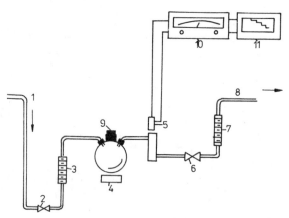

Fig. 4.27 — Device for the determination of traces of water in organic liquids: 1 — nitrogen supply, 2 — metering valve, 3 — flowmeter, 4 — magnetic stirrer, 5 — counting chamber, 6 — metering valve, 7 — flowmeter, 8 — exhaust, 9 — reaction vessel, 10 — ratemeter, 11 — recorder. (Reproduced from [217] by permission of the copyright holders, J. T. Baker Chemical Co.)

4.5.3.2 Determination of water pollutants

Polluted water can rarely be analysed satisfactorily by a single method; rather a group of methods is required. Emanometric methods have been successfully used in a number of areas.

Oxygen and pH. Dissolved oxygen has been a constituent of major concern in water quality studies and pollution control. It is vital in the protection of fish and other important aquatic species which must have an adequate oxygen supply in order to survive. The presence, or absence, of oxygen is a major determinant of water quality.

Dissolved oxygen can be measured by the use of radioactive thallium Kryptonate [230, 231]. The measurement is based on the reaction

$$4\,Tl[^{85}Kr] + O_2 + 2\,H_2O \quad \rightarrow \quad 4\,Tl^+ + 4\,OH^- + {}^{85}Kr\uparrow$$

The reaction is stoichiometric, therefore the amount of radioactive krypton released and the reduction of residual activity in the kryptonated thallium are proportional to the dissolved oxygen concentration in the sample under conditions of constant pH. The thallium carrier can be prepared by electrolytically coating a copper or platinum plate, or copper turnings with thallium. The carrier is then kryptonated by bombardment with ^{85}Kr ions or by diffusion.

Either the radioactive thallium Kryptonate is placed in the sample, or the sample is passed through a column of the turnings. Either the released ^{85}Kr or the decrease of activity of the thallium Kryptonate can be measured. The analysis can be automated.

With distilled water samples, the decrease in activity of the thallium Kryptonate was found to be linearly related to the concentration of oxygen down to the lowest concentration investigated (0.3 ppm).

The pH of a water sample can be measured by making use of the fact that the dissolution of thallium Kryptonate is pH-dependent (and therefore so is the release of ^{85}Kr) under conditions of constant temperature and pressure. These conditions are of course related to oxygen concentration.

In a study of this method, thallium carriers were prepared by electroplating thallium metal onto well polished discs of copper. The thallium carriers were kryptonated by ion bombardment.

To make a pH measurement, the thallium Kryptonate disc was submerged in 20 ml of solution for an arbitrary time. The disc was then dried in a desiccator under reduced pressure, and its residual activity measured. The pH was determined from an appropriate calibration chart [231, 232].

The optimal sample–Kryptonate contact time was found to be 60 sec and the residual activity of the Kryptonate linearly related to the pH under constant experimental conditions of reaction time, temperature and pressure.

Dichromate. Chromium(VI) is irritating and corrosive to mucous membranes, and has been shown to be toxic to test animals when administered systematically.

The Environmental Protection Agency of the United States has established maximum permissible levels of 50 µg/l. chromium for domestic water supplies, and 100 µg/l. for fresh-water aquatic life.

An emanometric method for the determination of dichromate makes use of radioactive silver Kryptonate according to the reaction

$$Cr_2O_7^{2-} + 6\,Ag[^{85}Kr] + 14\,H^+ \quad \rightarrow \quad 6\,Ag^+ + 2\,Cr^{3+} + 7\,H_2O + {}^{85}Kr \uparrow$$

The electrochemical potentials of the Ag/Ag^+ and $Cr^{3+}/Cr_2O_7^{2-}$ couples are such that the reaction is expected to be quantitative. The method was developed with neutral solutions of potassium dichromate in the concentration range 0.0125–$0.1 N$. Iron(III), chloride and nitrate interfere [231, 233].

Vanadium. This element is known to lower the free cholesterol and lipid content of man. At concentrations of 0.03–0.22 mg/l. in drinking water it has been reported to reduce dental caries in human teeth. Apparently the element can stimulate plant growth, and mitigate the deleterious effects of iron deficiency. On the other hand, excessive levels of vanadium have been demonstrated to be lethal to rats, and certain species of fish.

Vanadate can be measured in aqueous solutions by use of silver Kryptonate in a procedure similar to that reported above for dichromate. The reaction is carried out in fluoride media, and can be described by the equation

$$V(OH)_4^+ + 2\,H^+ + Ag[^{85}Kr] \quad \rightarrow \quad VO^{2+} + Ag^+ + 3\,H_2O + {}^{85}Kr \uparrow$$

To study the method, a series of solutions was made up as follows: 0.2 ml of NH_4F (300 mg/ml) and 0.15 ml of concentrated H_2SO_4 were added to 50 ml of ammonium vanadate solutions of varying concentrations. A $10^{-7}M$ solution gave a measurable decrease in the residual activity of the silver Kryptonate. It is possible that this sensitivity could be enhanced even more by increasing the time during which the silver Kryptonate disc is immersed in the sample [231, 233].

Hydrochloric acid. Silver Kryptonate may also be used to determine the concentration of hydrochloric acid, by use of a calibration curve as in the methods above. However, in this case, it is more convenient to plot residual activity of the silver Kryptonate against the logarithm of the acid concentration. The method might be useful at low acid concentrations, since a concentration of $10^{-5}M$ HCl still produced a decrease of about 15% in the residual activity of the silver Kryptonate [231, 233].

Sulphur dioxide. Chleck *et al.* [234] developed a method for the determination of sulphur dioxide in water by use of the radioactive Kryptonate of zinc fluoride. Sulphur dioxide forms sulphurous acid, which is allowed to react with $ZnF_2[^{85}Kr]$ suspended in the solution

$$H_2SO_3 + ZnF_2[^{85}Kr] \quad \rightarrow \quad ZnSO_3 + 2\,HF + {}^{85}Kr \uparrow$$

Attempts are being made to improve the sensitivity to better than 0.4 ppm of SO_2, and to reduce the measurement time [231].

Apparatus. Tölgyessy and co-workers [235, 236] have developed various types of analysers for the determination of dissolved oxygen in water (at constant pH) and the determination of pH (at constant oxygen concentration). The same analysers can be used for the determination of various pollutants with the aid of appropriate radioactive Kryptonates.

Analyser for the determination of dissolved oxygen in water, based on discontinuous measurement of the residual activity of thallium Kryptonate (Fig. 4.28). The lower part of the measuring cell (2) [with the radioactive thallium Kryptonate at (3)] is counted to determine the initial activity. The time necessary to accumulate 10000 counts is recorded as t_0. With the aid of suction device (1), the reaction cell (2) is filled with the solution to be analysed (2 ml). After 60 sec, the solution is discharged and the cell is dismantled. The dried thallium Kryptonate is again counted for the same time, t_0. This number of counts can be used to calculate the residual activity as a percentage of the original activity.

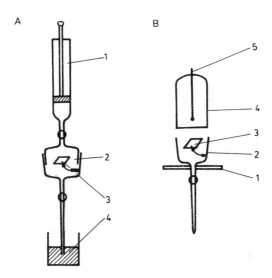

Fig. 4.28 — Analyser for the determination of dissolved oxygen in water based on discontinuous measurement of the residual activity of the thallium Kryptonate. A: 1 — suction device (5 ml syringe), 2 — reaction cell (2 ml), 3 — radioactive thallium Kryptonate, 4 — sample solution. B: 1 — reaction cell support, 2 — lower part of reaction cell, 3 — radioactive thallium Kryptonate, 4 — GM counter, 5 — connection to recording device. (Reproduced from [236] by permission of the copyright holders, Akademie-Verlag, Berlin.)

The oxygen concentration is read from an appropriate calibration chart [236].

Analyser for the determination of dissolved oxygen by measurement of the activity of the water after reaction with radioactive thallium Kryptonate (Fig. 4.29). The suction device (1) is used to pull the sample (7) into the reaction cell (2 ml volume) (4), which contains the radioactive thallium Kryptonate (5). After 60 sec contact time, the solution is passed to the measuring cell (10). The activity of the dissolved ^{85}Kr is measured with a scintillation detector, and the oxygen content is determined from the activity by means of a calibration curve. After the measurement, the sample is discharged from the measuring cell through outlet 12 [236].

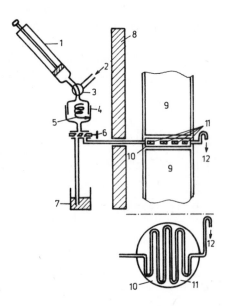

Fig. 4.29 — Analyser for the determination of dissolved oxygen in water by measurement of ^{85}Kr in the water: 1 — suction device (15 ml syringe), 2 — inlet for nitrogen or air, 3 — T-bore stopcock, 4 — reaction cell (2 ml volume), 5 — radioactive Kryptonate with homogeneous distribution (spiral form), 6 — two-way stopcock, 7 — sample, 8 — lead shielding, 9 — photomultiplier, 10 — scintillation measuring cell with a bored channel, 11 — bored channel (2 ml), 12 — outlet for analysed solutions.

Analyser for the determination of dissolved oxygen in water by measurement of the activity of ^{85}Kr released from thallium Kryptonate (Fig. 4.30). The suction device (1) is used to draw the sample into the reaction cell (4) (2 ml volume), which contains the thallium Kryptonate (5).

After 60 sec, the solution is passed to the purging cell (11) (2.5 ml volume), under positive pressure applied by the suction device. Stopcock (17) is closed and stopcock (18) is opened. Nitrogen is used [through inlet (2) and valve (3)] to purge the gaseous [85]Kr from the aqueous sample into the detection cell (13). After the purging, stopcock (18) is closed, and the activity of the gaseous [85]Kr in the detection chamber is measured. After the measurement, the aqueous sample is discharged from the purging cell (11), through stopcock (17). The observed activity is used to determine dissolved oxygen by means of a calibration curve [236].

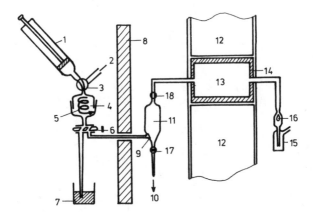

Fig. 4.30 — Analyser for the determination of dissolved oxygen in water by measurement of the activity of [85]Kr released from the aqueous solution. Items 1—8 as in Fig. 4.29; 9 — glass filter, 10 — outlet for analysed solutions, 11 — cell for purging of [85]Kr, 12 — photomultiplier, 13 — detection cell, 14 — scintillation detector, 15 — water seal, 16 — safety valve, 17, 18 — stopcocks.

REFERENCES

[1] G. Hevesy and F. Paneth, *Z. Anorg. Chem.*, 1913, **82,** 323.
[2] M. Haïssinsky, *La chimie nucléaire et ses applications*, Masson, Paris, 1957.
[3] R. Ehrenberg, *Biochem. Z.*, 1925, **164,** 183; 1928, **197,** 468.
[4] F. C. Henriques and C. Margnetti, *Ind. Eng. Chem., Anal. Ed.,* 1946, **18,** 477.
[5] J. Starý, J. Prášilová and T. Bouda, *J. Radioanal. Chem.*, 1975, **28,** 95.
[6] M. Kyrš and K. Přikrylová, *Chem. Listy*, 1981, **75,** 586.
[7] R. Becher, K. Buchtela and F. Grass, *Z. Anal. Chem.*, 1975, **274,** 1.
[8] G. Friedlander and J. W. Kennedy, *Lehrbuch der Kern- und Radiochemie*, Thiemig, Munich, 1962.
[9] S. A. Reynolds and G. W. Leddicotte, *Nucleonics,* 1963, **21,** 128.

[10] M. Kyrš and G. N. Bilimovich (eds.), *Novye metody radioanaliticheskoi khimii*, Energoizdat, Moscow, 1982.
[11] B. Mazière, *Produits Problèmes Pharm.*, 1972, **27**, 20.
[12] J. W. McMillan, *The Use of Tracers in Inorganic Analysis*, in *Radiochemical Methods in Analysis*, G. I. Coomber (ed.), Plenum Press, New York, 1975.
[13] M. I. Tousset, *Bull. Soc. Chim. France*, 1970, 2023.
[14] M. B. A. Crespi, *Pure Appl. Chem.*, 1971, **26**, 257.
[15] W. S. Lyon, *Radiochem. Radioanal. Lett.*, 1975, **20**, 233.
[16] W. Soaman, *Tracer Techniques*, in *Treatise on Analytical Chemistry*, I. M. Kolthoff and P. Elving (eds.), Part I, Vol. 9, Wiley, New York, 1971.
[17] V. I. Shamaev, *J. Radioanal. Chem.*, 1981, **63**, 301.
[18] V. I. Shamaev and A. M. Martynov, *Radiokhimiya*, 1972, **14**, 500.
[19] M. Kyrš, *Anal. Chim. Acta*, 1966, **33**, 245.
[20] K. Burger, *Acta Chim. Acad. Sci. Hung.*, 1964, **40**, 17.
[21] M. Kyrš and M. Pivoňková, *Chem. Listy*, 1967, **61**, 1502.
[22] H. Spitzy, *Mikrochim. Acta*, 1960, 789.
[23] S. Hirano, A. Mizuike and E. Nakai, *Radioisotopes*, 1964, **13**, 118.
[24] A. Langer, *Anal. Chem.*, 1950, **22**, 1288.
[25] V. Majer, *Základy jaderné chemie*, SNTL, Prague, 1961.
[26] M. H. Kurbatov and J. D. Kurbatov, *J. Am. Chem. Soc.*, 1947, **69**, 438.
[26a] A. R. Landgrebe, L. T. McClendon and J. R. DeVoe, *Chem. Abstr.*, 1965, **63**, 12310 g.
[27] V. I. Shamaev, G. K. Gaurbauskas and A. M. Martynov, *Radiokhimiya*, 1971, **13**, 94.
[28] J. Klas, J. Tölgyessy and E. H. Klehr, *Radiochem. Radioanal. Lett.*, 1974, **18**, 83.
[29] H. A. Das, H. Köhnemann and W. van der Mark, *J. Radioanal. Chem.*, 1975, **24**, 383.
[30] J. Tölgyessy, T. Braun and M. Kyrš, *Isotope Dilution Analysis*, Pergamon Press, Oxford, 1972.
[31] V. I. Shamaev, A. M. Martynov and M. A. Churilova, *Radiokhimiya*, 1972, **14**, 786.
[32] M. P. Menon, *J. Radioanal. Chem.*, 1973, **43**, 309.
[33] V. I. Shamaev, *J. Radioanal. Chem.*, 1979, **51**, 205.
[34] N. Suzuki and K. Kudo, *Anal. Chim. Acta*, 1965, **32**, 456.
[35] A. V. Pomerantseva, V. V. Atrashkevich and I. E. Zimakov, *Zh. Analit. Khim.*, 1971, **26**, 43.
[36] S. M. Grashchenko and E. V. Sobotovich, *Radiokhimiya*, 1967, **9**, 412.
[37] H. Koch, *Nukleonika*, 1980, **25**, 1181.
[38] J. Starý and K. Kratzer, *Radiochem. Radioanal. Lett.*, 1981, **46**, 249.
[39] N. D. Thaker and D. N. Patkar, *Radiochem. Radioanal. Lett.*, 1981, **48**, 185.
[40] E. V. Goode, *Analyst*, 1968, **93**, 663.
[41] P. Beneš and V. Majer, *Trace Chemistry of Aqueous Solutions*, Academia, Prague, 1980.
[42] V. I. Shamaev and M. E. Savkin, *Zh. Analit. Khim.*, 1973, **28**, 516.
[43] H. J. M. Bowen, *Radiochem. Radioanal. Lett.*, 1970, **3**, 339.
[44] J. Curry and M. P. Menon, *J. Radioanal. Chem.*, 1976, **34**, 319.
[45] V. J. Blaedel and T. J. Anderson, *Anal. Chem.*, 1971, **43**, 521.
[46] S. Mlinkó, *Mikrochim. Acta*, 1969, 517.
[47] M. Paikoff and J. E. Christian, *J. Am. Pharm. Assoc.*, 1956, **45**, 623.
[48] K. R. Kar and S. C. Jain, *Current Sci. (India)*, 1967, **36**, 542.
[49] R. M. Kniseley, *Nuclear India*, 1974, Nov.–Dec., 6.
[50] D. T. Baird, *J. Clin. Endocrinol.*, 1968, **28**, 244.

[51] R. W. Durham, G. R. Martin and H. C. Sutton, *Nature*, 1949, **164**, 1052.

[52] M. C. Stokes, F. C. Hickey and W. A. Fish, *J. Am. Chem. Soc.*, 1954, **76**, 5174.

[53] H. P. Kaufman and J. Budwig, *Fette, Seifen, Anstrichmittel*, 1951, **53**, 69.

[54] L. B. Gorbushina, L. Ya. Zhiltsova and E. N. Matreeva, *J. Radioanal. Chem.*, 1972, **10**, 165.

[55] W. J. McDowell and C. F. Coleman, *Proc. Int. Conf. Solvent Extraction*, held 8 September, 1974, Lyon, *Soc. Chem. Ind. (London)*, 1974, **3**, 2123.

[56] W. J. McDowell, D. R. Farrar and M. R. Billinge, *Talanta*, 1974, **21**, 1231.

[57] I. V. Melikhov, V. N. Rudin and V. F. Komarov. *Radiokhimiya*, 1973, **15**, 432.

[58] C. G. Taylor and J. Waters, *Analyst*, 1972, **97**, 533.

[59] C. W. Kao, E. R. Graham and K. W. Blanchar, *Soil Sci.*, 1971, **112**, 221.

[60] S. L. Hansard, quoted in [59].

[61] S. K. Das, A. K. Santikari and S. R. Upadhyay, *Technology (Sindri)*, 1973, **10**, 182.

[62] E. Rabinowicz, *Nature*, 1952, **170**, 1029.

[63] M. Burton, J. E. Ricci and T. W. Davies, *J. Am. Chem. Soc.*, 1940, **62**, 265.

[64] P. A. Leighton and R. A. Mortensen, *J. Am. Chem. Soc.*, 1936, **58**, 448.

[65] F. O. Rice and K. K. Rice, quoted in G. M. Harris, *J. Chem. Phys.*, 1947, **15**, 686.

[66] D. Dyrssen and B. Hok, *Svensk Kem. Tid.*, 1952, **64**, 80.

[67] H. P. Kaufmann and J. Budwig, *Fette, Seifen, Anstrichmittel*, 1951, **53**, 253.

[68] R. J. Ho, *Anal. Biochem.*, 1970, **36**, 105.

[69] C. McLean, A. D. Verner and D. Aminoff, *Anal. Biochem.*, 1973, **55**, 72.

[70] G. K. Koch and H. Van Tilborg, *J. Radioanal. Chem.*, 1977, **35**, 197.

[71] K. Nitzl, *Papier*, 1967, **21**, 393.

[72] D. E. Kramm, J. N. Lomonte and J. B. Moyer, *Anal. Chem.*, 1964, **36**, 2170.

[73] P. Valls, A. M. Venet and J. Pouradier, *Bull. Soc. Chim. France*, 1953, 106.

[74] V. G. Erwin and P. L. Pedersen, *Anal. Biochem.*, 1968, **25**, 477.

[75] D. K. Campbell and W. C. Warner, *Rubber Chem. Technol.*, 1969, **42**, 1.

[76] G. Ayrey, *Chem. Rev.*, 1963, **63**, 645.

[77] E. C. Hoffman and H. Hoberg, *Z. Elektrochem.*, 1954, **58**, 646.

[78] L. H. Howland, A. Nisonoff and L. E. Dannals, *Rubber Chem. Technol.*, 1959, **32**, 706.

[79] B. Hudson, J. Coghlan and A. Dulmanis, *Aust. J. Exp. Biol.*, 1963, **41**, 235.

[80] K. Bächmann, K. Büttner and J. Rudolph, *Z. Anal. Chem.*, 1976, **282**, 189.

[81] R. A. German, D. L. Hamilton and M. P. Menon, *Anal. Chem.*, 1975, **47**, 658.

[82] N. P. Rudenko, *Zh. Analit. Khim.*, 1967, **22**, 1736.

[83] H. J. M. Bowen, *Analyst*, 1974, **99**, 771.

[84] E. Banyai, F. Szabadváry and L. Erdey, *Mikrochim. Acta*, 1962, 427.

[85] J. E. Kenney and M. P. Menon, *Anal. Chem.*, 1972, **44**, 2093.

[86] A. Illaszewicz, K. Müller, D. Rabussay and H. Spitzy, *J. Radioanal. Chem.*, 1968, **1**, 39.

[87] J. Lengyel, J. Krtil and J. Večerník, *Radiochem. Radioanal. Lett.*, 1983, **58**, 253.

[88] J. Krtil, p. 45 in [10].

[89] L. E. Mattison and J. C. Wolford, *Anal. Chem.*, 1966, **38**, 1675.

[90] M. P. Menon, *Anal. Chim. Acta*, 1973, **64**, 151.

[91] C. G. Taylor and J. Waters, *Analyst*, 1972, **97**, 533.

[92] O. Gimesi, O. Weber and F. Szabadváry, *J. Radioanal. Chem.*, 1970, **6**, 457.

[93] N. Abdusalyamov, A. Ganiev and P. Kh. Nishamov, *Zh. Analit. Khim.*, 1969, **24**, 69.

[94] A. Zeman and J. Prášilová, *Radiochem. Radioanal. Lett.*, 1980, **43**, 329.

[95] R. B. Hahn and T. M. Schmitt, *Anal. Chem.*, 1969, **41**, 359.

[96] J. Starý, K. Kratzer and J. Burclová, *Radiochem. Radioanal. Lett.*, 1981, **46**, 191.

[97] M. P. Menon, *J. Radioanal. Chem.*, 1973, **14**, 63.

[98] J. Starý, *Moderní radioanalytické metody*, in *Nové směry v analytické chemii*, Vol. I, J. Zýka (ed.), SNTL, Prague, 1983.

[99] H. Sobczyk and J. Krawczyk, *Koks, Smola, Gaz*, 1981, **26**, 204.

[100] S. Sriman, V. Narayan and R. S. Rao, *Radiochem. Radioanal. Lett.*, 1982, **53**, 163.

[101] H. R. Schütte, *Radioaktive Isotope in der Organischen Chemie und Biochemie*, VEB Deutscher Verlag für Wissenschaften, Berlin, 1966.

[102] K. Schwabe and K. Thiemt, *Isotopenpraxis*, 1972, **8**, 139.

[103] Al. Cecal, C. Oniscu and E. Horoba, *Pharmazie*, 1983, **38**, 562.

[104] W. J. Driscoll, B. F. Scott and E. A. Huff, *Report, TID-11306*, 1961; see also [88].

[105] V. I. Shamaev, *J. Radioanal. Chem.*, 1980, **57**, 313.

[106] V. I. Shamaev and A. V. Eremina, *Zh. Analit. Khim.*, 1982, **37**, 1070.

[107] V. I. Shamaev, *Zh. Analit. Khim.*, 1984, **39**, 1961.

[108] V. I. Shamaev, *Zh. Analit. Khim.*, 1985, **40**, 49.

[109] G. A. Welford, E. L. Chiotis and R. S. Morse, *J. Chromatog.*, 1966, **25**, 144.

[110] P. C. Van Erkelens, *Anal. Chim. Acta*, 1961, **25**, 570.

[111] H. J. Bowen and J. A. Cook, *Radiochem. Radioanal. Lett.*, 1970, **5**, 103.

[112] G. W. Tang and C. J. Maletskos, *Science*, 1970, **167**, 52.

[113] W. Eckhart, G. Herman and H. D. Schüssler, *Z. Anal. Chem.*, 1967, **226**, 71.

[114] D. N. Sundermann and W. W. Meinke, *Anal. Chem.*, 1956, **29**, 1578.

[115] W. Zmijewska, *Radiochem. Radioanal. Lett.*, 1970, **3**, 285.

[116] I. H. Qureshi, *Talanta*, 1964, **11**, 1550.

[117] F. Girardi, R. Pietra and E. Sabioni, *J. Radioanal. Chem.*, 1970, **5**, 141.

[118] M. Csajka, *Radiochem. Radioanal. Lett.*, 1973, **13**, 151.

[119] M. Csajka, *Anal. Chim. Acta*, 1974, **68**, 31.

[120] M. Csajka, *Radiochim. Radioanal. Lett.*, 1974, **18**, 245.

[121] I. H. Qureshi, M. S. Shabir and S. M. Hasany, *Talanta*, 1967, **14**, 951.

[122] I. H. Qureshi and M. S. Shabir, *Talanta*, 1966, **13**, 847.

[123] C. Bigliocca and F. Girardi, *Anal. Chem.*, 1967, **39**, 1634.

[124] F. Girardi, *Activation Techniques in the Life Sciences*, IAEA, Vienna, 1967, p. 117.

[125] M. Csajka, *Talanta*, 1967, **14**, 1360.

[126] R. Kuroda and K. Oguma, *Anal. Chem.*, 1967, **39**, 1003.

[127] R. R. Ruch, J. R. DeVoe and W. W. Meinke, *Talanta*, 1962, **9**, 33.

[128] I. H. Qureshi and F. I. Nagi, *J. Inorg. Nucl. Chem.*, 1967, **29**, 2879.

[129] J. R. DeVoe, W. W. Meinke and H. W. Nass, *Anal. Chem.*, 1961, **33**, 1713.

[130] W. B. Silker, *Anal. Chem.*, 1961, **33**, 233.

[131] D. L. Love, *Anal. Chim. Acta*, 1958, **18**, 72.

[132] J. B. Luten, H. A. Das and C. Z. Ligny, *J. Radioanal. Chem.*, 1977, **35**, 147.

[133] M. Zaduban, N. Stollarova and S. Palágyi, *Radiochem. Radioanal. Lett.*, 1970, **3**, 129.

[134] T. Braun and A. B. Farag, *J. Radioanal. Chem.*, 1975, **25**, 5.

[135] M. Heurtebise and W. J. Ross, *Anal. Chem.*, 1971, **43**, 1438.

[136] J. Burclová, *Chem. Listy*, 1976, **70**, 561.

[137] G. N. Bilimovich, *Metod izotopnovo obmena*, p. 164 in ref. [10].

[138] G. N. Bilimovich and J. Starý, *Zh. Analit. Khim.*, 1979, **34**, 996.

[139] M. Kyrš and J. Krtil, *Chem. Listy*, 1968, **62**, 1409.

[140] J. Starý, J. Prášilová and T. Bouda, *J. Radioanal. Chem.*, 1975, **28**, 95.

[141] T. B. Pierce, *Analyst*, 1960, **85**, 166.

[142] T. B. Handley, *Anal. Chem.*, 1964, **36**, 153.

[143] H. G. Richter, *Anal. Chem.*, 1966, **38**, 772.

[144] N. Ikeda and S. Amano, *Radioisotopes*, 1967, **16**, 317.

[145] Y. Takahashi and N. Ikeda, *J. Radioanal. Nucl. Chem., Articles,* 1985, **89**, 365.

[146] A. Zeman and K. Kratzer, *Radiochem. Radioanal. Lett.,* 1976, **27**, 217.

[147] F. Vláčil, *J. Radioanal. Chem.,* 1980, **58**, 221.

[148] J. Starý and K. Kratzer, *Int. J. Environ. Anal. Chem.,* 1972, **2**, 79.

[149] A. Zeman and V. Mádr, *J. Radioanal. Nucl. Chem., Lett.,* 1984, **86**, 261.

[150] J. Starý and J. Prášilová, *Radiochem. Radioanal. Lett.,* 1976, **24**, 143.

[151] J. Starý and J. Prášilová, *Radiochem. Radioanal. Lett.,* 1976, **26**, 193.

[152] J. Starý, B. Valter, K. Kratzer and J. Prášilová, *Radiochem. Radioanal. Lett.,* 1977, **30**, 281.

[153] J. Starý, K. Kratzer and A. Zeman, *J. Radioanal. Chem.,* 1970, **5**, 71.

[154] J. Starý, K. Kratzer and A. Zeman, *Radiochem. Radioanal. Lett.,* 1971, **6**, 1.

[155] K. Kratzer, J. Starý, A. Zeman and V. Majer, *Collect. Czech. Chem. Commun.,* 1972, **37**, 3267.

[156] A. Zeman, J. Starý and K. Kratzer, *Radiochem. Radioanal. Lett.,* 1970, **4**, 1.

[157] K. Kratzer, *Radiochem. Radioanal. Lett.,* 1970, **5**, 69.

[158] N. Ikeda and Y. Takahashi, *J. Radioanal. Chem.,* 1977, **36**, 77.

[159] C. H. Gast and H. A. Das, *J. Radioanal. Chem.,* 1980, **57**, 401.

[160] G. Pfrepper and N. A. Malyshev, *Proceedings of "3. Tagung Nukleare Analysenverfahren",* held 11 April 1983 in Dresden, GDR, p. 246, Kernforschung Institut Rossendorf, 1983.

[161] J. L. Raven, P. L. Walker and P. Barkhan, *J. Clin. Pathol.,* 1966, **19**, 610.

[162] N. I. Polevaya and S. L. Mirkina, *Vsesoyuz. Nauch. Issled. Geol. Inst. Inform. Sbornik* 1955, No. 1, 123, *Chem. Abstr.,* 1958, **52**, 163c.

[163] H. Amano, *Sci. Repts., Research Inst. Tohoku Univ. A,* 1958, **11**, 367.

[164] K. B. Troitskii, *Trudy Komiss. po Analit. Khim.,* 1958, **9**, 187.

[165] V. L. Eggebraaten and L. E. Miller, *Int. J. Appl. Radiat. Isotopes,* 1964, **18**, 183.

[166] C. H. Wallace and J. E. Willard, *J. Am. Chem. Soc.,* 1950, **72**, 5275.

[167] R. Ehrenberg, *Biochem Z.,* 1926, **172**, 10.

[168] P. C. Van Erkelens, *Anal. Chim. Acta,* 1962, **26**, 46.

[169] K. Müller, *Anal. Chim. Acta,* 1966, **35**, 162.

[170] S. V. Bleskinskii, A. G. Nagaeva and V. F. Abramova, *Izv. Akad. Nauk. Kirg. SSR,* 1963, **5**, 27.

[171] A. R. Landgrebe, T. E. Gills and J. R. DeVoe, *Anal. Chem.,* 1966, **38**, 1265.

[172] P. C. Van Erkelens, *Anal. Chim. Acta,* 1962, **26**, 32.

[173] A. N. Busev and V. M. Byrko, *Vestn. Moskov. Univ.,* Ser. II, 1960, **3**, 46.

[174] V. I. Shamaev and A. V. Malysheva, *Zh. Analit. Khim.,* 1983, **38**, 1819.

[175] V. I. Shamaev and I. L. Karetnikova, *J. Radioanal. Chem.,* 1979, **50**, 33.

[176] M. Kyrš and J. Rais, *Chem. Listy,* 1967, **61**, 440.

[177] M. Kyrš, *Method of Concentration-Dependent Distribution,* in J. Tölgyessy and Š. Varga, *Nuclear Analytical Chemistry II,* University Park Press, Baltimore: Veda, Bratislava, 1972, p. 205.

[178] M. Kyrš and J. Rais, *Method kontsentratsionno-zavisimovo raspredeleniya,* in [10], p. 180.

[179] I. P. Alimarin, V. V. Atrashkevich, G. N. Bilimovich and A. V. Garanin, *J. Radioanal. Chem.,* 1980, **57**, 263.

[180] V. I. Shamaev, *J. Radioanal. Chem.,* 1979, **49**, 247.

[181] W. Van der Mark and H. A. Das, *J. Radioanal. Chem.,* 1974, **23**, 7.

[182] V. I. Shamaev, *J. Radioanal. Chem.,* 1981, **63**, 41.

[183] K. Přikrylová, E. Makrlík and M. Kyrš, *J. Radioanal. Nucl. Chem., Articles,* 1986, **97**, 13.

[184] V. I. Shamaev and T. V. Khudinovskikh, *Zh. Analit. Khim.,* 1981, **36**, 248.

[185] V. I. Shamaev and V. G. Dyachkova, *Zh. Analit. Khim.*, 1979, **34**, 1065.
[186] M. Kyrš and J. Hálová, *J. Radioanal. Chem.*, 1979, **52**, 53.
[187] M. Kyrš and J. Hálová, *J. Radioanal. Chem.*, 1983, **78**, 29.
[188] M. Kyrš and A. Moravec, *Chem. Listy*, 1984, **78**, 692.
[189] Z. Kolařík, *Collect. Czech. Chem. Commun.*, 1967, **32**, 311.
[190] T. Ishimori, E. Nakamura and H. Murakami, *J. At. Energy Soc. Japan*, 1961, **3**, 193.
[191] J. Rais, S. Podešva and M. Kyrš, *Anal. Chim. Acta*, 1966, **36**, 90.
[192] G. J. Cathers, *Actes de la première conférence atomique*, 1958, **7**, 569; in *Reactor Fuel Processing*, P/743.
[193] D. J. Salley, *Conf. At. Energy in Industry*, Nat. Ind. Conf. Board, New York, 1952.
[194] M. S. Merkulova, *Zh. Fiz. Khim.*, 1955, **29**, 1915.
[195] M. Kyrš, J. Rais and P. Pištěk, *Report ÚJV 1434/65*, Nuclear Research Institute, Řež, Czechoslovakia, 1965.
[196] S. Schoenherr, H. Goetz and G. Ackerman, *J. Radioanal. Chem.*, 1975, **26**, 73.
[197] K. B. Troitskii, *Radiokhimiya*, 1986, **10**, 598.
[198] V. Balek and J. Tölgyessy, *Emanation Thermal Analysis and other Radiometric Emanation Methods*, Elsevier, Amsterdam, 1984.
[199] D. J. Chleck, R. Maehl, O. Cucchiara and E. Carnevale, *Int. J. Appl. Radiat. Isotopes*, 1963, **14**, 581; D. J. Chleck and R. Maehl, *Int. J. Appl. Radiat. Isotopes*, 1963, **14**, 593.
[200] J. Tölgyessy, Š. Varga and E. H. Klehr, *Radioisotopy*, 1974, **15**, 623.
[201] A. M. Trofimov and Yu. N. Kazankin, *Radiokhimiya*, 1963, **7**, 288.
[202] V. Jesenák, Š. Varga and J. Tölgyessy, *Atompraxis*, 1968, **14**, 204.
[203] D. J. Chleck, *IAEA Symposium on Radiochemical Methods of Analysis*, 1965, **2**, 273.
[204] Š. Varga, J. Tölgyessy and V. Jesenák, *Sb. Ved. Prác Chemickotechn. Fak. SVŠT*, Bratislava, 1966, 117.
[205] J. Tölgyessy, M. Harangozó, P. Dillinger, Š. Varga and E. H. Klehr, *Radiochem. Radioanal. Lett.*, 1974, **19**, 147.
[206] R. Sizmann and W. Rupp, *Z. Naturforsch.* 1962, **170**, 179.
[207] P. J. Hidalgo and R. Sizmann, *Ber. Dtsch. Ker. Ges.*, 1986, **43**, 532.
[208] G. Carter, *Vacuum*, 1959, **9**, 190.
[209] C. Almén and G. Bruce, *Nucl. Inst. Meth.*, 1961, **11**, 257, 279.
[210] J. A. Davies, F. Brown and M. McCargo, *Can. J. Phys.*, 1963, **41**, 829.
[211] Č. Jech, *Int. J. Appl. Radiat. Isotopes*, 1960, **8**, 179.
[212] D. J. Chleck and O. Cucchiara, *Int. J. Appl. Radiat. Isotopes*, 1963, **14**, 599.
[213] D. J. Chleck and C. A. Ziegler, *Radioisotopes in the Physical Sciences and Industry*, IAEA, Vienna, 1962, p. 361.
[214] C. O. Hommel, F. J. Brousaides and R. L. Bersin, *Anal. Chem.*, 1962, **34**, 1608.
[215] Toxic Gas Monitor, Graf-8/65-10M, Tracerlab, Waltham, Mass., USA, 1965.
[216] V. Jesenák, M. M. Naoum and J. Tölgyessy, *Radiochem. Radioanal. Lett.*, 1973, **13**, 199.
[217] D. J. Chleck, R. Maehl and O. Cucchiara, *Chemist-Analyst*, 1965, **54**, 84.
[218] M. M. Naoum, J. Pružinec, J. Tölgyessy and E. H. Klehr, *Radiochem. Radioanal. Lett.*, 1974, **17**, 87.
[219] J. E. Carden, *US AEC Rept.*, ORN-IIC-18, July 1969.
[220] *Chem. Eng. News*, 1967, **35**, No. 50, 57.
[221] J. Pružinec, J. Tölgyessy and M. M. Naoum, *Radiochem. Radioanal. Lett.*, 1974, **20**, 73.
[222] R. L. Bersin, F. J. Brousaides and C. O. Hommel, *J. Air Pollution Control Assoc.*, 1962, **12**, 129, 146.

[223] J. Pružinec and J. Tölgyessy, *Sb. Ved. Prác Chemickotechn. Fak. SVŠT*, Bratislava, 1977, 123.

[224] J. Pružinec and J. Tölgyessy, *Jad. Energie*, 1973, **19**, 345.

[225] O. Cucchiara and P. Goodman, *Development of Krypton-85 as a Universal Tracer*, NYO-2757-6, Panametrics, Inc., 1966.

[226] M. M. Naoum, J. Tölgyessy, V. Jesenák and E. H. Klehr, *J. Radioanal. Chem.*, 1975, **25**, 129.

[227] M. M. Naoum, V. Jesenák, J. Tölgyessy and J. Lesný, *Radiochem. Radioanal. Lett.*, 1973, **14**, 265.

[228] M. M. Naoum, V. Jesenák and J. Tölgyessy, *Radiochem. Radioanal. Lett.*, 1973, **14**, 373.

[229] P. Goodman, *Int. Lab.*, 1971, Nov./Dec., 32.

[230] J. Tölgyessy and Š. Varga, *Talanta*, 1969, **16**, 625.

[231] J. Tölgyessy, Š. Varga and V. Jesenák, *Peaceful Uses of Atomic Energy*, IAEA, Vienna, 1972, **14**, 257.

[232] J. Tölgyessy, Š. Varga, V. Jesenák, P. Lukáč and P. Dillinger, *Radiochem. Radioanal. Lett.*, 1970, **5**, 331.

[233] J. Tölgyessy, V. Jesenák and E. Koval, *J. Radioanal. Chem.*, 1970, **4**, 13.

[234] D. J. Chleck, R. Maehl and O. Cucchiara, *Development of Krypton-85 as a Universal Tracer*, NYO-2757-3, Panametrics, Inc., Nov., 1963.

[235] P. Dillinger and J. Tölgyessy, *Jad. Energie*, 1973, **19**, 94.

[236] J. Tölgyessy, E. H. Klehr, P. Dillinger and M. Harangozó, *Isotopenpraxis*, 1980, **16**, 268.

5

Radiometric titrations

Methods of titrimetric analyses are based on the relations between some physical parameter characterizing the solution to be analysed, and the volume of an added titrant. In titrations based on reactions in which the product is in a phase other than the reacting medium (for example, by precipitation or extraction), radioactive nuclides or a sealed source of radiation can be used to indicate the end-point. These so-called radiometric titrations can be divided into groups according to:

(1) the nature of the distribution of the titration components between the individual phases; (a) precipitate–solution, (b) ion exchanger–solution, (c) solid indicator–solution, (d) two immiscible solutions (water–organic solvent);

(2) the reaction type used; (a) precipitation, (b) neutralization, (c) redox, (d) complexometry;

(3) the radiometric principle of determination of the end-point; (a) indicator methods (the titrant, the test solution or both contain a radioactive indicator), (b) methods based on the interaction of nuclear radiation with the system to be analysed;

(4) the technique of titration; (a) discontinuous titration in one solution, (b) discontinuous titration in a series of solutions, (c) continuous titrations;

(5) the type of radiation (gamma and hard beta);

(6) the type of detector used (Geiger–Müller and scintillation counter).

The necessity of separation is one of the greatest drawbacks of radiometric titrations, compared to other instrumental titrations. The titration is both lengthy and cumbersome because the titration and activity measurements have to be done alternately and repeatedly. A continuous titration is possible only if the separation can be done continuously, which is practicable only in a few instances. On the other hand, radiometric titrations with use of radioactive indicators excel by their sensitivity. This follows from the nature of radioactive labelling. Even minute amounts of substances can be monitored if the specific activity is high enough. The intensity of the radioactivity monitored during the titration can be adjusted by the addition of carrier-free isotopes (i.e. by varying the specific activity). This possibility is obviously advantageous for low concentrations. With other instrumental end-point detection methods, such as photometric or conductometric techniques, where the measured property is a linear function of concentration, both the measured property and its changes during titration become smaller as the range of low concentrations is approached, and accurate location of the end-point on the titration curve becomes more difficult. In contrast, the radioactivity and its change can be adjusted as desired for use in radiometric titrations. For precipitation radiometric titrations, where the amounts of the substances to be determined are about 1 mg, the relative error is generally $\pm$ 3–4 %. In titrations with dithizone (with phase separation by solvent extraction) where higher specific activities are generally employed, the relative errors are less than $\pm$ 1 % when µg amounts of substances are determined.

Radiometric titrations can be done in heterogeneous, coloured, turbid, or non-aqueous systems, or under artificial lighting, for none of these factors affects in any way the measurement of radioactivity.

5.1 RADIOMETRIC TITRATIONS BASED ON PRECIPITATION

In radiometric titrations based on precipitate formation the radioactive indicator passes from a liquid to a solid phase. The end-point can be detected by following the changes in the radioactivity of one of the phases (the liquid phase in all the published methods). An appropriate isotope of the element to be determined (or of a component element of the compound to be determined) or of an element in the active component of the titrant is generally used, and this technique is known as *isotopic labelling*. The choice of radionuclide to be used depends on the nature and conditions of the determination. From a purely practical aspect, only radionuclides with a sufficiently long half-life and high radiation energy are suitable.

At their simplest, radiometric precipitation titrations may yield three types of titration curve. The shape of the curve will depend on whether the titrant, the test solution, or both are labelled (Fig. 5.1).

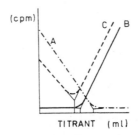

Fig. 5.1 — Radiometric precipitation titration curves: A — test solution active, titrant inactive; B — test solution inactive, titrant active; C — both test solution and titrant active.

There are many instances where suitable radioisotopes of the element to be determined are not available (e.g. copper, nickel, aluminium, magnesium, titanium). In such situations, the radioisotope of another element can often be used as the indicator. This technique is called the *non-isotopic method of labelling*. Non-isotopic indicators can be non-isomorphous, or isomorphous. *Non-isomorphous indicators* must yield a more soluble compound with the titrant than the analyte does. During the titration, the analyte is precipitated first, and the activity of the supernatant liquid is constant; after the end-point, the indicator is precipitated and the activity of the supernatant liquid decreases. The sharp break in the curve indicates the end-point. Non-isotopic *isomorphous indicators* are co-precipitated with the analyte throughout the titration and the end-point is indicated by constant activity of the supernatant liquid.

The accuracy of radiometric titrations based on precipitate formation is affected by factors similar to those which play a role in all precipitation titrations: the stoichiometry of the reaction, the adsorption of foreign ions by the precipitate, the value of the solubility product, the concentration of the initial solutions, the occurrence of side-reactions, etc.

The determination limit of radiometric titrations based on precipitate formation is generally about 1 mg. This limiting value follows from the nature of the process and is determined by the solubility product of the precipitate. Lower limits can be achieved only by the use of special titration techniques, e.g. ion exchange, use of collectors and titrations in non-aqueous media.

5.1.1 Theoretical basis of the method

In precipitation radiometric titrations the function $I = f(V)$ is monitored, where I is the activity of the supernatant liquid and V the volume of titrant added. When a solution containing A^+ is titrated with a titrant containing B^-, to form precipitate AB (solubility product L_{AB}), the relation between the concentration $[A^+]$ and the volume of titrant added can be expressed by the equation

$$\gamma^2 - \left(\frac{\gamma_0 V_0 - c_0 V}{V_0 + V}\right)\gamma - L_{AB} = 0 \qquad (5.1)$$

where γ_0 is the initial concentration of A^+, γ the concentration of A^+ during the titration, c_0 the titrant concentration, and V_0 the initial volume of titrant [1, 2]. Various cases arise.

(1) *Titrand active, titrant inactive*. Because the radioactivity of the solution is directly proportional to the titrand concentration, i.e. $\gamma_0 = kI_0$ (where I is the activity in cpm per unit volume and $\gamma = kI$, then substitution into (5.1), gives

$$I^2 - \left[\frac{kI_0 V_0 - c_0 V}{k(V_0 + V)}\right]I - \frac{L_{AB}}{k^2} = 0 \qquad (5.2)$$

(2) *Titrand inactive, titrant active*. The progress of titration is described by

$$I^2 + \left[\frac{\gamma_0 V_0 - c_0 V}{k(V_0 + V)}\right]I - \frac{L_{AB}}{k^2} = 0 \qquad (5.3)$$

(3) *Titrand and titrant both active*. The end-point is determined by the minimum on the titration curve. If $\gamma_0 = k_1 I_{1,0}$, $\gamma = k_1 I_1$, $c_0 = k_2 I_2$ then the activity (I) of the supernatant, measured during the titration, will be given by

$$2I = R\left(1 - \frac{k_1}{k_2}\right) + \left(R^2 + \frac{4L_{AB}}{k_1^2}\right)^{1/2} + \left[\left(\frac{k_1 R}{k_2}\right)^2 + \frac{4L_{AB}}{k_2^2}\right]^{1/2} \qquad (5.4)$$

where

$$R = \frac{k_1 I_{1,0} V_0 - c_0 V}{k_1 (V_0 + V)} \tag{5.5}$$

In certain instances it is possible to simplify (5.4) as follows:
(1) When $k_1 = k_2 = k$, i.e. when the ratio of the activities of the titrand and titrant corresponds to the ratio of their concentrations, (5.4) simplifies to

$$I = \left(R^2 + \frac{4 L_{AB}}{k^2} \right)^{1/2} \tag{5.6}$$

(2) When k_1 is negligible compared to k_2, (5.4) becomes

$$2I \approx R + \left(R^2 + \frac{4 L_{AB}}{k_1^2} \right)^{1/2} \tag{5.7}$$

(3) When k_2 is negligible compared to k_1, (5.4) simplifies to

$$2I \approx -\frac{k_1 R}{k_2} + \left[\left(\frac{k_1 R}{k_2} \right)^2 + \frac{4 L_{AB}}{k_2^2} \right]^{1/2} \tag{5.8}$$

The titration curves obtained most frequently in precipitation radiometric titrations are characterized by these correlations [1–3].

The end-point may be determined graphically by plotting the volume of titrant added against the activity of the supernatant liquid (usually in cpm). The point of intersection of extrapolations of the two branches of the curve marks the end-point.

The determination of the end-point with the aid of the titration curve, for the construction of which at least six data-points are necessary, requires 40–50 min. The time needed depends on the nature and sedimentation rate of the precipitate.

The end-point can also be determined by extrapolation from two points on the curve. In *titration of active titrand with inactive titrant* the end-point volume V_{ep} is determined from the initial activity I_0, and the activity (I) of the supernatant liquid when a volume V of titrant $(V < V_{ep})$ has been added:

$$V_{ep} = \frac{I_0 V}{I_0 - I} \tag{5.9}$$

For *inactive titrand and a labelled titrant*,

$$V_{ep} = \frac{V_1 I_2 - V_2 I_1}{I_2 - I_1} \tag{5.10}$$

where I_1 and I_2 are the activities of the supernatant liquid when volumes V_1 and V_2 of titrant ($V_2 > V_1 > V_{ep}$) have been added.

The following method may also be used. Equal volumes of sample and distilled water are taken, and to each is added the same known volume ($V > V_{ep}$) of titrant, and the activity of both solutions is measured. If the activity of the blank is I_b and that of the test solution I_t, then

$$V_{ex} = V(I_b - I_t)/I_t \qquad (5.11)$$

For *titration of labelled titrand with labelled titrant*, the end-point can be calculated by using either (5.9) or (5.10). However, the error in the calculated values will be significant. In the practical applications of this type of titration, it is in general satisfactory to label either the test solution or the titrant. Though both solutions can be labelled, the evaluation is more cumbersome. Since the correlation between the activity and the volume of added titrant is not linear, the end-point determined by extrapolation will embody a certain error; the greater the solubility product of the precipitate and the higher the ratio V/V_0, the greater the error. Therefore, some workers suggest a relatively complex calculation, or the use of nomograms for determining the end-point. In practical applications, however, this appears a less convenient procedure, though it may be useful for repetitive routine analysis.

5.1.2 Methods of precipitate separation and technique of titration

A fundamental prerequisite for radiometric titrations is separation of the radioactive reagent(s) from the radioactive reaction product(s). This is possible when the reactants are in a different phase from the products, as in precipitation reactions, where filtration, centrifugation or flotation may be used for phase separation.

5.1.2.1 Filtration method

Filtration is the method usually used, and can be carried out discontinuously or continuously.

In the simplest case, the precipitate is removed on filter paper of suitable porosity. Filtration apparatus devised for radiometric titrations is shown in Fig. 5.2. A known amount of titrant is added to the test solution, then the supernatant solution is sucked through the glass filter into the glass mantle, and the activity of the liquid measured by a liquid GM counter. On discarding the solution, another portion of titrant is added to the test solution, and the process repeated [4]. Special

GM counting tubes for use in radiometric titrations are shown in Fig. 5.3 [4].

If the indicator emits gamma radiation, a scintillation detector is used. A scheme of an instrument of this type is presented in Fig. 5.4. The supernatant solution is sucked into the test-tube placed in the well of the scintillation counter, where its radioactivity is measured, and is then returned to the titration vessel [5].

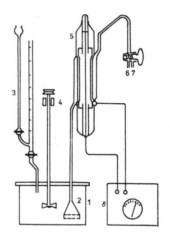

Fig. 5.2 — Radiometric titration apparatus: 1 — titration flask, 2 — suction device, 3 — burette, 4 — stirrer, 5 — GM counter, 6 — aperture to air, 7 — to the pump, 8 — ratemeter. (Reproduced from A. Langer, *J. Phys. Chem.*, 1941, **45**, 639, by permission. Copyright 1941, American Chemical Society.)

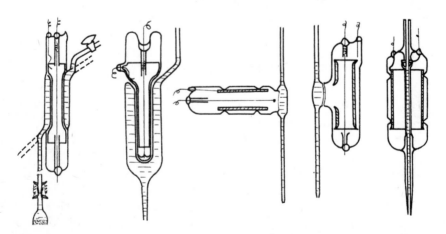

Fig. 5.3 — GM counting tubes for liquids. (Reproduced from A. Langer, *J. Phys. Chem.*, 1941, **45**, 639, by permission. Copyright 1941, American Chemical Society.)

Volume changes occur on addition of the titrant when these instruments are used. Consequently, the measured counts must be corrected for the volume change by multiplication by $(V_0 + V)/V_0$.

A great drawback of the filtration method is that the filter often gets blocked. This lengthens the duration of filtration, and requires a frequent change of filter plates.

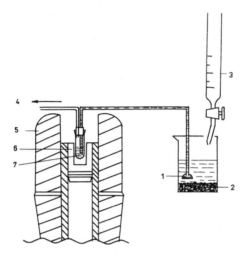

Fig. 5.4 — Well-type scintillation detector [5] applied in radiometric titrations: 1 — filter-tube, 2 — titration flask, 3 — burette, 4 — to the pump, 5 — lead column, 6 — scintillator, 7 — supernatant solution. (Reproduced by courtesy of Nuclear Chicago Corp.)

The scheme of an apparatus for automatic precipitation radiometric titration [6] is shown in Fig. 5.5. The titrant is automatically added to the test solution at a constant rate. A micropump maintains continuous circulation of the supernatant liquid from over the precipitate, through the filter plate into the glass mantle of the GM counter, and hence back into the titration vessel. The counter is coupled with a ratemeter, the output from which is continuously recorded.

The recorded titration curves can be evaluated from the rate of titrant addition and the time taken to reach the end-point, or by comparing the titration times for the sample and a standard. The time is measured from the start of addition of titrant to the time corresponding to the intersection of the extrapolated linear sections of the curve. Evaluation by comparison with a standard is more frequently employed, and accurate knowledge of the titrant concentration is not necessary, since the apparatus is standardized with a sample of accurately known concentration.

The method of automatic radiometric titration up to the end-point
[7] is used for serial determinations of approximately equal quantities
of a given substance. Before the actual determination, it is necessary to
plot the entire titration curve for one typical sample. This is required in
order to establish the radioactivity of the supernatant liquid at the
end-point. This technique is essentially similar to the previous method

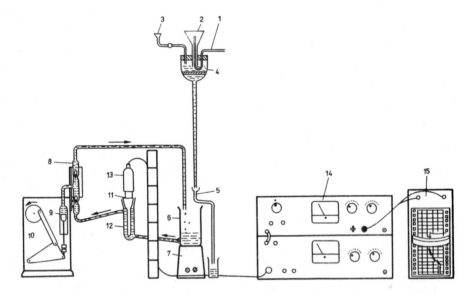

Fig. 5.5 — Scheme of an automatic radiometric titration device: 1 — auxiliary tube
connected with suction line for filling reservoir flask, 2 — reservoir flask, 3 — funnel,
4 — storage flask, 5 — outflow pipe, 6 — titration flask with fritted glass insert, 7 —
electromagnetic stirrer, 8 — mercury valve, 9 — syringe, 10 — electric motor and hinged
transmission, 11 — GM counter, 12 — glass mantle, 13 — amplifier head, 14 — ratemeter
with high-voltage current supply, 15 — recording device. (Reproduced from [6] by
permission of the copyright holder, Akadémiai Kiadó, Budapest.)

in which the titration curves are automatically recorded. The only
differences are in the measurement of radioactivity and the detection of
the end-point. The time elapsed between starting the addition of titrant
and the end-point being reached is measured with an electric stop-
watch which is automatically started and stopped by relay from a
ratemeter (which also stops and starts addition of the titrant). When the
radioactivity of the supernatant liquid has decreased to the level present
on the ratemeter (usually the activity at the end-point of the preliminary
titration), the stop-watch and the addition of titrant are both stopped.
The automatic titrator is shown diagrammatically in Fig. 5.6.

Provided that the titrant is added at a constant rate, the volume of titrant consumed can be calculated from the time elapsed. When an automatic burette or other calibrated vessel is employed for the titration, no stop-watch is necessary, because the volume of titrant consumed can be read directly.

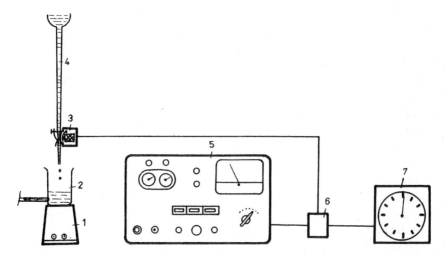

Fig. 5.6 — Scheme of the automatic apparatus for radiometric titration up to the end-point: 1 — titration flask with fritted glass disc, 2 — electromagnetic stirrer, 3 — electromagnetic valve, 4 — polyethylene tube, 5 — TISS ratemeter, 6 — relays, 7 — electric timer. (Reproduced from [7] by permission of the copyright holders, Veda, Bratislava.)

Use of the automatic instrument should eliminate subjective errors and increase the accuracy of the determination, and make the titration quicker. A single determination does not require more than 8 min. Frequent blockage of the filter is still a disadvantage, however.

5.1.2.2 Centrifugation method

In this method identical volumes of test solution are transferred into several centrifuge tubes, increasing amounts of titrant are added, and the solutions are made up to identical volume with distilled water. The precipitates are sedimented by centrifugation, then from each supernatant liquid a 0.5-ml sample is transferred dropwise to a filter paper and dried. The activity is then measured, and a titration curve plotted.

The counting and recording instruments combined with automatic sample-changers manufactured by various firms lend themselves well to carrying out the centrifugation method rapidly.

Centrifuging offers some advantages over filtration. It avoids block-ed filters, a common drawback in the filtration methods. This is par-ticularly relevant for fine precipitates, which would either cause a blockage, or not be retained by the filter plates. As the operations are carried out with a constant volume of liquid, no corrections for volume change are necessary, thus making the method more rapid. When about six readings are used for plotting the titration curves, a determination requires about 25 min.

5.1.2.3 Flotation method
Flotation has also been proposed for phase separation in radiometric titration based on precipitation [8]. In the usual flotation technique, particles suspended in water are separated from the liquid by air bubbled vigorously through the system. The solid particles are carried upwards by the ascending air bubbles. The same mechanism operates when flotation is used in radiometric titrations, except that the gas is replaced by an organic liquid that is immiscible with water. When this system is shaken or stirred, the two liquid phases are converted into an emulsion, which is unstable in water. The precipitate particles adhere to the droplets of the organic compound, and are finally collected at the interface of the two liquids. It is easier to measure the activity when the labelled component is present in the upper phase. As it is the activity of the aqueous layer that is measured in the titration, it is advisable to use an organic compound with a density greater than that of water.

5.1.3 Determination of components of mixtures
One advantage of radiometric titrations is the possibility of simul-taneously determining two components by one titration, or two or more components by two titrations. This can be achieved by the forma-tion of precipitates with markedly different solubility products, suitable choice of radioactive (isotopic or non-isotopic) indicators, proper ad-justment of pH, etc. We present some examples of the simultaneous determination of two or three components by precipitation radiometric titration.

For analysis of a mixture of zinc and lead, an aliquot of test solution is titrated with ^{59}Fe-labelled potassium hexacyanoferrate(II). This gives the total amount of zinc and lead. In another aliquot the lead is precipitated with sulphuric acid and the zinc is titrated with potassium hexacyanoferrate(II). The lead content is obtained by difference [9].

To analyse a mixture of iodide and iodine, in one aliquot, the iodide is determined radiometrically with ^{204}Tl-labelled thallium(I) sulphate as titrant. In another aliquot the iodine is reduced to iodide with sodium thiosulphate, and the total iodide concentration is similarly determined [10].

Two components can be determined simultaneously with the same labelled titrant provided that they form precipitates with the titrant at different pH values. This method has been used for the determination of Tl(III) and In(III) by titration with sodium 1-dithiocarboxy-3-methyl-5-phenylpyrazolinate labelled with ^{35}S [11, 12]. First, the thallium is precipitated quantitatively at pH 14 by the titrant. After the end-point, further addition of the titrant increases the activity of the supernatant liquid (Fig. 5.7, curve 1). When the pH of the solution is then adjusted to 7, the indium is precipitated by the excess and further portions of the titrant, and the activity of the solution remains constant at a level determined by the solubilities of the precipitates. Once all the indium has been precipitated, the activity of the solution rises again (Fig. 5.7, curve 2).

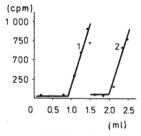

Fig. 5.7 — Radiometric titration of Tl(III) and In(III) in the presence of each other, using as titrant a sodium 1-dithiocarboxy-3-methyl-5-phenylpyrazolinate solution, labelled with ^{35}S: 1 — titration of Tl, 2 — titration of In. (Reproduced from [11] by permission of the copyright holders, Pergamon Press, Oxford.)

A different type of titration curve is obtained if the ion precipitated at the higher pH is labelled with a radionuclide. This can be used for the determination of Tl(III) (labelled with ^{204}Tl) in the presence of zinc, by titration with sodium 1-dithiocarboxy-3-methyl-5-phenyl-pyrazolinate labelled with ^{35}S [11, 12]. Addition of the titrant decreases the activity of the solution because of precipitation of the thallium. After the precipitation is complete, the activity of the solution is increased by addition of further portions of active titrant (Fig. 5.8, curve 1). When the pH is then decreased the excess of titrant reacts with the zinc and the activity of the solution is determined by the solubilities of the two precipitates. After the zinc precipitation is complete, further addition of titrant increases the activity of the supernatant liquid (Fig. 5.8, curve 2).

Titration curves of quite different type are obtained when a mixture of zinc (labelled with ^{65}Zn) and cadmium is titrated with sodium 1-di-thiocarboxy-3-methyl-5-phenylpyrazolinate labelled with ^{35}S [11, 12]. As shown by Fig. 5.9, curve 1, during the precipitation of cadmium the activity of the solution is constant and is that of the ^{65}Zn. After quantitative precipitation of the cadmium, and adjustment of the pH to 7, a small excess of titrant precipitates some zinc, and the activity of the solution decreases until all the zinc is precipitated. Then addition of excess of titrant increases the activity of the solution (the beta activity of the ^{35}S is being measured). Thus the end-point on curve 2 in Fig. 5.9 corresponds to the combined amounts of cadmium and zinc.

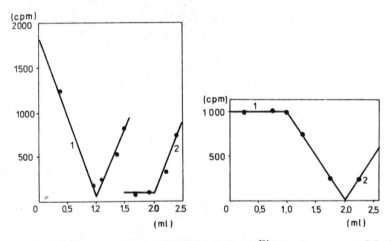

Fig. 5.8 — Radiometric titration of Tl(III) labelled with ^{204}Tl, in the presence of Zn(II), with sodium 1-dithiocarboxy-3-methyl-5-phenylpyrazolinate solution: 1 — titration of Tl, 2 — titration of Zn. (Reproduced from [11] by permission of the copyright holders, Pergamon Press, Oxford.)

▶

Fig. 5.9 — Radiometric titration of Zn labelled with ^{65}Zn, in the presence of Cd, with sodium 1-dithiocarboxy-3-methyl-5-phenylpyrazolinate solution: 1 — titration of Cd, 2 — titration of Zn. (Reproduced from [11] by permission of the copyright holders, Pergamon Press, Oxford.)

Differences in the solubility products of the compounds formed during the titration have been utilized for the determination of halides [13]. In this procedure, the halide which forms the least soluble pre-cipitate (i.e. the precipitate that forms first) is labelled, and the solution is titrated with ^{110m}Ag-labelled silver nitrate solution. In this way, iodide, labelled with ^{131}I, and chloride or bromide have been determined. The titration curves are presented in Fig. 5.10. During the titration, the iodide is precipitated first, and the activity of the solution decreases up to the iodide end-point. The activity then remains steady during the precipitation of the chloride or bromide. After the second end-point, the addition of excess of titrant results in an increase in activity.

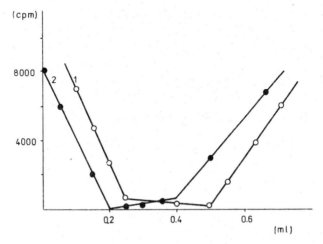

Fig. 5.10 — Radiometric titration of mixture of halide ions, using as titrant a silver (^{110m}Ag) nitrate solution: 1 — *I + Cl, 2 — *I + Br. (Reproduced from [13] by permission of the copyright holders, Japan Society for Analytical Chemistry.)

Three components can be determined by radiometric titrations, provided that all three react with the titrant to form precipitates with solubility products that are appreciably different from each other. Suitable radioisotopes of the two components which form the most and least soluble precipitates are used as indicators.

Obviously, these examples do not cover all possibilities. Selective titrations can also be carried out by employing auxiliary complexing or demasking agents, the combined use of several radioactive indicators, varying the pH, etc. Theoretical considerations indicate that up to five components may be determined in the same solution [14]. However, the errors are greatly multiplied in systems of this type. At present, therefore, methods for the simultaneous determination of more than three components have no practical significance.

5.1.4 Radiometric titrations in non-aqueous media

The determination limits of precipitation titrations are governed by the solubility products of the precipitates concerned. Because the solubility product is constant at a given temperature, the sensitivity of the method cannot be increased when aqueous media are used. However, if the titration medium is another solvent, in which the precipitate is less soluble, it is possible to increase the sensitivity and accuracy of the titration.

The solubility product (L) generally changes with the dielectric constant of the solvent (ε) according to

$$\ln L = A - \frac{C}{\varepsilon} \qquad (5.12)$$

where A and C are constants. Consequently, in radiometric precipitation titrations, the lower the dielectric constant of the solvent, the better the results. The degree of dissociation of electrolytes also changes according to the solvent used:

$$\ln K = a - \frac{c}{\varepsilon} \qquad (5.13)$$

where a and c are constants.

There are interesting possibilities for the radiometric titration of mixtures of several salts by precipitation in non-aqueous media. Thus, a mixture of salts which yield compounds with different solubility products can be analysed, e.g.

$$M_1A_1 + M_2A_2 + 2\,RS \;\rightleftharpoons\; RA_1\downarrow + RA_2\downarrow + M_1S + M_2S \quad (5.14)$$

In addition, compound yielding the same precipitate can be determined in the presence of each other, e.g.

$$M_1A + M_2A + 2\,RS \;\rightleftharpoons\; 2\,RA\downarrow + M_1S + M_2S \qquad (5.15)$$

Titrations of this type can be carried out in differentiating solvents, i.e. those capable of differentiating the degree of dissociation of salts, such as acetone or methyl ethyl ketone.

It is rather more difficult to choose suitable solvents for the titration of acids or bases. This is because the dissociation of acids and bases is determined not only by the solvation energy of the ions, but also by the energy of proton exchange with the solvent, and by other factors which depend on the nature of the acid (or base) and the solvent.

When the particular acid or base is a strong electrolyte in both aqueous and non-aqueous media, the solvent is chosen in the same way as for the titration of salts, i.e. solvents with low dielectric constants must be employed. For weak or only moderately strong acids or bases, incomplete dissociation must be taken into account.

In radiometric precipitation titration of acids, good results may be expected with mixed solvents of low basicity and dielectric constant, such as ethanol and benzene or ethanol and carbon tetrachloride. Organic acids and their salts have been titrated in methanol, or in a mixture of methanol and acetone or dioxan, with a silver nitrate solution, labelled with ^{110m}Ag, as titrant. The methanol–dioxan mixture was used to reduce the dielectric constant, whereas use of the methanol–acetone mixture increased the solubility of the titrant in the solvent mixture [15].

Calcium butyrate, sodium acetate and sodium benzoate have been titrated with silver nitrate in a methanolic medium [15]. The solubility products of the corresponding silver salts in water are about 10^{-3}–10^{-4}. The concentrations of the solutions investigated were about $0.01M$ and the relative error was 1–2%. To obtain results of this quality by aqueous titration the solubility product in water would have to be about 10^{-8}.

5.1.5 Miscellaneous applications of radiometric precipitation titrations

Precipitation radiometric titrations have been applied with success in fields other than direct analysis, for example to determine the composition of compounds, investigate co-precipitation, determine solubility products and measure the specific activity of radioactive preparations.

Radiometric titration has been applied to elucidate the composition of the triphenylselenonium tetraiodobismuthate(III) complexes [16]. In centrifuge tubes, 25 ml of $1.09 \times 10^{-3}M$ BiCl$_3$ solution (acidified with $0.5M$ hydrochloric acid), 3.0 ml of $0.1M$ potassium iodide labelled with ^{131}I, and various volumes of $1.00 \times 10^{-2}M$ triphenylselenonium chloride were mixed. After measurement of the activity of the supernatant solutions, titration curves were plotted (Fig. 5.11). From the end-point it was concluded that $[(C_6H_5)Se]^+$ $[BiI_4]^-$ was precipitated.

When no triphenylselenonium chloride solution was added, the ratio $Bi^{3+} : I^-$ was $1 : 11.1$, and no precipitation occurred; the radioactive iodine all remained in solution. The activity at the start of the titration curve shown in Fig. 5.11 was 5841 cpm. Consequently, the activity per mole of iodide was $5841/3 \times 10^{-4} = 1.947 \times 10^7$ cpm. The activity at the end-point, obtained from the titration curve, was 3744 cpm; thus the activity of iodide corresponding to the amount held in the precipitate was $5841 - 3744 = 2097$. Hence, $2097 : 526 = 3.99$ is the number of moles of iodide involved per mole of bismuth, i.e. the precipitate is $(C_6H_5)_3SeBiI_4$.

Radiometric titration has been used for determination of the composition of the precipitate formed in the reaction of beryllium and phosphate [17], zirconyl chloride and phosphoric acid [18], the composition of mercury hexacyanoferrate(II) species [19], etc.

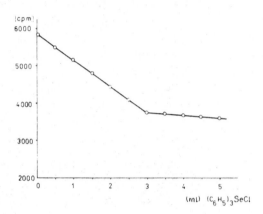

Fig. 5.11 — Investigation of the composition of triphenylselenonium iodobismuthate by radiometric titration. (Reproduced from [16] by permission of the copyright holders, Japan Society for Analytical Chemistry.)

The study of *co-precipitation* is of extreme importance, particularly in nuclear chemistry, where it plays an essential role. Radiometric titrations can provide answers to certain questions concerning the co-precipitation of microquantities of various elements in rapid precipitation. Radiometric titration enables the concentration of micro- and macrocomponents to be checked at every measurement of the solution activity. The co-precipitation of silver, thallium and lead with iodide, chloride, thiocyanate, chromate and sulphide has been studied in this way by use of ^{204}Tl, ^{110m}Ag and ^{212}Pb [20]. It was shown that the co-precipitation of thallium with silver iodide is adsorptive in nature. It begins when a small excess of iodide appears in the solution, i.e. after reaching the end-point. Owing to the adsorption of iodide on the silver iodide, the surface becomes negatively charged. Consequently, during titration, the less soluble silver iodide precipitate is formed first, followed by the thallium iodide precipitate. The amount of adsorbed ^{204}Tl did not change when radioactive thallium was added to the already precipitated silver iodide.

Precipitation radiometric titrations have also been suggested for the determination of the *solubility* and *solubility product* of sparingly soluble substances. The solubility product can be determined from individual points of the radiometric titration curve, or can be calculated from the relations deduced for the simplified experimental conditions of determination. This method was used to determine the solubility product of lead iodide (PbI_2) [21] and caesium hexachloroplatinate (Cs_2PtCl_6) [22].

The determination of specific activity by radiometric titration is relatively simple. No specific indicator or special equipment for radioactivity measurement is necessary to establish the concentration of the substance. The radioactivity of the substance serves as the indicator in the titration with an inactive titrant. The starting point of the titration curve gives the actual radioactivity of the preparation and the concentration of the preparation can be calculated from the end-point. The specific activities of ^{65}Zn and ^{237}U preparations have been determined in this way [23]. For zinc, a solution of potassium hexacyanoferrate(II) was used as titrant, and solutions of sodium hydroxide or disodium hydrogen phosphate were used as titrants for uranium. Microamounts of zinc were also determined by radiometric titration based on solvent extraction, with dithizone as titrant and carbon tetrachloride as solvent. The measuring instrument was calibrated in both instances with standard ^{65}Zn and ^{237}U preparations.

5.1.6 Radiocoulometric titration

In radiocoulometric titrations, the active component of the titrant is not added from a burette, but is generated directly in the solution by electrolysis at constant current. This technique has the following advantages over the radiometric titration methods described so far:
— the accuracy is increased,
— difficulties connected with the accurate and continuous addition of titrant are eliminated,
— problems connected with the preparation, standardization and storage of unstable titrants are similarly eliminated,
— volume correction for dilution is not required.

Three types of radiocoulometric precipitation titrations have been devised: the test solution or the working electrode or both may be labelled with radioactive tracers.

The titration curve obtained in titrations of labelled ions by use of an inactive electrode is described by

$$I^2 - I\left(I_0 - \frac{At}{kV}\right) - \frac{L}{k^2} = 0 \qquad (5.16)$$

where I is the activity of a constant volume of the solution during titration, I_0 the initial activity of the same volume of labelled test solution, V the volume of the titrated solution, k the ratio of the concentration of labelled species to its radioactivity (i.e. $[A^+] = kI$), L the solubility product of the precipitate, A a constant which includes the electrochemical equivalent, and t time.

When the last term in Eq. (5.16) is negligible, the titration curve can be separated into two straight lines:

$$I = I_0 - \frac{At}{kV} \quad \text{and} \quad I = 0 \qquad (5.17)$$

which are the asymptotes of the original hyperbolic curve. The end-point is given by the time corresponding to the point of intersection.

Radiocoulometric titrations can be done either incrementally or continuously.

Incremental titration has been used for the determination of iodide, labelled with ^{131}I [24], with silver ions generated electrochemically from a silver electrode: 100 ml of labelled test solution containing 40 ml of acetate buffer (pH 6.0) and 40 ml of $2M$ potassium nitrate were placed in 200-ml beaker and two silver electrodes were inserted. A direct current of 10 mA, supplied by a 12-V battery, was used. Portions (0.2 ml) of the supernatant liquid were withdrawn and their activity was measured. The values obtained were plotted against time. The activity decreases during the titration, becoming constant once the end-point is attained.

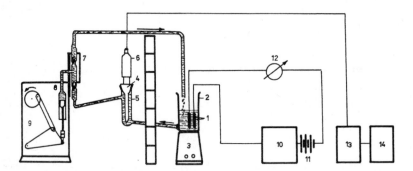

Fig. 5.12 — Automatic radiocoulometric titrimeter: 1 — silver electrodes, 2 — titration beaker with inserted filter disc, 3 — electromagnetic stirrer, 4 — GM counter, 5 — glass mantle, 6 — amplifier head, 7 — mercury valve, 8 — syringe, 9 — electric motor with transmission, 10 — amperostat, 11 — D.C. supply, 12 — ammeter, 13 — ratemeter, 14 — recording device. (Reproduced from [24] by permission of the copyright holders, IAEA, Vienna.)

An automatic titrimeter has been constructed for continuous radio-coulometric titrations [24]. A diagram is shown in Fig. 5.12. Iodide labelled with ^{131}I was titrated in this apparatus with silver ions generated from a silver electrode. The device works on the same principle as the automatic titrimeters already referred to (see Section 5.1.2.1). The titration curve of activity *vs.* time is recorded automatically and can be evaluated in two ways:

(a) the amount of electricity (coulombs) required to attain the end-point (and hence the amount of analyte) can be calculated from the time and current;

(b) the amount of analyte can be calculated by simple proportion from its titration time and that for a standard solution.

Two or three components can be titrated radiocoulometrically in the presence of each other provided they form with the generated ions precipitates having sufficiently different solubility products.

5.1.7 Radioactive Kryptonates as end-point indicators

In the application of radioactive Kryptonates (for details Chapter 4) to end-point indication, an auxiliary reaction of the radioactive Kryptonate with an excess of the titrant is used, in which radioactive krypton (^{85}Kr) is released. Since the reaction product is a gas, it can be relatively easily separated from the other components of the reaction, which are in the solid (Kryptonate indicator) and liquid (titrant and test solution) phases [25, 26].

Radioactive Kryptonates can only be used as end-point indicators if the kryptonated solid does not react with the solution being titrated. Substances reacting with the Kryptonate indicator could interfere with the end-point determination by releasing ^{85}Kr prematurely. If this interfering reaction is slow, it will increase the background count, but the end-point will still be detectable. If the interfering reaction is so fast that it is impossible to determine the exact start of ^{85}Kr release during the titration, either the interfering ion must be removed, or a different Kryptonate must be used.

Radiometric titrations using radioactive Kryptonates as end-point indicators may be performed incrementally or continuously. During *incremental titrations,* the radioactivity of the solid indicator or of the krypton released is measured after each addition of titrant. The titration curve is plotted as count rate *vs.* volume of titrant. The radioactivity of the krypton released can be measured if the radioactive Kryptonate has a high specific activity. In *continuous titrations* the titrant is added continuously to the test solution and the released ^{85}Kr is carried by nitrogen or other suitable carrier-gas into a flow-through detector connected to a ratemeter and recorder.

Several types of radioactive Kryptonates have been used as indicators in precipitation titrations. Use is made of the dissolution of the surface of a radioactive Kryptonate of a metal, or of disturbance of the surface of a radioactive Kryptonate of glass, or of redox reactions on the surface of the radioactive Kryptonate, caused by the titrant after the end-point.

Fluoride has been determined by titration with thorium, with kryptonated zinc foil as the indicator [27]. After titration of the fluoride is complete, the excess of thorium hydrolyses and the hydrogen ions formed react with $Zn[^{85}Kr]$, releasing ^{85}Kr.

Cadmium and calcium have been titrated with $0.2M$ sodium fluoride, with radioactive fused-silica Kryptonate as indicator [28]. For calcium, the best results were obtained at pH 7.0; for cadmium, at pH 6.0. After the end-point, the first excess of fluoride attacks the silica, with release of ^{85}Kr and decrease in the radioactivity of the indicator. Thorium has been determined by incremental titration with $0.2M$ sodium fluoride, and radioactive glass Kryptonate as indicator [29]. The thorium solution was acidified to pH < 1 with $1M$ sulphuric acid before titration.

Barium has been titrated with $0.002M$ $K_2Cr_2O_7$ with radioactive silver Kryptonate as the indicator [30]. Up to the end-point, the precipitation reaction is:

$$2\,Ba^{2+} + Cr_2O_7^{2-} + H_2O \; \rightleftharpoons \; 2\,BaCrO_4 + 2\,H^+ \qquad (5.18)$$

After the end-point, the redox reaction

$$Cr_2O_7^{2-} + 6\,Ag[^{85}Kr] + 14\,H^+ \; \rightarrow$$
$$6\,Ag^+ + 2\,Cr^{3+} + 7\,H_2O + {}^{85}Kr\uparrow \qquad (5.19)$$

occurs. The decrease in activity of the silver Kryptonate is measured after each addition of titrant.

5.2 RADIOMETRIC TITRATIONS BASED ON COMPLEX FORMATION

In radiometric detection of the end-point of complex-forming reactions, the separation of the components is the most difficult operation. In these reactions, the reacting species and the reaction products are in the same phase, before and after the reaction. Thus, titrations of this type can only be carried out by using an auxiliary method of separation: the success of the titration thus depends on the success of the separation. Several methods of separation have been described. Besides solvent extraction, the earliest method applied, methods based on ion exchange, paper chromatography, on focused paper electrophoresis, solid indicators and radioactive Kryptonates are used.

The sensitivity of radiometric titrations based on complex formation depends on the factors known to govern complexometry. Determinations can be carried out successfully even on 10^{-6}–$10^{-5}M$ solutions.

5.2.1 Extraction radiometric titration

In extraction radiometric titrations, one of the reaction components, usually the reaction product, is extracted with a solvent immiscible with the system. Organic reagents usually serve as titrants for the determination of metal ions. The resulting complex, sparingly soluble in water, is extracted into an organic phase. Thus, the end-point is detected by following the changes in radioactivity of the aqueous or organic phase.

5.2.1.1 *Titration of one, two or three components in the presence of each other*

In the simplest case, three types of titration curves are obtained, depending on whether the test solution, the titrant, or both, are radioactive [31].

(1) *Determination of an active component with an inactive titrant.* The element or compound to be determined is labelled with one of its radioactive isotopes. It is most practical and convenient to employ a carrier-free radionuclide for labelling. As the amounts of labelling substances used are minute, no correction is needed when calculating the results. When radionuclides diluted with a carrier are used, the amounts added to the titrated sample must obviously be known and subtracted from the result.

When the changes in radioactivity of the aqueous and organic phases during titration are followed, curves 1 and 2, respectively, of Fig. 5.13A are obtained. This method is used for titration of zinc, mercury, lead, etc., with dithizone, or of zinc and mercury with oxine, the resulting complexes being extracted into chloroform, carbon tetrachloride, etc.

(2) *Determination of an inactive component with an active titrant.* Curves 1 and 2 of Fig. 5.13B are obtained. An example is titration of potassium thiocyanate with cobalt(II) sulphate labelled with ^{60}Co (water–isoamyl alcohol system).

(3) *Determination of an active component with an active titrant.* Titration curves are shown in Fig. 5.13C, curves 1 and 2.

As procedure (3) does not offer any advantages over (1) and (2), only these two are widely applied in practice.

When the component to be determined is labelled with small or large amounts of its radioactive isotope, the method is called isotopic labelling, as in radiometric precipitation titrations. When no suitable radioactive isotope of the element to be determined is available, *non-isotopic labelling* may be employed [31, 32]. This is labelling by an isotope of an element different from the element being determined. Two alternatives of this labelling technique are known. In the first, the labelling element also forms an extractable complex with the titrant, with an extraction constant of the same order of magnitude as that of the analyte complex. Both complexes are extracted simultaneously, and in both the extraction and the titration, the isotope of the labelling element behaves in just the same way as the element to be determined.

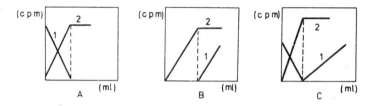

Fig. 5.13 — Extraction radiometric direct titration curves: 1 — aqueous phase, 2 — organic phase. A — titration of radioactive test solution with inactive titrant, B — titration of inactive test solution with active titrant, C — titration of active solution with active titrant.

In the second alternative, the labelling element forms a complex with the titrant with an extraction constant lower by several orders of magnitude than that of the element to be determined. Only the analyte reacts initially with the titrant, and the measured radioactivity is constant (and high or low, depending on whether the radioactivity is measured in the aqueous or the organic phase). When all the analyte has reacted, the labelling element begins to react. This is shown on the titration curve by a decrease or increase of activity. The end-point is thus denoted by a break in the titration curve. An example is the titration of zinc with dithizone, with ^{60}Co as indicator (Fig. 5.14) [32].

If the reaction rate between the analyte and the titrant is too low, the *indirect extraction radiometric titration method* should be applied. Here, the reaction rate with excess of titrant solution is generally sufficiently high. An excess of titrant is added to the sample, followed by excess of a labelled solution of an element with which the titrant reacts immediately. The excess of labelled solution is then determined by extraction radiometric titration. The quantity of the analyte can then be calculated from the amounts of reactants consumed. This method has been used for the determination of palladium by use of Ag$^+$ labelled with ^{110m}Ag, and dithizone as titrant [33].

For radiometric determinations based on extraction and done by direct titration of several components in the presence of each other, quite a number of methods are available. Of these, the application of non-isotopic labelling in systems containing two or three components to be determined, the use of labelled titrants of various specific activities, titrations done at various pH values or in the presence of masking agents, etc., are most frequently used.

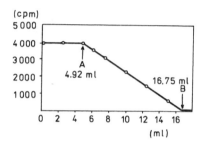

Fig. 5.14 — Extraction radiometric titration, with a non-isotopic indicator. Determination of zinc as dithizonate, with ^{60}Co as indicator. Point A corresponds to the end-point for zinc, point B to that for zinc plus cobalt (added: 10. 18 µg of Co^{2+} and 6.90 µg of Zn^{2+}; found: 10.27 µg of Co^{2+} and 6.89 µg of Zn^{2+}). (Reproduced from [32] by permission of the copyright holders, Pergamon Press, Oxford.)

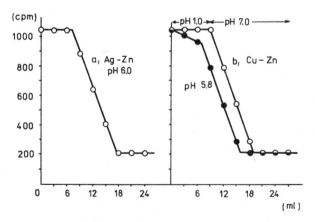

Fig. 5.15 — Titration of Ag and Zn, and of Cu and Zn with dithizone as titrant, with and without varying the pH (non-isotopic indicator ^{65}Zn). (Reproduced from [33] by permission of the copyright holders, Springer-Verlag, Vienna.)

For the titration of two or three components in the presence of each other, non-isotopic labelling has mainly been employed. For example, traces of cobalt and zinc in the presence of each other have been titrated with dithizone, with ^{60}Co as non-isotopic indicator (see Fig. 5.14) [32]. Traces of silver and zinc have been titrated with dithizone, with ^{65}Zn as non-isotopic indicator [33]. However, the determination of traces of copper and zinc was found possible only by performing two titrations

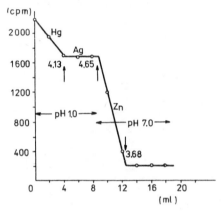

Fig. 5.16 — Titration of Hg, Ag and Zn with dithizone as titrant (non-isotopic indicators ^{203}Hg and ^{65}Zn) by varying the pH. (Reproduced from [33] by permission of the copyright holders, Springer-Verlag, Vienna.)

at different pH values [33]. Copper was titrated at pH 1.0, then zinc at pH 7.0. The titration curves are presented in Fig. 5.15. It can be seen that erroneous end-points are obtained when the titrations are done at pH 5.8. Traces of mercury, silver and zinc were also titrated with dithizone (^{203}Hg and ^{65}Zn as non-isotopic indicators) [33]. Mercury and silver were titrated at pH 1.0, and zinc at pH 7.0. The titration curve is shown in Fig. 5.16. Traces of mercury, copper and zinc (^{203}Hg and ^{65}Zn as indicators) and silver, copper and cadmium (^{110m}Ag and ^{115}Cd as indicators) have also been titrated with dithizone [34, 35].

5.2.1.2 Apparatus and technique of titration
Radiometric titrations based on complex formation carried out by solvent extraction are, in general, conducted incrementally, i.e. the addition of each portion of titrant is followed by an extraction and measurement of the radioactivity of the aqueous or the organic phase. The cycle is repeated as many times as required to establish the titration curve.

Equal portions of test solution and of buffer to adjust the pH to the required value are placed in a series of test-tubes, to which increasing quantities of the titrant (chelating agent dissolved in organic solvent) are added. The contents of all the tubes are then made up to the same volume with the organic solvent used to dissolve the chelating agent. It is expedient to mark all the tubes at the appropriate volume and dilute up to the mark. It is also advisable to use equal phase volumes. The test-tubes are then shaken automatically until equilibrium is reached (approximately 15 min). The test-tubes are then centrifuged for 1–2 min. The activities of the two layers are measured with a GM counter or with a scintillation counter after evaporation of an aliquot. Since it is possible to measure the activities of both layers simultaneously, two determinations can be made simultaneously on the same sample. The values obtained serve for plotting the titration curve.

An apparatus for extraction radiometric titration is shown in Fig. 5.17. The NaI(Tl) scintillator is housed in a lead shield into which the titration test-tube is inserted. Besides reducing the background, the lead shields the detector from the activity of the aqueous phase. The increase of activity in the organic phase is monitored [36].

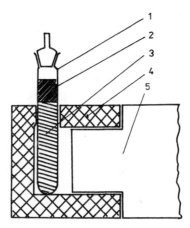

Fig. 5.17 — Apparatus for extraction radiometric titration: 1 — test-tube, 2 — aqueous phase, 3 — organic phase, 4 — lead shield, 5 — scintillation head with NaI(Tl) crystal. (Reproduced from [36] by permission of the copyright holders, ÚVVVR, Prague.)

Only a few attempts have been made to make the titration continuous. The main obstacle is the intermittent nature of the liquid extraction. Continuous titration would require continuous separation of the phases, and attempts to achieve this have all failed. Similar difficulties are met in attempts to automate the titration.

Semi-automatic instruments for radiometric titration based on solvent extraction have been described, however [37, 38]. Though these instruments do not completely solve the problem of continuous titration, they do facilitate and accelerate the determination.

An apparatus for semi-automatic extraction radiometric titration is shown in Fig. 5.18. The lighter layer is circulated by suction or gas pressure through the spiral glass tube fixed below a GM counter. An electric stirrer is used for mixing, and the activity of the aqueous phase is measured after the addition of each portion of the titrant [37]. The scheme of a similar apparatus is shown in Fig. 5.19. The lighter phase is circulated below the detector by suction [38].

A fully automatic apparatus is shown in Fig. 5.20 [39, 40]. It consists essentially of three parts.

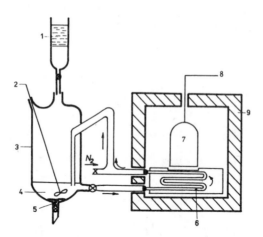

Fig. 5.18 — Semi-automatic apparatus for extraction radiometric titration: 1 — burette, 2 — stirrer, 3 — titration flask, 4 — aqueous phase, 5 — CCl$_4$ layer, 6 — glass spiral, 7 — GM counter, 8 — electrical connection to counter, 9 — lead shield. (Reproduced from [37].)

(1) The titration flask (1) has arbitrary volume and its lower part is a test-tube fitted with a stopcock. The stirrer (13) is driven by a motor (12). The test-tube part of the flask is inserted into the lead shield (2) housing a scintillation crystal (3). If the stopcock is omitted, the test-tube can be inserted into a suitably screened well-type crystal.

(2) The detection system consists of a scintillation head (or GM tube) with voltage supply (4) and amplifier (5), integrator (6) and

recorder (7). The scintillation head can also be connected to a multichannel pulse-height analyser and recorder. If desired, the pH can be simultaneously measured [pH-meter (15) and electrodes (14)] and recorded.

(3) Addition of titrant and mixing are periodic and alternate. The principal control element is switch (8), adjustable to close for periods between 0.5 and 5 sec at intervals of 1–3000 sec, to operate the electromagnetic valve (10) for addition of titrant (11) and simultaneously operate switch (9) controlling motor (12). Switch 9 may be set for 0–3 min intervals. Use of this device considerably reduces the titration time.

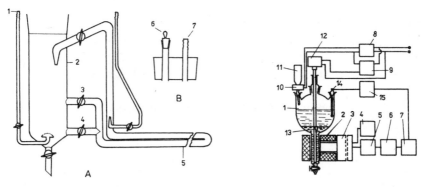

Fig. 5.19 — Semi-automatic apparatus for extraction radiometric titration. A: Titration instrument: 1 — mixing of titration liquid by air-bubbles, 2 — reaction flask (when stopcock 3 is opened the lighter phase circulates and is measured; when stopcock 4 is opened the heavier phase circulates), 5 — part of the tube in the lead shield, under the detector (GM or scintillation counter); B: Reaction flask seal: 6 — inlet for titrant, 7 — suction connection. (Reproduced from [38] by permission of the copyright holders, IAEA, Vienna.)

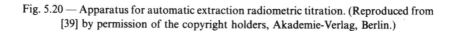

Fig. 5.20 — Apparatus for automatic extraction radiometric titration. (Reproduced from [39] by permission of the copyright holders, Akademie-Verlag, Berlin.)

5.2.2 Ion-exchange radiometric titration

Titration products can be separated from the initial components by the use of ion exchangers. The works published so far [41–47], use EDTA (H_2Y^{2-}) and DCTA (H_2Z^{2-}), as titrants.*

---

* Some of these works, particularly those dealing with the determination of very low concentrations, are more conveniently assigned to the group of analytical methods designated as 'saturation analysis' (see p. 213).

5.2.2.1 Essence of the method

In the titration of a bivalent metal ion Me^{2+} with EDTA or DCTA, an anionic chelate complex is formed:

$$Me^{2+} + H_2Y^{2-} \rightleftharpoons MeY^{2-} + 2H^+ \qquad (5.20)$$

The unreacted Me^{2+} can be separated by removal of Me^{2+} with a cation exchanger or of the chelate with an anion exchanger.

The activity of either the solution or the ion exchanger is monitored and the titration curve plotted.

When isotopic indication is used, in an idealized case the titration curve is linear up to the end-point. Since conditions are not ideal in practice, the titration curves will not be linear, and difficulties may arise in locating the end-points. Thus, the use of calibration curves is advisable. These show the correlation between the measured radioactivity and the amount of substance to be determined, on addition of a fixed substoichiometric (constant) volume of titrant. Theoretical titration curves for measurement of the radioactivity of a cation exchanger used to remove free indicator ion are shown in Fig. 5.21. From the theoretical titration curves the calibration curves are plotted (cf. Fig. 5.22) [46].

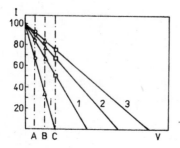 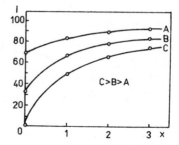

Fig. 5.21 — Correlation between the concentration of analyte (proportional to cpm) (ordinate) and the volume of titrant (abscissa): i — titration of the indicator, 1, 2 and 3 — titration curves for various amounts of inactive components to be determined.

Fig. 5.22 — Correlation between the radioactivity measured and the amount of component to be determined, on the addition of a constant substoichiometric amount of the titrant. The curves are for three different amounts of titrant (A < B < C).

The calibration curves also offer the advantage that the concentration of the titrant need not be accurately known. This will often make the work much easier in the case of low concentrations. However, for plotting the calibration curve, standard solutions of accurately known concentrations are required.

When non-isotopic indication is used, the calibration curves (Fig. 5.23) obtained from the theoretical titration curves (Fig. 5.24) will be linear. The prerequisite is that the conditional stability constant of the analyte complex (K'_{ML}) should be much higher than that of the indicator complex (K'_{M_iL}). The sensitivity of the determination will theoretically be lower with isotopic indication than with non-isotopic indication [46].

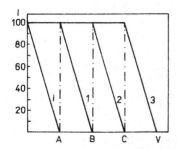 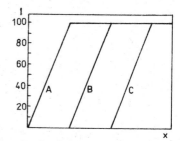

Fig. 5.23 — Correlation between the concentration of metal ions to be determined and the volume of titrant: i — titration curve of the indicator, 1, 2 and 3 — titration curves for various amounts of component to be determined.

Fig. 5.24 — Correlation between the radioactivity measured and the amount of inactive component to be determined, on the addition of equal volumes (A, B, C) of titrant.

5.2.2.2 Titration techniques

Ion exchanger added to the labelled titrant. About 100–200 mg amounts of ion exchanger are added to the aliquots of the isotopically labelled sample solution to which various volumes of titrant have been added. The exchanger is filtered off and the specific activity of the liquid is measured and plotted against volume of titrant. On the microscale the determination is done with 0.5–1.5 ml aliquots of test solution and 50–100 mg of ion exchanger in 2.5-ml vials. Various volumes of titrant are added and the vials are shaken until equilibrium is attained. The ion exchange can also be done on microcolumns, or the exchanger separated by centrifugation (Fig. 5.25). Figure 5.26 illustrates the titration of zinc (^{65}Zn as indicator) with $10^{-3}M$ EDTA. The separation was done with Dowex 50W-X2 cation exchanger [47].

Titration of one aliquot by means of a single ion-exchange membrane. The sample, a buffer solution, and a radioactive indicator are added to a 2.5-ml titration flask and an ion-exchange membrane (1 × 1 cm) in a polyethylene holder connected to an electric stirrer is inserted into the

solution. During stirring the membrane acquires a certain activity
which is then measured. The titration curve is constructed from the
equilibrium activity values attained by the membrane after each addition of titrant.

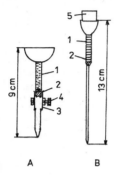 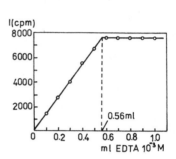

Fig. 5.25 — Micro-columns for granular ion-exchange resin: 1 — ion exchanger,
2 — cotton-wool plug, 3 — PVC tubing, 4 — flow-control, 5 — stopper.

▶

Fig. 5.26 — Titration curve of ^{65}Zn indicator solution.

Batchwise titration with one type of exchanger. Several aliquots of
sample are treated individually, with buffer and increasing amounts of
titrant. Ion-exchange membranes (1 × 1 cm) are added to the solutions,
and the vessels are sealed and then shaken until equilibrium is attained.
The activity of the individual membranes is measured and plotted
against titrant volume.

Batchwise titration with two types of exchanger. The procedure is the
same as for one type of exchanger except that both anion-exchange and
cation-exchange membranes (1 × 1 cm) are inserted into the solutions.
The labelled cation is thus sorbed on the cation exchanger and the
labelled negatively-charged chelate complex on the anion exchanger.

Use of ion-exchange membranes has the advantage that the activity
of the membranes can be measured directly in a well-type scintillator,
or with a Geiger–Müller counter. Furthermore, the membranes can be
regenerated by a simple technique for use in series of analyses. The
analysis time is shorter with ion-exchange membranes than with granular ion exchangers.

A titration assembly for use of phase separation by ion exchange is
shown in Fig. 5.27 [48]. The activity is measured with a single-channel
analyser coupled to an NaI(Tl) scintillation crystal, with a hole, perpen-

dicular to the axis, made in its middle, to carry a glass tube (with a
fritted disc attached to the lower end), dipping into the solution. When
equilibrium had been attained after each addition of the titrant, the
solution is drawn up into the tube with a pump, and its activity is
measured.

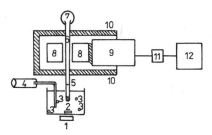

Fig. 5.27 — Titration assembly for ion-exchange titration: 1 — stirrer motor, 2 — stirrer
bar, 3 — anion exchanger, 4 — microburette, 5 — fritted disc, 6 — glass tube, 7 — pump,
8 — NaI(Tl) crystal, 9 — photomultiplier, 10 — lead shield, 11 — preamplifier, 12 —
single-channel analyser. (Reproduced from [48] by permission of the copyright holders,
Akadémiai Kiadó, Budapest.)

5.2.3 End-point detection by paper chromatographic separation
A labelled solution of a metal ion is titrated with a complexing agent.
After the addition of each portion of titrant, a small aliquot of the
solution is transferred to give an individual spot on the start-line of a
chromatogram. After the titration, the chromatogram is run, and the
radioactivity along its length is measured; two spots appear for each
aliquot added, one for the metal ion, the other for the anionic metal
complex. As the titration progresses, the activity of the anion spot
increases, while that of the cation spot decreases. The titration curve is
obtained by plotting one of the sets of activities against the volume of
titrant added [41].

5.2.4 End-point detection by focusing paper electrophoresis
Focusing paper electrophoresis has been successfully applied for detecting the end-points of radiometric titrations based on complex formation [49].

Figure 5.28 gives a schematic presentation of the apparatus used.
There are two containers linked by a strip of chromatographic paper.
A drop of a dilute solution of the analyte cation, M^{x+}, is placed in the
centre of this paper strip, which is held in a carbon tetrachloride bath
(2) as coolant. The cathode compartment (3) contains a solution of a
complexing agent L^{y-} which forms an anionic complex with M^{x+} (e.g.

acetate). The anode compartment (1) contains a mineral acid (e.g. hydrochloric). Figure 5.29 shows schematically the processes that occur in the system. When through capillary migration, these solutions have arrived at the spot of M^{x+}, a potential of about 300–500 V is applied between the electrodes. The protons and the complexing anions then

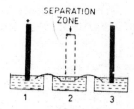

Fig. 5.28 — Diagram of apparatus for separation by focusing paper electrophoresis. (Reproduced from [50] by permission of the copyright holders, the Swiss Chemical Society.)

migrate towards each other and react on meeting, which leads to a pH–pL gradient along the paper strip. Depending on the concentration of the free ligand, L^{y-}, the element M will be present in this gradient either as the cation M^{x+} (anodic region) or as a complex anion (cathodic region). Consequently, in the electric field these ions will also migrate towards each other. Thus, in a part of this field, the element M will be enriched just as at an electrophoretic focus. It has been proved

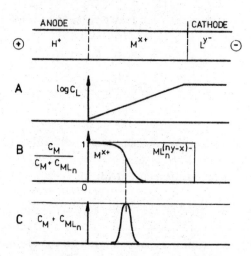

Fig. 5.29 — Schematic diagram of separation by focusing paper electrophoresis: A — ligand gradient, B — metal distribution at start, C — metal profile at end of electrophoresis. (Adapted from [51].)

that the position of this focus depends mainly on the stability constant of the complex $ML_n^{(ny-x)-}$ [51].

Double focusing may also occur, in which two separate lines ensue for the same metal. This occurs when the metal can form two complexes of different stability (a) with different ligands, (b) with one ligand, the metal being in two oxidation states.

Case (a) can be used in titrations. In these determinations, the test solution is titrated with EDTA or some similar complexant. After addition of each portion of titrant, a small aliquot is withdrawn from the solution and focused in the electrophoresis apparatus. There is now competitive focusing of the two types of metal complex and two foci appear, one corresponding to the free metal ion (and its complex with the reagent in the cathode compartment) and the other to the metal–titrant complex. The activity of the EDTA complex focus increases during the course of titration, and at the end-point there is only the focus corresponding to the EDTA complex. The radioactivity of the focus of this complex (A_c) and the total activity (A_t) are measured after each focusing. The titration curve is obtained by plotting the ratio A_c/A_t against the volume of titrant added. For example, a $1mM$ solution of yttrium was titrated in this way with 0.1 or $0.01M$ EDTA [49], with an acetate buffer in the cathode compartment. Figure 5.30 shows the activity of the two foci of yttrium at various stages of the titration [49].

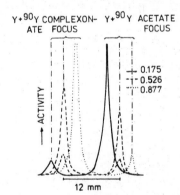

Fig. 5.30 — Radiograms of foci obtained in the titration of yttrium at various degrees of titration (0.175, 0.526, 0.877). (Reproduced from [49] by permission of the copyright holders, the Swiss Chemical Society.)

5.2.5 End-point detection with the use of solid indicators (precipitates and radioactive Kryptonates)

The end-point of complexometric titrations can also be detected with the aid of a labelled precipitate [38, 52–54].

Cation M is titrated with complexing anion L according to the reaction

$$M + L \quad \rightleftharpoons \quad ML \tag{5.21}$$

The end-point is indicated with the aid of a labelled indicator cation $*M_i$ originally present as a precipitate $*M_iB$. On completion of reaction (5.21) L reacts with the precipitate to give a soluble compound of $*M_i$:

$$*M_iB \downarrow + L \quad \rightarrow \quad *M_iL + B \tag{5.22}$$

Such titrations require the following conditions to be fulfilled:
(a) $pK_{ML} > pK_{*M_iL}$,
(b) the conditional stability constant of M_iL and solubility product of $*M_iB$ should be such that the complexing agent can dissolve the precipitate under the conditions used.

During the titration reaction (5.21) proceeds first, so the radioactivity of the solution remains constant and depends on the solubility product of $*M_iB$. After complexation of M is complete, the increasing excess of L will progressively dissolve $*M_iB$, and the radioactivity of the solution will increase in proportion to the excess of titrant added.

The titration instruments for use with solid indicators do not differ much from those generally used for radiometric titrations. When the solid indicator is applied as a suspension, the necessary phase separation can be done by filtration, centrifugation or flotation. Figure 5.31 shows the apparatus for complexometric radiometric titration with the aid of an indicator membrane (radioactive indicator precipitated in the pores of a filter paper) [38].

Figure 5.32 shows the complexometric radiometric titration curves obtained for various amounts of Ca^{2+}, with EDTA as titrant and silver iodate labelled with ^{110m}Ag as non-isotopic indicator [52].

Solid indicators have also been used in complexometric radiocoulometric titrations (for radiocoulometric titrations see p. 271) of nickel and other cations [55] with cyanide ions coulometrically generated in a solution of dicyanoargentate(I), according to the electrode reaction

$$Ag(CN)_2^- + e^- \quad \rightleftharpoons \quad Ag + 2\,CN^- \tag{5.23}$$

The cyanide ions react with the Ni^{2+} ions:

$$Ni^{2+} + 4\,CN^- \quad \rightleftharpoons \quad Ni(CN)_4^{2-} \tag{5.24}$$

The indicator used was AgI labelled with ^{110m}Ag. After the end-point, the excess of CN^- ions dissolves the indicator:

$$*AgI + 2\,CN^- \quad \rightleftharpoons \quad *Ag(CN)_2^- + I^- \tag{5.25}$$

and the activity of the solution increases proportionally with generation time.

A drawback of these titrations is the cumbersome and dangerous handling of radioactive precipitates and solutions and difficulties with setting up an automated titration procedure. These drawbacks are eliminated by the use of radioactive Kryptonates (see p. 217) as solid indicators. For example, iron has been determined [56] by adding the radioactive Kryptonate $Y_2(C_2O_4)_3[^{85}Kr]$ to the sample and titrating with $10^{-2}M$ EDTA. Up to the end-point the reaction is:

$$Fe^{3+} + EDTA \quad \rightarrow \quad Fe(EDTA) \qquad (5.26)$$

and after the end-point:

$$Y_2(C_2O_4)_3[^{85}Kr] + 2\,EDTA \quad \rightarrow$$
$$2\,Y(EDTA) + 3\,C_2O_4^{2-} + {}^{85}Kr\uparrow \qquad (5.27)$$

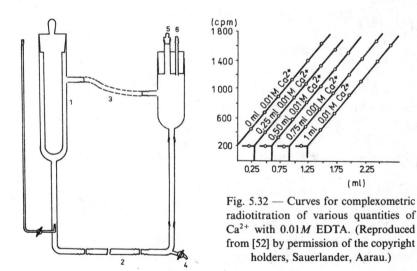

Fig. 5.32 — Curves for complexometric radiotitration of various quantities of Ca^{2+} with $0.01M$ EDTA. (Reproduced from [52] by permission of the copyright holders, Sauerlander, Aarau.)

Fig. 5.31 — Circulating titrimeter with a precipitate membrane: 1 — membrane holder, 2 — portion of tube in the lead shield under the detector, 3 — flexible connecting tube, 4 — outflow to waste on completion of titration, 5 — inlet for addition of titrant, 6 — to suction device. (Reproduced from [38] by permission of the copyright holders, IAEA, Vienna.)

For the determination of nickel, the radioactive Kryptonate $AgI[^{85}Kr]$ has been used as the indicator and $0.1M$ potassium cyanide as the titrant [57]. The titration was done in ammoniacal medium by either the incremental or the continuous method. Up to the end-point, reaction (5.24) takes place. After the end-point, the reaction which takes place is:

$$AgI[^{85}Kr] + 2\,CN^- \quad \rightarrow \quad Ag(CN)_2^- + I^- + {}^{85}Kr\uparrow \qquad (5.28)$$

The solid kryptonated indicator dissolves and its activity decreases with increasing volume of titrant added.

5.3 RADIOMETRIC TITRATIONS BASED ON REDOX REACTIONS

5.3.1 End-point detection with indicator precipitates

The use of indicator precipitates (solid indicators) in complexometric titrations has already been dealt with (p. 288), and these indicators can also be used for radiometric end-point detection in redox titrations [38, 52, 58]. In titration of an oxidant Ox_2, with a reductant, Red_1, in the presence of a solid indicator, $*M_iB$, radiometric detection of the end-point is possible if Red_1 can form a complex with the ion $*M_i$ after quantitative reduction of Ox_2. Excess of Red_1 added to the system dissolves some $*M_iB$ by the complexing reaction:

$$*M_iB + Red_1 \quad \rightleftharpoons \quad *M_iRed_1 + B \qquad (5.29)$$

The prerequisites for this to occur are as follows:
(a) the standard potentials of the systems Ox_1/Red_1 and Ox_2/Red_2, and the stability constant of the complex $*M_iRed_1$, must be such that Ox_2 is completely reduced by Red_1 before $*M_i$ is complexed by Red_1,
(b) the stability constant of $*M_iRed_1$ and the solubility product of $*M_iB$ should be such that the excess of titrant is able to dissolve the precipitate by complex formation.

Thus iodine has been titrated with thiosulphate, with $^{110m}AgSCN$ as solid indicator [38]. The following reaction takes place up to the end-point:

$$I_2 + 2\,S_2O_3^{2-} \quad \rightleftharpoons \quad 2\,I^- + S_4O_6^{2-} \qquad (5.30)$$

and the activity of the solution is constant and low. After quantitative reduction of the iodine, the excess of thiosulphate reacts with the indicator:

$$^{110m}AgSCN + 2\,S_2O_3^{2-} \quad \rightleftharpoons \quad ^{110m}Ag(S_2O_3)_2^{3-} + SCN^- \qquad (5.31)$$

and the activity of the solution increases in proportion to the excess of titrant added.

In this determination the radioactive Kryptonate AgSCN[^{85}Kr] can also be used as solid indicator [59].

5.3.2 End-point detection with radioactive metals

Radioactive metals have also been suggested as indicators in redox titrations [60]. For instance, neutron-activated silver has been used as indicator in titrating arsenious acid with ceric sulphate. In these titrations the oxidizing agent must always possess a higher redox potential than the indicator used. The ion to be determined must have the lowest redox potential. For the given case, we have

$$Ce^{4+} + e^- \rightleftharpoons Ce^{3+} \qquad + 1.500 \text{ V}$$
$$Ag^+ + e^- \rightleftharpoons Ag \qquad + 0.799 \text{ V}$$
$$H_2AsO_4^- + 2e^- + 3H^+ \rightleftharpoons H_3AsO_3 + H_2O \qquad + 0.574 \text{ V}$$

5.3.3 End-point detection by use of amalgams

Radioactive amalgams can be used in certain instances for radiometric end-point detection in redox titrations [38, 52]. If a liquid amalgam with a labelled metal atom, suitable for reducing the oxidizing titrant Ox_2, is present in the system

$$Red_1 + Ox_2 \rightleftharpoons Ox_1 + Red_2 \qquad (5.32)$$

then after oxidation of all the Red_2, the excess of oxidizing titrant will react with the amalgam and liberate from it an equivalent quantity of radioactive ion *Me according to the reaction

$$*Me(Hg) + Ox_2 \rightleftharpoons Hg + *Me + Red_2 \qquad (5.33)$$

During the titration, the radioactivity of the solution remains at a low level until the end-point. After the end-point, reaction (5.33) proceeds and the radioactivity of the solution increases. This procedure has been applied, for example, to the titration of ascorbic acid with iron(III) sulphate in the presence of zinc amalgam labelled with ^{65}Zn.

5.3.4 End-point detection by solvent extraction

Solvent extraction can be employed for end-point detection in radio-iodometric titration [38, 52], e.g. thiosulphate has been titrated with iodine solution labelled with ^{131}I, with carbon tetrachloride as solvent. The radioactivity of both phases was measured. There was practically no radioactivity in the organic phase while thiosulphate was still present in the solution. An excess of iodine solution, however, resulted in an increase in radioactivity in the organic phase.

5.4 NEUTRALIZATION RADIOMETRIC TITRATIONS

Neutralization radiometric titrations have so far been done only with the use of radioactive Kryptonates (for more details, see p. 217).

The radioactive Kryptonates of magnesium and zinc have been used as indicators in the titration of a strong base (sodium hydroxide) with a strong acid (hydrochloric acid). The radioactivity of the released ^{85}Kr was determined after each addition of titrant [27].

Radioactive glass Kryptonate is a suitable indicator in the titration of acids with strong bases [61]. After the end-point, the excess of base decomposes the surface layers of the glass, with the consequent release of ^{85}Kr and decrease in the radioactivity of the glass. Two types of titrations using a radioactive glass Kryptonate have been used.

(a) The sample and the glass Kryptonate are placed in a titration vessel fitted with an inlet tube for passage of nitrogen, which carries off the released ^{85}Kr. After each addition of the titrant, the radioactive Kryptonate is left in the solution for 3 min and then removed and dried for a predetermined time. Its radioactivity is measured under constant geometrical conditions with an end-window counter.

(b) The radioactivity of the glass Kryptonate is measured directly in the solution with a cylindrical beta counter fitted with a Plexiglas ring. The radioactive Kryptonate is placed in a special holder in the ring (Fig. 5.33) [26]. Three minutes after each addition of titrant, the Kryptonate activity is measured for 2 min; the next increment of titrant is added during 1 min. Continuous titration is also possible with this method.

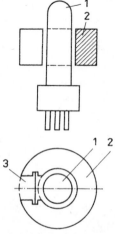

Fig. 5.33 — Measuring head of titration assembly: 1 — GM counter, 2 — Plexiglas holder, 3 — kryptonated cover glass or kryptonated silver plate.

5.5 THERMORADIOMETRIC TITRATIONS

In classical thermometric titrations, the interconversion of chemical energy and heat is monitored as a function of the amount of titrant added. Constancy or only a slight change of temperature indicates that the end-point has been passed and that it can be located by extrapolation.

In thermoradiometric titrations the conversion of chemical energy into heat is followed by means of an inert radioactive gas; the determination is based on the fact that the solubility of an inert gas decreases with increasing temperature [62–64].

The sample is saturated with a radioactive inert gas (e.g. ^{85}Kr) and the titrant is added continuously or discontinuously. If the enthalpy change of the reaction is negative, heat is evolved, the temperature of the sample will increase and radioactive gas will be released during the titration. The amount of ^{85}Kr released, or alternatively the decrease in the activity of the sample, may be measured as a function of the amount of the titrant added. The apparatus developed for radiometric titrations with Kryptonates as indicators may be used. The titration curve can be plotted manually, or recorded automatically.

The method has been applied to different types of reactions, including neutralization (HCl with NaOH), precipitation (AgNO$_3$ with HCl) and redox reactions (As$_2$O$_3$ with KMnO$_4$). The end-point can be detected sensitively and precisely, and the method should be applicable to neutralization, precipitation and complexometric titrations, but not those for which ΔH is positive (e.g. Mg^{2+} with EDTA, for which the temperature falls during the titration).

A typical procedure for a neutralization titration is as follows: 1 ml of distilled water saturated with ^{85}Kr is added to 9 ml of the acid sample, and the mixture is brought to a given temperature. The solution of base (concentration at least 10 times that of the acid, to minimize dilution problems), is added from a burette kept at the selected temperature. After each addition of base the sample is stirred for 2 min and the residual activity of the sample is measured [64]. The titrant can also be added continuously to the sample, and the released ^{85}Kr flushed with nitrogen through a flow-detector connected to a ratemeter and recorder.

REFERENCES

[1] V. Jesenák and J. Tölgyessy, *Acta Chim. Acad. Sci. Hung.*, 1962, **32,** 397.
[2] V. Jesenák and J. Tölgyessy, *Chem. Zvesti*, 1963, **17,** 161.

[3] J. Tölgyessy, *Rádiometrické titrácie*, Publ. House of the Slovak Academy of Sciences, Bratislava, 1966.

[4] A. Langer, *J. Phys. Chem.*, 1941, **45**, 639.

[5] Nuclear Chicago Corp., *Techn. Bull.*, No. 6, 1959.

[6] J. Tölgyessy and V. Šajter, *Acta Chim. Acad. Sci. Hung.*, 1961, **26**, 179.

[7] J. Tölgyessy and V. Šajter, *Chem. Zvesti*, 1962, **16**, 217.

[8] T. Braun, *Chim. Anal. (Paris)*, 1964, **46**, 61.

[9] J. Tölgyessy, *Sb. Chemickej Fakulty SVŠT*, Bratislava, 1960, p. 3.

[10] J. Majer, M. Šaršúnová and J. Tölgyessy, *Schweiz. Apoth. Ztg.*, 1960, **98**, 631.

[11] A. I. Busev and V. M. Byrko, *Talanta*, 1961, **8**, 492.

[12] A. I. Busev and V. M. Byrko, *Tr. Komis. po Analit. Khim., Akad. Nauk SSSR*, 1958, **9**, 200.

[13] K. Kimura and F. Kametani, *Bunseki Kagaku*, 1961, **10**, 1293.

[14] T. Braun and J. Tölgyessy, *Radiometric Titrations*, Pergamon Press, Oxford, 1967.

[15] V. S. Chernyi, E. I. Vail and N. A. Izmailov, *Tr. Komis. po Analit. Khim., Akad. Nauk SSSR*, 1963, **13**, 445.

[16] M. Shinagawa, H. Matsuo and M. Yoshida, *Bunseki Kagaku*, 1955, **4**, 139.

[17] I. P. Alimarin and I. M. Gibalo, *Zavodsk. Lab.*, 1957, **23**, 412.

[18] I. P. Alimarin and I. M. Gibalo, *Zavodsk. Lab.*, 1956, **22**, 635.

[19] A. Basiński, W. Szymański, A. Krygier and G. Zapalowska, *Roczniki Chem.*, 1963, **37**, 1345.

[20] I. P. Alimarin and I. A. Sirotina, *Zh. Neorgan. Khim.*, 1958, **3**, 1709.

[21] J. F. Duncan, *J. Inorg. Nucl. Chem.*, 1959, **11**, 161.

[22] A. T. Casey and W. Robb, *Nature*, 1963, **198**, 581.

[23] F. Kametani, K. Kimura and A. Kataoka, *J. At. Energy Soc. Japan*, 1962, **4**, 373.

[24] J. Tölgyessy, V. Jesenák and T. Braun, *Radiochemical Methods of Analysis*, Vol. 2, IAEA, Vienna, 1965, p. 99.

[25] J. Tölgyessy, *Chem. Listy*, 1969, **63**, 513.

[26] J. Tölgyessy, *Isotopenpraxis*, 1971, **7**, 208.

[27] R. Maehl, O. Cucchiara and D. J. Chleck, *Chemist-Analyst*, 1965, **54**, 83.

[28] J. Tölgyessy, B. Síleš, V. Jesenák and Š. Varga, *Isotopenpraxis*, 1968, **4**, 383.

[29] J. Tölgyessy, B. Síleš, V. Jesenák and Š. Varga, *Isotopenpraxis*, 1968, **4**, 429.

[30] J. Tölgyessy and V. Jesenák, *Isotopenpraxis*, 1969, **5**, 186.

[31] I. M. Korenman, F. R. Sheyanova, H. M. Mezina and M. I. Ostasheva, *Zh. Analit. Khim.*, 1957, **12**, 48.

[32] J. F. Duncan and F. G. Thomas, *J. Inorg. Nucl. Chem.*, 1957, **4**, 376.

[33] H. Spitzy, *Mikrochim. Acta*, 1960, 789.

[34] I. M. Korenman and F. R. Sheyanova, *Tr. Komis. po Analit. Khim., Akad. Nauk SSSR*, 1958, **9**, 205.

[35] I. M. Korenman, F. R. Sheyanova, T. N. Miktrokhina and T. N. Trapeznikova, *Tr. Khim. i Khim. Tekhnol.*, 1958, **1**, 109.

[36] F. Kukula and M. Křivánek, *Radioisotopy*, 1964, **5**, 211.

[37] J. F. Duncan and F. G. Thomas, *Australian Atomic Energy Symp., Radio-isotope–Physical Sciences*, 1958, 637.

[38] T. Braun and E. Körös, *Radiochemical Methods of Analysis*, Vol. 2, IAEA, Vienna, 1965, p. 213.

[39] F. Kukula, *Isotopenpraxis*, 1967, **3**, 17.

[40] J. Tölgyessy, *Radiometric Titrations*, in J. Tölgyessy and Š. Varga, *Nuclear Analytical Chemistry*, Vol. 2, University Park Press, Baltimore; Veda, Bratislava, 1972.

[41] E. Schumacher and W. Friedli, *Helv. Chim. Acta*, 1960, **43**, 1013.

[42] J. Konečný, J. Tölgyessy and T. Braun, *Acta Chim. Acad. Sci. Hung.*, 1967, **51**, 245.

[43] J. Konečný, J. Tölgyessy and T. Braun, *Proc. Anal. Chem. Conf. Budapest*, Vol. 2, 1966, p. 471.

[44] J. Konečný, J. Tölgyessy and M. Šaršúnová, *Z. Anal. Chem.*, 1967, **232**, 343.

[45] J. Konečný and J. Tölgyessy, *Chem. Zvesti*, 1968, **22**, 521.

[46] J. Tölgyessy, J. Konečný and T. Braun, *Nuclear Applications*, 1967, **3**, 383.

[47] J. Konečný and J. Tölgyessy, *Isotopenpraxis*, 1967, **3**, 479.

[48] A. Heijink and H. L. Polak, *J. Radioanal. Chem.*, 1969, **3**, 359.

[49] E. Schumacher and H. J. Streiff, *Helv. Chim. Acta*, 1958, **41**, 1771.

[50] E. Schumacher and H. J. Streiff, *Helv. Chim. Acta*, 1957, **40**, 228.

[51] E. Schumacher, *Helv. Chim. Acta*, 1957, **40**, 221.

[52] T. Braun, *Chimia (Aarau)*, 1967, **21**, 16.

[53] T. Braun, I. Maxim and I. Galateanu, *Nature*, 1958, **182**, 936.

[54] T. Braun and J. Tölgyessy, *Talanta*, 1964, **11**, 1543.

[55] J. Tölgyessy, V. Jesenák, T. Braun and M. Hradil, *Chem. Zvesti*, 1965, **19**, 465.

[56] J. Tölgyessy, Š. Varga and V. Jesenák, *Isotopenpraxis*, 1968, **4**, 223.

[57] J. Tölgyessy, V. Jesenák, Š. Varga and B. Síleš, *Isotopenpraxis*, 1968, **4**, 259.

[58] T. Braun, *Acta Chim. Acad. Sci. Hung.*, 1964, **41**, 199.

[59] J. Tölgyessy, P. Dillinger, J. Pružinec and Š. Varga, *Radiochem. Radioanal. Lett.*, 1970, **4**, 231.

[60] A. Illaszewicz, K. Müller and H. Spitzy, *Microchem. J.*, 1966, **10**, 1.

[61] J. Tölgyessy, B. Síleš, Š. Varga and V. Jesenák, *Isotopenpraxis*, 1968, **4**, 368.

[62] J. Tölgyessy and J. Lesný, *Jad. Energie*, 1973, **19**, 28.

[63] J. Tölgyessy, P. Dillinger and J. Lesný, *Radioisotopy*, 1973, **14**, 245.

[64] J. Tölgyessy, P. Dillinger and J. Lesný, *Uspekhi Analiticheskoi Khimii*, Izd. Nauka, Moscow, 1974, p. 97.

SELECTED BIBLIOGRAPHY

T. Braun and J. Tölgyessy, *Talanta*, 1964, **11**, 1277.

T. Braun and J. Tölgyessy, *Radiometrische Titrationen*, Hirzel Verlag, Stuttgart, 1968.

V. Balek and J. Tölgyessy, *Emanation Thermal Analysis and other Radiometric Emanation Methods*, Elsevier, Amsterdam; Akadémiai Kiadó, Budapest, 1984.

J. Tölgyessy, Š. Varga and E. H. Klehr, *Radioisotopy*, 1974, **15**, 623.

B. M. Maryanov, *Radiometricheskoe titrovanie*, Atomizdat, Moscow, 1971.

M. Kyrš, J. Tölgyessy, G. Bilimovich, J. Krtil, L. Kuča, P. Dillinger and I. Rais, *Novye metody radioanaliticheskoi khimii*, Energoizdat, Moscow, 1982.

T. Braun and J. Tölgyessy, *Radiometric Titrations*, Pergamon Press, Oxford, 1967.

6

Radioimmunoassay*

6.1 INTRODUCTION

6.1.1 Characteristics of the method

In the classification on p. 158 this group of determinations should belong to type 3.1.2, but the term *radioimmunoassay* (RIA) does not appear as a separate group there. The reason is that the method differs from the others in the group in the nature or character of the analyte and matrix substances, which are in most cases organic compounds. As our classification was based on more general criteria, RIA does not appear in it as a separate group. On the other hand, a book devoted to new methods in radioanalytical chemistry cannot neglect RIA, which is much more widely used than the method of radioactive reagents in inorganic chemistry.

To illustrate the 'explosion' in the number of papers devoted to RIA and related methods, it may be noted that the number of papers in English devoted to these methods since their introduction in 1959 [1] was about 3000 in the period 1964–1973 [2]. The total number of papers

---

* By M. Kyrš and L. Kronrád.

in the period 1973–1975 was around 4000. In 1977, 60–70 papers in this field appeared monthly. Thus, it is little wonder that it became almost impossible, even for a specialist, to keep abreast of developments in this field. The number of symposia on RIA methods organized is also evidence of the rapid development of this field (1973 — Istanbul, 1974 — Lyon [3], 1974 — Oxford, 1977 — West Berlin). In 1974, a new monthly journal *Radioassay News* appeared, devoted to the field of RIA and similar methods.

The number of reviews appearing during recent years shows the need felt for a summary of the experimental results. As the methods of RIA are mainly applied in the specialized fields of biochemistry and medicine, however, it is not reasonable to give a detailed review in this book.

Thus, only some principles and the characteristic features of RIA will be described, the knowledge of which will be useful to the reader with a wider range of interests in radioanalytical chemistry. Generally, the principle of the method does not differ from the principles of several variants of the method of radioactive reagents (Chapter 4) but distinguishing features are the type of analyte and the type of reaction used.

6.1.2 Basic terms

6.1.2.1 Antibody and antigen

If a substance penetrates into an organism which recognizes it as foreign (such a substance is called an *antigen*), the organism responds by the formation of *antibodies* in the globulin fractions of blood (gamma globulins of various classes, IgM, IgG, etc.). Antibodies should be regarded as defensive weapons created by the organism for protection against invasion by 'undesirable guests' (mostly pathogenic microorganisms). It appears that an antibody has an unusual ability to bind very selectively an antigen which has stimulated its formation.

The ability of an antibody to discriminate between its 'own' antigen and the great many other substances, of widely diverse structure, that are found in biological fluids (blood, urine, etc.) is the basis of its application as an analytical tool.

At the same time, an antibody is also able to bind selectively a group which is only a constituent part of the antigen. It should be noted that the term 'antigen' has a firm place in the nomenclature of biochemistry and many other fields of science. One of the most important properties of an antigen is the ability to create immunity, i.e. its *immunogeneity*. Proteins are simultaneously immunogens and antigens. On the other hand, most drugs are antigens but not immunogens, and in order to become immunogens, they must first be bound with a large molecule, for example, with a natural protein (such as albumin or bovine serum) or with a synthetic protein.

6.1.2.2 Haptens
An antigen that must be coupled to a large molecule in order to provoke an antibody response is called a *hapten* (Greek; 'to capture'). The term was first introduced by Landsteiner in 1921 to designate small organic residues that have specifically reacted with antibodies produced by immunization compounds where these residues are covalently bound to large carrier molecules. Haptens do not usually have molecular weights greater than about 2000. Substances with molecular weight > 2000 can on their own promote formation of determinable concentrations of antibodies. However, a larger amount of antibodies is formed after the conjunction of a hapten with a molecule of appreciably higher molecular weight [4].

6.1.2.3 Antibody titre
The *antibody titre* is the degree of dilution at which an antiserum will bind a particular fraction (often chosen as 50 %) of antigen under standard conditions (fixed incubation volume, fixed amount of antigen).

6.1.2.4 Antigenic determinant
This is the part (or parts) of the antigen molecule entering into combination with the binding site of an antibody. In the case of a peptide or protein, each determinant consists of a sequence of about 3–5 amino-acids. In the case of drugs, it comprises only part of the molecule, usually that most exposed on the *conjugate* (compound of the hapten with the carrier) used for *immunization* (process of antibody formation).

6.1.2.5 Cross reactivity
Cross reactivity is the extent of reaction between an antibody and other substances which do not produce the given immunization response (e.g. accompanying substances and admixtures similar in structure to the analyte substance). It is usually expressed by the formula

$$\text{cross reactivity } = 100A_{50\%}/X_{50\%} \quad (\%)$$

where $A_{50\%}$ is the amount of non-labelled antigen A that replaces 50 % of the labelled antigen A at the binding site with the antibody, $X_{50\%}$ the amount of an accompanying substance X that also replaces 50 % of the labelled antigen A at the binding site of the same antibody.

An illustrative example of the cross reaction of human luteinizing hormone (HLH), human follicle-stimulating hormone (HFSH) and human chorionic gonadotropine (HCG) in the radioimmunoassay for human thyroid-stimulating hormone (HTSH) is given in Fig. 6.1 [5].

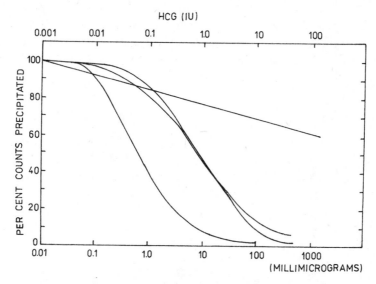

Fig. 6.1 — Cross-reaction of HLH, HFSH and HCG in the radioimmunoassay of HTSH with a non-specific antiserum.

6.1.3 Principles of the methods

6.1.3.1 Basic principle
There are several variants of RIA, but the principle can be illustrated by one typical general case. The reactions on which the determination is based can be written

$$*Ag + Ab \; \rightleftharpoons \; *Ag\text{--}Ab \tag{6.1}$$

$$Ag + *Ag\text{--}Ab \; \rightleftharpoons \; Ag\text{--}Ab + *Ag \tag{6.2}$$

Equation (6.1) characterizes the equilibrium state of the reaction of radioactively labelled free antigen with antibody. If this equilibrium is disturbed by the addition of an unknown or standard amount of non-labelled antigen, then the competition converts a certain part of the labelled antigen from the bound state Ag–Ab into the free state. The higher the amount of non-labelled antigen added, the lower the ratio

(S) of the amounts of bound and free labelled antigen obtained. The dependence of S (measured from the radioactivity of the bound and free labelled antigen, after decomposition) or a suitable function of it (at constant concentration of antibody and labelled antigen) serves as a calibration graph for determination of the antigen investigated. In other words, the concentration of antigen in the sample is determined by comparing the degree of inhibition (retardation) of binding of the labelled antigen, caused by the presence of the sample, with that caused by standards. The principle of the determination is shown in Fig. 6.2.

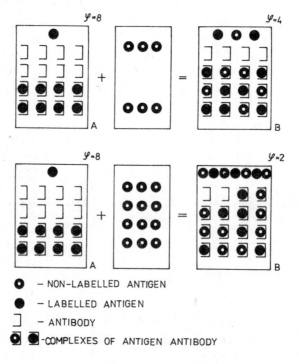

Fig. 6.2 — The principle of radioimmunoassay.

State A corresponds to Eq. (6.1) and represents one extreme of the calibration curve (zero concentration of non-labelled antigen). State **B** arises on addition of 6 particles of non-radioactive antigen, and state C of 12 particles. The figure demonstrates the principle of competition between labelled and non-labelled antigens (dilution of isotopic compounds) to form compounds with antibodies, and also illustrates the principle of the equilibrium state (part of the antibody can remain free even in the presence of excess of antigen). The hypothetical case shown in Fig. 6.2 corresponds to a value of 1 for the equilibrium constant defined as

$$K = [AgAb]/([Ag][Ab]) = [*AgAb]/[*Ag][Ab] =$$
$$([*AgAb] + [AgAb])/([*Ag] + [Ag])[Ab]$$

where Ag and *Ag are labelled and non-labelled antigens and Ab is antibody. For simplicity, $[*Ag]$, $[Ab]$, $[AgAb]$ and $[*AgAb]$ are presented as integers. However, the condition $K = 1$ may lead to fractional values of these quantities.

It can be seen from the figure why RIA is called 'saturation analysis' by some authors.

Recently, thermodynamic analysis of the immunological reaction has been presented, for example, by Kalmakoff *et al.* [6]. A simple theoretical model has been proposed for the antigen–antibody reaction in connection with solid-phase radioimmunological determination. A unified theory based on the law of mass action has been worked out by Trautman [7] for radioimmunology and neutralization of viruses.

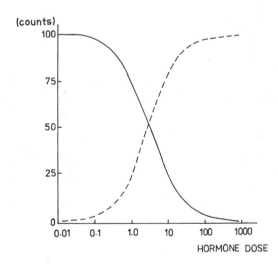

Fig. 6.3 — Semilog plot of data from a radioimmunoassay. *Y*-axis — the fraction of counts attributed to bound antibody (solid line) or free antibody (dashed line); 100 % is defined as maximum free or bound antibody.

Many methods for evaluating radioimmunoassay results are given in the literature and routinely used. Figure 6.3 shows the dose–response curves plotted as log (dose of antigen) against percentage of free labelled antigen and percentage of bound labelled antigen. If increasing amounts of antigen are added, the percentage of bound antigen falls within the *dose–response area* of the curve. There is a region of the curves corresponding to additions of amounts of antigen that are so small that the system is incapable of detecting the change in amount of labelled antigen bound. Similarly, when no further labelled antigen is bound to the protein, an additional amount of unlabelled antigen produces no further change in the system. 'Per cent counts free' is the reciprocal of the 'per cent counts bound', and this curve increases with increasing amount of non-radioactive antigen added.

Two other ways of plotting the dose–response curves are given in Figs. 6.4 and 6.5. Many authors used Scatchard plots to determine the apparent equilibrium constants of the corresponding binding reactions of both components [8–10]. The graph presents the dependence of

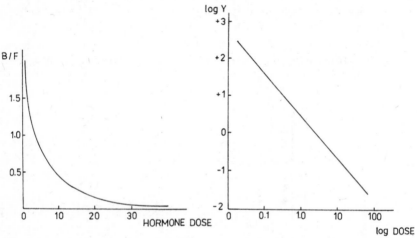

Fig. 6.4 — Arithmetic plot of radioim-munoassay data: B/F — bound/free anti-body ratio.

Fig. 6.5 — Plot of radioimmunoassay data as log $[Y/(100 − Y)]$ *vs.* log dose.

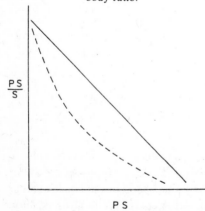

Fig. 6.6 — Scatchard plot of radioim-munoassay data. The solid line indicates theoretical plot of data from a reaction be-tween a binding protein having one type of binding site with a single affinity, the dashed line indicates type of data commonly ob-tained from antigen–antibody reactions.

C_{bound}/C_{free} on C_{bound}, where C is the concentration of radioactive antigen, C_{bound} is related to the part bound with antibody, and C_{free} refers to the unbound part. Linear character of this dependence is considered as evidence for acceptance of the law of mass action in its simple form, i.e. evidence for the existence of only one value of K, or evidence of the homogeneity of the corresponding antigen. A Scatchard plot of the radioimmunoassay data is schematically given in Fig. 6.6

6.1.3.2 Generalized procedure for radioimmunoassay

This is not intended as a detailed description to enable the user to perform a radioimmunoassay, but as a brief account to give the reader some idea of the general method used. The reagents (e.g. hormones) are usually added to small test-tubes (10–15 ml) in the following order:
(a) enough buffer to give a final total volume of 1 ml;
(b) serum sample to be analysed, 100–300 µl;
(c) radioactive hormone (100 µl) with sufficiently high specific activity to produce 10–200000 cpm;
(d) antiserum ($\sim$ 100 µl) appropriately diluted to give a starting bound fraction of unity;
(e) in the case of double antibody 100 µl of $0.01 M$ EDTA.

Various standard or reference amounts of hormone are placed in a series of tubes to give a dose–response curve covering the entire range of the assay. This usually requires 15–20 tubes; the remaining tubes are used for assay samples. All samples are analysed at least in duplicite, and frequently as a series of different volumes to produce the equivalent of a dose–response curve. The hormone in the unknown serum samples must have a dose–response curve identical in shape to that of the reference preparation, for the assay system to be valid. The tubes are incubated for several hours or days at defined temperature. At the end of this time, the bound hormone is separated from the free hormone by any of the techniques described later and the amount of radioactivity of each fraction is determined.

In the double antibody techniques, at the end of the incubation, 50–150 µl (depending on the particular antiserum) of sheep or goat anti-rabbit gamma globulin is added to each tube. The tubes are further incubated for several hours, then centrifuged, and the bound hormone, along with the carrier rabbit gamma globulin present in the buffer, is precipitated. The total activity is measured (and, of course, should be constant throughout the assay) for each tube; the supernatant liquid is discarded, and the activity of the precipitated fraction is assessed in an automatic gamma spectrometer. This generalized procedure applies, with minor variations, to all the polypeptide radioimmunoassays. An example of a dose–response curve for human luteinizing hormone (HLH) is given in Fig. 6.7 [11].

The procedure for steroid radioimmunoassays differs slightly from that for peptide radioimmunoassays. Generally the sample to be assayed is extracted with an organic solvent (e.g. ether extraction is used for oestradiol). The solvent extract is then evaporated to dryness and the assay buffer added (at the very low molar concentrations present in biological fluids, steroids are sufficiently water-soluble to permit their assay). The radioimmunoassay reagents are then added as for peptide assays. If the antiserum is not sufficiently specific to permit assay of the crude extract (containing all steroids present that are soluble in the solvent used), then an additional purification step is required, such as chromatography.

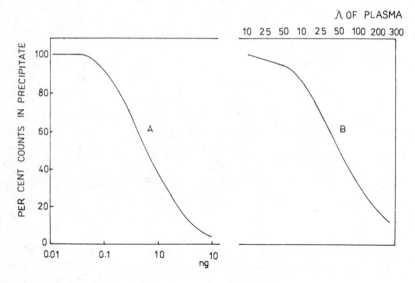

Fig. 6.7 — Semilog dose–response curves for purified human luteinizing hormone (A) and for plasma from a postmenopausal woman (B).

In addition, the steroid antiserum is partially purified prior to routine use. This is necessary because such antisera contain albumin and alpha globulins which bind steroid hormones. If untreated antisera are used (containing albumin, alpha and beta globulins), the competitive displacement curves are complex resultants of the effects of all these binding proteins. Albumin in particular has a high capacity and low affinity, and decreases the steepness of the dose–response curve.

6.1.4 Reaction of antigen with antibody

The values of the equilibrium constant (K) of the reaction between an antigen and antibody often lie in the range 10^7–10^9 [12]. This means that 50 % of the antigen is bound when $[Ab] = 10^{-9}$–$10^{-7} M$. As C_{Ab} and C_{Ag} should be commensurable (according to the principle of saturation), this concentration range corresponds approximately to the total concentration of antibodies in the individual determinations.

Numerous cases have been described for which K is $> 10^{10}$. However, in practice, the K value does not seem suitable for characterizing the reaction of an antibody with a given antigen, because antibodies may not be quite homogeneous, and antigens can appear in various forms. For example, the interaction of lactoferrin glycoprotein with rabbit antiserum antibodies proceeds in such a way that two values of the reaction constant are obtained (1.8×10^{11} and 1×10^9) [13]. Hence to characterize the affinity between the two substances, the quantities *antibody titre* or *antiserum titre* are mostly used. The connection between K and the titre is shown by the following simple calculation. Let the initial concentration of antigen be β and its final free concentration b. Then the concentration of bound antigen will be $\beta - b$ and the initial and bound antibody concentrations can be designated α and $\alpha - a$. If a 1 : 1 complex is assumed to be formed, we can write $\beta - b = \alpha - a$. The association constant is equal to

$$K = (\beta - b)/ab$$

It follows from its definition that the titre is equal to α/α_0, where α_0 is the concentration of antibody in the initial serum (binding half of the antigen), so $b = \beta/2$, and $K = 1/a$. The quantity a can be expressed as

$$a = \alpha - (\alpha - a) = \alpha - (\beta - b) = \alpha - \beta/2 = 1/K$$

Then the titre is equal to

$$[(1/K) + (\beta/2)]/\alpha_0$$

Hence the titre is determined by the initial concentration of antibody and generally depends on the concentration of antigen and the association constant. In the case of a high association constant and high specific activity of antigen ($1/K \ll \beta/2$), the titre is given by the ratio of the initial concentrations α_0 and β. In the case of low values of K and β, $1/K \gg \beta/2$, the titre is given by the product $\alpha_0 K$. Simplifications made in these considerations are analogous to those made by Scatchard in the determination of K.

In practice, the desired dilution of the antiserum (i.e. the titre) is

determined empirically. A series of solutions is prepared, each after the
first having twice the dilution of the one preceding it. The calibration
curve is obtained by finding the dilutions leading to the binding of
20–60 % of labelled antigen. A substantial hindrance to the wide application of RIA is the difficulty of preparing antisera with the desired
values of K and sufficient selectivity. It is often necessary to immunize
a great number of animals in order to obtain the necessary antibodies
in sufficient amount. The costs of keeping these animals during several
months as well as of the investigation of a large number of sera can be
rather high for many laboratories.

6.2 IMMUNIZATION — THE PREPARATION
   OF ANTIBODIES

Antibodies are obtained from laboratory animals (young animals are
preferred), usually rabbits or guinea-pigs, after repeated injections of
antigen. Chickens, sheep [14], goats [15], mice and rats are also used.
About 4–6 animals of one type, selected at random, are immunized. The
time necessary to form the antibodies usually varies from 6 weeks to 4–6
months.

   An adjuvant is introduced together with the antigen into the animals. An adjuvant is a substance enhancing the response of the organism to an antigen, for example, Freund's adjuvant, which consists of a
neutral detergent, paraffin oil and killed mycobacteria.

   The procedure depends on the nature of the animal used, the dose
and frequency of immunization, and on accidental circumstances.
There are several practical recommendations for immunization of individual types of animals [16].

   It has been calculated that one animal has the ability to produce
antibodies able to distinguish 10^7–10^8 various antigen determinants.

   It should be emphasized that the antigen used for the preparation of
antibodies need not be pure, provided that pure antigens are used as
standards or as labelled substances. At the same time, if the preparative
antigen Ag used for the immunization also contains a certain antigen
type Ag', the antibodies obtained (a mixture of Ab and Ag') are also
able to bind mixtures of this type present in the test sample. When
antibodies Ab' partially react with the antigen to be determined, Ag, the
presence of Ag' in various concentrations in the sample leads to different results for determination of one and the same amount of Ag. In
such a case, an excess of Ab' can be used in the incubation of samples
and standards. The interfering effect of variations in the Ab' concentration on the determination can be eliminated in this way.

The selectivity of the antibody obtained, for the substance to be determined, can sometimes be increased by chemical purification. This can be illustrated by the example of antibodies produced by haemoglobin C [17]. The antibodies obtained cannot adequately distinguish between haemoglobins A and C. To improve the selectivity, affinity chromatography has been used to isolate from the mixture of antibodies the fraction with increased affinity for haemoglobin C. The chromatographic column is filled with a synthetic peptide, the composition of which corresponds to the first 13 amino-acid residues of the β-chain of haemoglobin C. The peptide is covalently bound to Sepharose.

Antibodies against polypeptides with molecular weights larger than approximately 15000 are routinely prepared. The majority of anterior pituitary hormones are larger polypeptides and antibodies against them may be easily prepared. It is very difficult to prepare antisera against some of the smaller polypeptide hormones (e.g. gastrin with molecular weight M_r 2096, ACTH with M_r 4566, vasopressin with M_r 1080) and these present a special problem. These hormones may result in antibody formation in some instances if large numbers of animals are immunized for prolonged periods. Such hormones (haptens) may also be conjugated to larger proteins and antibodies raised against the conjugates. With the use of such conjugation techniques, antibodies can be produced against almost any organic substance. For example, antisera are now routinely prepared against steroid–protein conjugates; these antibodies are used in immunoassay for steroid hormones. The size of the reactant sites on antibodies is not known accurately, but is probably equivalent to M_r 600–800. Thus steroids with molecular weights of around 300 may be smaller than the reactive sites. Generally, 10–40 steroid molecules are conjugated to each protein molecule to produce antibodies against steroids. Antibodies may react against the steroids and adjoining areas of the protein.

6.3 LABELLED ANTIGENS

When choosing the labelled atom, it should be taken into account that if a certain number of atoms of the radionuclide ^{14}C will provide a certain counting rate (cpm), the same counting rate is obtained with only about a hundredth of this number of ^{3}H atoms and about a ten-thousandth of this number in the case of ^{125}I. Substances labelled with ^{14}C are rarely used, owing to their expensive synthesis. On the other hand, ^{14}C has the advantage that the molecule can be labelled more homogeneously than, for example, with iodine ^{125}I.

6.3.1 Antigens labelled with tritium

The advantages of ^{3}H as the labelling atom are the suitable half-life (12.3 yr), similarity of the properties of the labelled and non-labelled substances, low radiation damage during prolonged storage of the preparation, and the low irradiation dose to personnel. The disadvantage is, as a rule, the relatively low specific activity, high cost of the liquid scintillator and, in the case of measurement of the complex AgAb, the necessity to decompose this complex before the measurement because the presence of antibody can substantially lower the counting efficiency of the preparations. The disadvantage is also the possibility of the 'isotope effect' decreasing the affinity of the labelled antigen for the antibody. The difficulty of obtaining, for example, steroids labelled with ^{3}H should also be taken into account [18]. ^{3}H atoms can be especially recommended when the number of samples to be analysed is not too large.

6.3.2 Antigens labelled with iodine

The most commonly used substances are compounds labelled with radioactive iodine. Recently, compounds labelled with ^{125}I have been used in many cases instead of the compounds labelled with ^{3}H used in earlier methods [19, 20]. An interesting example has been published by Baehr *et al.* [21] who found (when determining insect juvenile hormone) that the titre (for 40 % binding) is 1/6000 for the tritium-labelled preparation and 1/20000 for that labelled with ^{125}I. The application of iodine-labelled compounds is based on the fact that the difference in the K values for the antigen and its slightly modified iodine-labelled molecule is low. The stable isotope of iodine is not present in the initial antigen.

In principle, suitable derivatives of antigens can be labelled with either ^{125}I or ^{131}I, the former having substantial advantages over the latter (higher counting efficiency and isotopic abundance, longer half-life and less danger for personnel). The radio-iodination can be done immediately after preparation of a suitable derivative (e.g. compounds with a tyrosine residue). Another way is first to iodinate a suitable auxiliary substance, which is then conjoined with the antigen [22]. It is important to use this method when the iodine is simultaneously joined to the molecule of the initial antigen. This leads to a K value lower than that with the substance to be determined. For example, the free amine groups of proteins can be allowed to interact with iodine-labelled 3-(4-hydroxyphenyl)propionic acid and N-hydroxysuccinimide. It has been shown that antigens labelled in this way exhibit the same immunoreactivity as the non-modified proteins. At the same time, if iodine is introduced into the tyrosine groups instead, the immunoreactivity of the antigen with some antisera decreases [23].

If an appreciable part of the antigen activity is lost in the course of iodination of small quantities of antigens purified by the chloramine-T method, it is necessary to use a tenfold higher total concentration of the antigen proteins. The required antigen fraction is isolated after the iodination, e.g. by the method of isoelectrofocusing [24]. Another method consists in iodination of the tyrosine residues of a protein covalently bound with the antigen. Recently, it has been shown that the iodination of antibodies bound with an adsorbent is more efficient than iodination in a homogeneous solution.

Modification of the molecule for preparation of the ^{125}I-labelled derivative should be different from that for preparation of the immunogen. Some populations of antibodies have higher K values than the molecules of the carrier or bridge. In that case, the iodinated substance containing bridges identical with the immunogen would have an appreciably higher value of the K constant than that of the substance determined. This would decrease the sensitivity of determination.

6.3.3 Antigens labelled with other radionuclides

Knight et al. [25] have worked out a method for bromination of proteins with ^{77}Br with a high yield. Products which are more stable in vitro than iodinated products prepared under the same conditions are obtained by this method. Antigens have also sometimes been labelled with ^{75}Se.

6.3.4 Quality of the labelled antigens

The labelled antigen should not react with a component of plasma, serum or blood when the antigen to be determined is present. If such reactions are not taken into account, lower or higher results (depending on the method of calculation) can be obtained. Accidental adsorption of the labelled substance on the walls of the vessels or on some components of the biological fluid of the samples can also lead to erroneous results.

The quality of the labelled antigen can be examined not only by the usual chromatographic procedures (appearance of only one peak) but also by an indirect method using two control experiments: (a) without non-labelled antigen, (b) without antiserum. In case (a), the amount of bound antigen should be maximal and decrease with increasing concentration of the inactive antigen. In case (b), serum of an animal which has not undergone immunization is used in the same dilution. The relative amount of the bound radioactive antigen should not exceed 5 % in that case.

6.4 PREPARATION OF THE ANTIGEN

As the resulting antibodies usually react with the most active part of the molecule, it is necessary (when finding a suitable reaction of hapten with the molecule of the carrier) to choose a reaction for the immunogen preparation that leaves untouched the most characteristic part of the hapten, i.e. the part which can distinguish the hapten from its main metabolites and natural or synthetic analogues.

This has been shown by Park *et al.* [26] for equilin and oestrone, where it was necessary to link the equilin to protein through the 17-position to leave the C_7–C_8 double bond of equilin free for differentiation of the equilin from oestrone.

Another example is the synthesis of immunogen for the determination of nicotine in the presence of its basic metabolite cotinine and its by-products such as 4-(3-pyridyl)-4-oxo-*N*-methylbutyramide. Nicotine cannot be directly joined with the carrier molecule, so it is necessary to introduce, for example, carboxyl groups into the nicotine molecule, to react with the amino groups of human serum albumin. Use of *trans*-3′-succinylmethylnicotine has been found to give adequate specificity [27].

However, the scope for changing the specificity of the reaction of hapten with antibodies by choice of the site of conjunction should not be overestimated. For example, in the case of testosterone, androstenedione and dihydrotestosterone, it is rather narrow [27a].

A supplementary requirement for choice of the conjunction site is that the corresponding bond should not be attacked by the metabolic processes in the human or animal organisms.

In addition to the conjunction site, an important role is also played by auxiliary substances forming special 'bridges' between the macromolecules and haptens [28]. For example, a conjugate containing tuftsin (Tu) [29] exhibits a structure which is schematically illustrated by the formula

This formula shows that 36 units of tuftsin join with one macromolecule of bovine serum albumin (BSA). It appears, in the binding of testosterone with albumin, that the titre, capacity and affinity (but not the specificity) of the antibodies obtained are dependent on the type of the conjunction agent (mixed anhydride or carbodi-imide) even if the molar ratio of steroid to albumin is the same (30 in the case cited) [30].

Various methods for covalent binding of drugs with the carrier-protein, and their dependence on the functional grouping of the drug have been reviewed by London and Moffat [16].

6.5 SEPARATION OF FREE AND BOUND ANTIGENS

This step in the analytical procedure should be carried out after the antigen binding reaction and before the radioactivity measurement. It should be mentioned that methods not using radioactivity have also been worked out, and do not require separation of the bound and free parts of the antigen [31]; this is their most important advantage.

6.5.1 Requirements for an ideal separation method
The separation method should
(a) not disturb the equilibrium (or pseudoequilibrium) separation of both parts of the antigen;
(b) not be affected by any foreign substances present;
(c) be sufficiently reproducible without excessive demands on the operator;
(d) use reagents and apparatus that are readily available and cheap;
(e) not be excessively labour-consuming, complicated and prolonged;
(f) be suitable for the treatment of large series of samples.

If the equilibrium is disturbed by the separation method, the sensitivity decreases. The reproducibility also decreases if the equilibrium is affected by factors that are difficult to control. The total error of the RIA method is often determined by the error in the separation methods.

6.5.2 Routinely used separation methods
We shall classify the separation methods according to the principle on which they are based. The main methods are as follows.

(1) *Chromatographic methods*. The separation is based on differential velocity of migration of free and bound particles through a column of stationary phase, as a consequence of differences in charge or size, etc. The methods include electrophoresis, radioelectrocomplexing [32], chromatography [33], electrochromatography, gel filtration, etc.

(2) *Precipitation methods*. The bound substance is precipitated with salts, acids or organic solvents, e.g. $(NH_4)_2SO_4$, Na_2SO_4, CCl_3COOH, dioxan, ethanol, acetone, polyethyleneglycol.

(3) *Adsorption methods*. Free antigen is adsorbed on granular sorbents, e.g. talc, ultrafine silica gel, Florisil, activated carbon, activated carbon coated with dextran (allowing penetration only by particles below a certain size), activated carbon bound with albumin, kaolin, cellulose, anion exchangers, activated glass powder, bentonite [34], ion-exchange membranes or porous gels, under equilibrium conditions.

(4) *Double antibody method*. The complex of antigen and antibody is precipitated with another antibody [35–37].

(5) *Solid-phase techniques*. In 'solid-phase' RIA, the carrier (solid phase) is already present in the system during incubation. If the sorbent is added after the incubation, the methods should not be considered as solid-phase RIA even if particles fixed, for example, on a disc, have been used as a sorbent.

A number of different techniques are described in the literature:

(a) antibody is immobilized in plastic test-tubes [38–40];
(b) the antibody is covalently bound with Sephadex activated with Sepharose [41];
(c) the second antibody is bound with the solid matrix [42];
(d) the antibody itself forms an insoluble polymer;
(e) antibody is bound by a silane binding agent with magnetically-sensitive inorganic particles; this permits a magnetic separation [43];
(f) the sorbent is formed by the reaction of gamma globulin, antibody (or labelled antibody) and polyethyleneglycol with M_r 1000–5000;
(g) solid-phase RIA with an antigen bound on plastic spheres [44] or a membrane surface or on cells adhering to plastic [45].

It must be stressed that separation of the free and bound parts of the antigen is one of the most important stages of the determination [46]. It has been shown that the separation method can appreciably affect the accuracy of the results of RIA. For example, higher results have been obtained in insulin determination by the double antibody method than by the adsorption technique [47]. Another example is the use of activated carbon containing substances which can inhibit the reactions of some hormones with antibodies. Thus, pretreatment of the serum with activated carbon for the purpose of preparation of standard samples can lead to false results. It is interesting that a change in the method of separation of the bound and free parts of the hormone–antibody combination can cause enhancement or inhibition of the reaction [48].

6.5.3 Evaluation of the different separation methods

None of the methods available seems to be ideal. When working out a new RIA method, it is best to compare two, or better three, rather different methods and to choose the most suitable [16]. The chromatographic and migration methods are rarely used, because of their complexity and unsuitability for the analysis of large numbers of samples. Adsorption on 'coated' activated carbon is one of the most popular methods owing to its experimental simplicity, speed and convenience for dealing with large series of samples. The sedimentation methods usually need careful choice of the salt concentration or solvent and other reaction conditions in order to obtain complete precipitation of the bound part while preventing precipitation of the free particles. Further, slight fluctuations in the quality of solvents can appreciably change the course and degree of precipitation. Nevertheless, sedimentation with polyethyleneglycol (final concentration 10–12 %, in the presence of serum from a non-immune animal) is commonly used.

The double antibody method can be considered a universal method. However, the precipitating antibodies are rather expensive because of their specifications. Recently, variants of solid-phase RIA have become the most widely used methods. They were disregarded earlier on account of insufficient accuracy and a relatively high consumption of antibodies. Further development of this method has shown that these disadvantages can be overcome. Carefully chosen specific antisera, when used under optimum conditions, might provide a universal means of separation.

Practical recommendations for the choice and application of separation methods have been given by many authors [49]. It is important to mention that the sorbents and precipitating antibodies often vary in quality. It is therefore necessary to have a sufficient supply of homogeneous material before starting the often laborious operation of optimizing the separation method. If the system is not fully insensitive to the concentration of plasma or serum in the given range of their dilution, it is necessary to equalize the concentrations by addition of a suitable plasma or serum not containing an antigen.

6.6 MODIFICATIONS OF RIA

6.6.1 Competitive binding with albumin

This method is characterized by the use of natural albumins (usually of non-immune globulin type) exhibiting selective affinity towards the analyte hormone or class of hormones (e.g. steroid hormones, thyroxine, vitamin D_3 and its metabolites, corticosterone, oestradione) as radioactive reagents [50]. Typical applications are determination of progesterone in plasma, testosterone, folic acid in serum [51], vitamin B_{12} [52, 53], cyclic adenosine monophosphate [54], and corticosteroids [55]. These methods are usually less sensitive and selective than other RIA methods. Thus, a preliminary chromatographic separation of the analyte(s) is necessary.

6.6.2 Radioreceptor method
Cellular receptors prepared from certain tissues (kidney, liver, ovaries, fat tissue, lymphocytes, etc.) are used instead of antibodies for binding with radioactive or non-labelled ligands [56]. In general, for steroid hormones such receptors exist in the cytosol and nucleus; for polypeptide hormones they exist in the cell membrane. Such receptor proteins usually have a high affinity and may be used to develop sensitive assays. Such assays have been described for oestrogens with the uterus receptor proteins, for testosterone with the prostate cytosol receptor protein, for progesterone with chick oviduct or uterus receptor protein, and for a number of peptide hormones [57].

This method has won wide recognition but its full potential has not yet been realized.

6.6.3 Immunoradiometric method
A labelled antibody can be used as a radioactive reagent before removal of the reaction product of the labelled antibody with the antigen to be determined. A reaction with the solid-phase antigen is used [16]. This method is especially suitable for compounds for which tissue receptors have been identified and isolated. There is no need to run blank experiments (which are normally required in RIA methods). However, the method is more expensive and its development is rather difficult [31, 58].

6.6.4 Two-site method
Labelled antibodies are also used in this method. The difference from the preceding method is that the antigen to be determined is extracted by the solid-phase non-labelled antibody before the reaction with labelled antibody. The term 'two-site' refers to the antigen having two sites which are able to bind with the antibody.

One part of the antigen is bound with radioactive antibody (the amine determination reaction) and the second part is bound with the solid-phase non-labelled antibody. The features of this method are similar to those of the preceding method. However, the method uses less antigen. It also gives lower blanks as well as higher selectivity. The terms 'sandwich' or 'junction' test are sometimes used for this method [16, 59].

Methods using a labelled antibody should be employed in the following cases:

(a) if the labelled antigen is not stable;
(b) if the labelled antibody can be prepared with higher specific activity than labelled antigen;
(c) the separation of bound and free antibody is simpler than that of free and bound antigen.

6.6.5 Method of successive saturation

The standard or sample is incubated with antiserum in the absence of a labelled substance. Then secondary incubation is done with a labelled substance. The aim is to increase the sensitivity [60, 61].

6.6.6 Method of multivariable standard curves

This method has recently been proposed for the determination of testosterone and 5-α-dihydrotestosterone in the same sample. The sample is incubated with two antisera, the first exhibiting higher affinity for the testosterone, the second for 5-α-dihydrotestosterone. From the two values for inhibition of binding, the concentrations of both substances can be found by means of a nomogram [62].

6.6.7 Determination of antibodies

It follows from its basic principle that RIA can be used to determine not only antigens but also antibodies [63]. In the latter case, the amount of antigen (serving as an auxiliary substance) should be kept strictly constant. Methods such as successive saturation [60] or the 'two-site' method can be used simultaneously for determination of antibodies.

6.6.8 Method of radioimmunological precipitation with polyethyleneglycol

The acronym RIPEGA (radioimmunoprecipitation-PEG-assay) can be found in the literature. Repeated precipitation with polyethyleneglycol ($M \sim 6000$) is used for isolation of the desired fraction of antibodies from an antiserum (at the beginning of the analysis) and for separation of the free and bound antigen (at the end of the analysis) [64]. The RIPEGA method is considered more sensitive than classical RIA.

6.6.9 Heterologous RIA

In homologous RIA, the same substance (e.g. a steroid) is used both as antigen and labelled substance. In the heterologous system, different substances are used for the two purposes (for example, polypodin B for the immunization and ecdisone as a label) [56].

6.6.10 Method of radioactivity release from thrombocytes
The separation of the free and bound complex of antigen and antibody is not necessary in this method, which is based on the fact that thrombocytes labelled with ^{3}H-serotonin release the label when they undergo reaction with the complex. Hence, the thrombocytes can act as indicating cellules for finding the equivalence point in the titration of antiserum with the corresponding antigen. The sensitivity of this method is comparable with that of RIA [65].

To conclude this section on modifications of RIA, we shall briefly discuss the characteristics of the RIA method. It has often been stated that the RIA method is based on the principle of isotope dilution [66, 67]. According to our classification, however, this is not the case, since the radioactive substance often differs from the analyte not only in the radioactivity of some atoms (for example, testosterone is determined with ^{125}I-testosterone-3-carboxymethyloximotyramide) introduced into the substance, but also in the molecular size of the derivative formed; the amount of antigen bound with the antibody usually changes with the concentration of antigen and it is therefore not possible to use simple formulae of the type

$$C_0/C = i(I_0) + 1$$

for the calculations, where C_0 is the radioactivity of the bound antigen in the absence of non-labelled antigen, C the radioactivity in the presence of non-labelled antigen, and $i(I_0)$ the concentration of non-labelled (labelled) antigen [68]. Some determinations by the method of competitive binding of proteins (for example, the determination of vitamin B_{12} [53]) can be considered as an example of isotope dilution, however, because the amount of vitamin bound does not depend on its concentration. Some authors consider this conclusion as unjustified. A detailed review has been given by Wunderwald et al. [54].

6.7 ACCURACY AND PRECISION OF THE RESULTS

6.7.1 Accuracy
By its nature, the RIA method is prone to systematic as well as random errors. Hence great attention is paid to evaluation and minimization of these errors. It should be noted that similar errors also arise in the determination of hormones and similar substances by biological methods.

The sources of error in RIA can be classified as follows [69]:
(a) the reaction system;
(b) the apparatus;

(c) the reagents;

(d) the quality of the sample.

To obtain high-quality results it is necessary to make regular checks of the errors and accuracy. Various systems have been proposed [70, 71]. The coefficient of variation ($c_v = 100\,\sigma/(\bar{x})$ or the accuracy index ($\lambda = \sigma/b$) is generally used (σ = standard deviation, $\bar{x}$ = mean result, b = slope of the calibration curve). The c_v and λ values for a given substance are usually lower (i.e. more advantageous) for the RIA method than the biological methods.

6.7.1.1 Sources of errors in radioimmunoassay

One source of error is underestimation of the time effect on reagent and samples. It has been shown that antibodies or labelled antigens undergo degradation which is sometimes significant even in a month. For example, radioactive testosterone undergoes rapid change on storage. Its radiochemical purity decreases by up to 80 % during six weeks, and therefore it is recommended to check its purity and apply a supplementary chromatographic purification before its use [72]. It is evident that if these operations are omitted, and a certain degree of standardization is not attained, the results of analyses done in different laboratories with the same type of kit may not be comparable.

In determination of the concentration of drugs in an organism, a prerequisite for obtaining accurate results is attainment of equilibrium between the concentration of a substance in the blood and in tissues [73]. The content of various hormones (testosterone, FSH, LH) in the biological fluids of a patient varies appreciably with time. Thus, it is recommended to take three samples in a period of 6–8 min and analyse for the hormone in the mixture of the three samples. In some cases (e.g. insulin secretion in acute radiation sickness), the period of variation may be several days [74].

A source of serious error is the so-called 'hook effect', i.e. increase of the degree of binding of radioactive antigen with increasing concentration of inactive antigen over a certain range [75]. This effect can usually be avoided by use of another antiserum or careful choice of analytical reagents.

A serious source of errors (usually systematic) is the presence of matrix substances which (despite high selectivity of the method) interfere in the determination.

6.7.1.2 Recommendations for avoiding errors in RIA

An obvious precaution is preliminary separation of the analyte(s) from the sample, e.g. by extraction. Undesirable reactions of some antigens with proteins can be suppressed by adding inhibitors. The use of 8-aniline-1-naphthalenesulphonic acid as an inhibitor has proved successful in many cases (e.g. determination of thyroxine, 1-tri-iodothyronine, thrombin, etc.). Serum proteins that interfere in the determination can sometimes be enzymatically denatured. In some diseases, antibodies formed in the organism may also be able to bind the substances to be determined. This leads to erroneous results in the determination of thyroxine and tri-iodothyronine. Hence in the screening of body fluids of patients suffering from Hashimoto's disease, cancer of the thyroid gland, precautions should be taken to avoid wrong results.

The accuracy of results should always be checked by use of more than one method. A necessary (but not sufficient) condition for validation of the accuracy of an RIA method is that the measured concentration of the antigen (hormone) in an unknown sample should not depend on the degree of dilution at which it is measured; for example, a twofold dilution of the sample should lead to the concentration found being half that found before dilution. This is tested by a series of experiments at different dilutions and comparison of the experimental curve with that obtained by measuring the diluted standard. If the curves do not coincide, the results of the analysis cannot be considered trustworthy.

An important criterion of the accuracy of determination of hormones is correlation of the RIA results with those obtained by chemical or biochemical methods. However, it has to be taken into account that the biological and immunological activities of hormones can differ appreciably. In drug analysis, the RIA results should be compared with those of chemical analysis (mass spectrometry or gas–liquid chromatography). If it is known that a certain disease leads to increase or decrease of the hormone concentration in the organism, then it is necessary to correlate the results of RIA obtained with normal and diseased subjects. The same applies when the hormone concentration depends on physiological factors or the presence of certain reagents.

Especially thorough verification of the accuracy of results should be made when a human hormone is determined with antibodies prepared by immunization of an animal hormone. A stricter procedure should be applied when an animal hormone, e.g. from a bull or a cow, serves as the standard for determination of the hormone in human plasma.

The effect of the surface of glassware on the results of analysis has also been studied [76]. It has been found that the results can sometimes be improved by treatment of the glass vessels in a 'self-purification' furnace. It has also been recommended to make the glass water-repellent by treatment with 0.05 % Triton X-405 solution.

We can give no general recommendation for choice of calibration curve function (per cent of bound substance as a function of concentration, semilogarithmic or logarithmic plots, polynomial functions, spline functions, etc.). Each case must be treated on its merits. A linear plot is often obtained by using the 'logit' function $\log [Y/(1 - Y)] = f(\log X)$, where $Y = (B - N)/(B_0 - N)$, X = concentration of non-labelled antigen or hapten in the solution, B is the measured radioactivity of antigen bound with antibodies in the sample, B_0 the radioactivity in the absence of non-labelled antigen, N the radioactivity of the part of the antigen bound with the serum components in the absence of specific antibodies (non-specifically bound) [77]. However, the importance of this transformation should not be overestimated. The experience of some laboratories has shown that only about 50 % of all calibration RIA curves can be linearized by using this function [78].

Calibration by adding known amounts of hormone to the patient's own pre-treatment serum is not recommended. It is better to use a pooled serum from several patients as the medium for the calibration [79]. Even if the experimental points can be fitted into a smooth calibration curve, it is still necessary to verify whether the curve is the best for the purpose and to decide whether it is based on assumptions which may not be true (e.g. that the maximum binding is an error-free quantity). There should be an adequate number of data-points, and several curve-fitting equations tested. A programmable calculator or a computer is recommended for ease of calculation. A comparative test of published equations has resulted in recommendation of use of overlapping quadratic equations [80].

The computer treatment of experimental data for the determination of insulin by RIA has been described [68]. It was found that use of different programs could give differences as large as hundreds of per cent, for interpretation of the same experimental data. The reproducibility can generally be considered adequate if the coefficient of variation is less than 10 %. If the concentration of antigen is sufficiently high, a coefficient of variation of about 5 % can be obtained provided that the count used is at least 10^4, the samples are accurately diluted (preferably automatically or semi-automatically) and the bound and unbound parts of the radioactive substance are separated efficiently (e.g. with adequate washing).

The reproducibility of the RIA methods is generally lower than that of chemical methods, but higher than that of biological methods, where irreproducibility in sampling of the tissues is the main source of error.

6.7.2 Precision

The precision of radioimmunoassay systems is generally superior to that of most biological methods. This statement, however, must be qualified. Within a single assay comparing one serum sample with another, the relative concentrations are determined fairly accurately. Usually the 95 % confidence limits of a determination, calculated from single assay data, are the value determined $\pm$ 5 %. However, this is the relative potency of two or more samples. If the same sample of pituitary or urinary preparation is assayed repeatedly in completely separate assays, and the mean potency is calculated from ten or more assays, the confidence limits often increase to as much as the mean $\pm$ 30 %. Stated in another way, the interassay variability is usually considerably greater than the intra-assay variability. Thus, true potency estimates derived from a variety of assays are usually about as precisely determined as they are by several bioassays of the same preparation. Frequently, both for experimental and clinical purposes, the exact hormonal value is not as important as the relative changes in a series of blood samples obtained from the same animal or human subject. Thus for determining growth hormone, a single value is seldom of any use, but a series of serum samples obtained after or during insulin hypoglycaemia or after arginine treatment will show increases from low levels to very high levels in normal humans. The character of this response is what is important, and whether the growth hormone response is say 30 or 40 µg/ml is less important.

We would like to stress that the RIA method sometimes needs an extreme effort by analytical chemists for accurate results to be obtained. Even in the relatively well investigated determination of insulin, an interlaboratory comparison made by 36 laboratories in Italy has shown that though the sequence of samples (according to the insulin content) practically agreed in all cases, there were appreciable deviations in the absolute values of the concentrations.

6.8 SENSITIVITY OF DETERMINATION

The sensitivity (limit of determination) of the RIA method is a function of the reproducibility of determination. A simple equation is given in the literature [81] for the sensitivity of determination of steroids:

$$S = 200 \, \sigma / Rf$$

where S is the lowest amount of the substance which can be measured in an aliquot of the biological fluid, σ the standard deviation of the blank (4 replicates), R the chemical yield of the purification (%), f the fraction of the purified substance used in the determination. If the blank value is practically zero, then σ should be replaced by $*\sigma/2$, where $*\sigma$ is the standard deviation of the slope of the calibration curve, i.e. the lowest amount of the standard substance which can be differentiated from zero with a statistical likelihood of 95 %.

The sensitivity of determination by RIA is limited, as a rule, either by the K value or the specific activity of the radioactive antigen. It also depends, in some cases, on the blank value. An approximate rule which is valid when the K value is the limiting factor is that the lower limit of determination of hapten is of the same order of magnitude as the K value of the antibody used. For a number of drugs, the sensitivity is several pg/l. and superior to that of chromatographic and spectral methods if they exist. For antipyrine, for example, the classical colorimetric method has a determination limit three orders of magnitude higher than that of RIA [22]. High specific radioactivity, though generally giving high sensitivity, can lead to radiation damage of the labelled antigen, which must therefore be periodically checked for quality.

The limiting factor in the determination of low concentrations is often the decrease in equilibration rate of the reactions, with decreasing concentration of the reagents. This leads to a prolonged incubation time. Thus an increase in sensitivity can sometimes be obtained by optimizing the incubation conditions.

The high sensitivity of RIA often allows interferences to be eliminated by dilution of the sample to bring the interferent concentration below the threshold value.

There are several methods for enhancing sensitivity. Preliminary concentration of separation of the analyte obviously increases the sensitivity, as does increasing the ratio of sample volume to total volume for the incubation.

An interesting method for increasing the sensitivity is based on the kinetic factor. The standard or sample is first incubated with an antiserum and then the labelled antigen is added. Thus, the analyte is preferentially bound and the labelled antigen occupies only the remaining free active sites.

RIA methods are often more sensitive than methods using two radionuclides [82]. If a fluorescent label is used instead of a radioactive label for the immunological determinations, the sensitivity is impoverished. The same applies to binding the hapten with molecules containing a free radical — the 'spin label' method [16]. On the other hand, the 'ferment label' or 'bacteriophage label' methods often reach the same level of sensitivity as RIA [83].

It should be mentioned that the current sensitivity of RIA is insufficient for its wide application in clinical practice. Thyroid stimulating hormone can be mentioned as an example [84].

6.9 KITS FOR RIA

During the last few years, 'kits' containing all the reagents, additives, test-tubes, etc., necessary for standard determinations of one substance or a limited number of substances have been produced by numerous firms. They are especially suitable for small laboratories. We shall mention only some examples, as many new types of kit appear on the market every year, and an attempt to give a complete list would be unsuccessful and quickly out-of-date.

Table 6.1 — Antigens most frequently estimated by radioimmunoassay.

(A) Protein hormones

Insulin	Calcitonin
Adrenocorticotropin (ACTH)	Angiotensin
	Renin
Human growth hormone (HGH)	Prolactin
Glucagon	Gonadotropin
Thyrotropin (TSH)	Gastrin
Follicle-stimulating hormone	Oxytocin
Luteinizing hormone (LH)	Vasopressin

(B) Steroid hormones

Aldosterone	Oestradiol
Androstendione	Oestriol
Cortisol	Oestrone
Corticosterone	17-α-Hydroxyprogesterone
Dehydroepiandrosterone	Progesterone
Dehydrotestosterone	Testosterone
Desoxycorticosterone	11-Desoxycortisol

(C) Hormones of thyroidal glands

Thyroxine	Tri-iodothyronine

(D) Other compounds

Digoxin	Transferrin
Digitoxin	Thyroxine binding globulin
Immunoglobulins	Cyclic AMP
Australia-antigen	Morphine and derivatives
α-Foetalprotein	Vitamin B_{12}
Prostaglandins	Folic acid

A short survey of the antigens most frequently and routinely estimated by radioimmunoassay is given in Table 6.1.

A critical comparison of kits (from British suppliers) for the determination of steroid hormones has been given by Jackson *et al.* [85].

At present, many kits contain reagents in a form which needs no
treatment before the determination. For example, all necessary reagents
are placed in one test-tube, in which all operations of the analytical
procedure then take place. The determination requires only addition of
the sample of biological fluid. Antibodies are present in the form of a
compound adsorbed (bound) on the surface of the test-tube, where the
free labelled antigen is also present. The test-tubes can be stored for a
prolonged period in the absence of air and humidity. In another type
of kit, small cylinders containing immobilized antibody are used. A
solution containing the antigen of unknown concentration is passed
through the cylindrical column. The columns can be stored for a
prolonged period before use [86].

The kits make the work of clinical laboratories easier, especially by
supplying them with antibodies, labelled antigens and other reagents.
However, time and labour can also be saved by the automation of RIA.
This will be further discussed in Section 6.10. Here we mention, for
example, the fully automated RIA system 'Gammaflow' which gives
results in minutes (as compared with hours or days in the case of
manual methods) [87].

Parallel incubation and separation of a large number of samples
under non-equilibrium conditions can be achieved in less than 30 min
by using the 'Centria' system [88].

6.10 GENERAL EVALUATION OF RIA

The RIA method, which was first proposed for determination of protein and polypeptide hormones, is now also used for substances which
are not themselves immunogenic, for example steroid hormones, or for
determination of drugs which can induce formation of antibodies after
suitable conjugation with a carrier molecule. The principal advantages
of the radioimmunoassay methods are

    (a) simplicity and ease of assay, once the reagent have been de-
        veloped;

    (b) increased sensitivity.

The smallest amount of hormone determinable by radioimmunoassay systems may be smaller by a factor of $10-10^4$ than that determinable
by bioassay systems. For example, the smallest amount of human
growth hormone (HGH) detectable in a radioimmunoassay system is

about a thousandth of that detectable in bioassay systems. Owing to this simplicity, sensitivity and wide applicability, the RIA method has become one of the most important and rapidly developing methods of application of labelled substances in the fields of biochemical, biological and medical science and practice (clinical chemistry, virology, microbiology, physiology, pathophysiology, endocrinology).

An important advantage of RIA is the possibility of choosing a highly specific antiserum in order to secure the most advantageous conditions for analysis. The specificity and activity of an antiserum are advantageous not only from the point of view of selectivity and rate of reaction (shorter incubation) but also with regard to any shift of the equilibrium between the free and bound antigen, arising from separation of these two forms.

Another important feature of RIA is the possibility of its relatively simple automation, owing to the relatively high decay energy of the radioactive labelling atoms. Automation and the use of computers have led to the possibility of a throughput of several thousands of analyses per week [89]. This was impossible for determination of hormones without the use of radionuclides.

At present, no fully automated apparatus exists for RIA and other modifications of saturation analysis. Nevertheless, substantial progress has been made in the development of programs for computer control of the analysis as well as in the development of technical equipment. A combination of these two aspects should lead to the development of fully automated systems. Another advantage of the RIA method is shown by the determination of synthetic anabolic steroids in the urine and plasma of athletes. It is possible to prepare group-selective antibodies, which can distinguish certain structural features characteristic for a given group. Thus, only 4–6 antigens and corresponding antibodies are necessary to detect tens of potentially used anabolic steroids. In this case, the advantage of RIA over chromatographic methods lies not in its greater sensitivity, but in its ability to handle the complexity and number of samples for analysis.

The basic advantage of RIA over biological methods is that it does not depend on the character of the tissues investigated [66, 67]. It also has appreciably higher specificity and reproducibility.

The introduction and wide application of RIA has considerably affected the development of some fields of science and contributed to solution of a number of serious problems, especially in the field of endocrinological physiology [90].

Radioimmunological techniques, in combination with the methods of filtration and chromatography on Sephadex or polyacrylamide gel, have led to understanding of the inhomogeneity of the circulating forms of hormones. Such forms include 'precursors' (prohormones), isohormones, polymers, half-units, metabolites, analogues of hormones formed in tumours, etc. The application of RIA together with the method of competitive binding of proteins in the determination of ng and pg quantities of hormones in the plasma of the peripheral blood of large domestic animals has substantially extended our knowledge of the reproductive physiology of these animals. The method has firmly gained a place in numerous diagnostic laboratories [91]. It is also used in the investigation of the mechanism of disease transmission [92].

However, RIA has not yet matured. At present, it is applied in clinical practice for diagnostic purposes and routine determination of drugs, etc., but such methods as determination of the LG-releasing factor and of other hypophysiotropic hormones of the hypothalamus are still available in only a few laboratories [93].

In the future, RIA will find new applications, for example in the field of the determination of immunoglobulins, the synthesis of which is induced by bacterial antigens [94]. It can also be expected that new methods of preliminary separation of some antigens before the direct application of RIA will be developed. This is especially related to steroids, as most antibodies do not exhibit sufficient selectivity. Thus, the steroids cannot be determined without a preliminary separation [81].

Problems existing at present in determination of important hormones will also be solved. In the near future, the application of RIA is expected to be extended in nephrology [95] and in prophylaxis and control of cancer treatment [14].

The cheapness, automation and speed of radioimmunoassay may lead to its use for screening of a whole population. One example is the determination of thyroxine in new-born children by RIA. When hypothyreosis is established early enough, appropriate treatment can eliminate the brain damage which would otherwise occur several years later.

Thus, RIA is one of the most important and rapidly developing applications of radioactive indicators in *in vitro* biological and biochemical studies.

REFERENCES

[1] S. A. Berson and R. S. Yalow, *Ann. N. Y. Acad. Sci.,* 1959, **82,** 338.
[2] C. D. Hawker, *Anal. Chem.,* 1973, **45,** 878A.
[3] *Pathol.-Biol.,* 1975, **23.**
[4] G. Court, *Pathol.-Biol.,* 1975, **23,** 859.

[5] R. L. Hayes, F. A. Goswitz and B. E. P. Murphy (eds.), USAEC Symposium Series, No. 13 (CONF-671111), Oak Ridge, Tenn., 1968.

[6] J. Kalmakoff, A. J. Parkinson, A. M. Crawford and B. R. G. Williams, *J. Immunol. Methods*, 1977, **14**, 73.

[7] R. Trautman, *Scand. J. Immunol.*, 1976, **5**, 609.

[8] J. Binoux, J. G. Pierce and W. D. Odell, *J. Clin. Endocrinol. Metab.*, 1974, **38**, 674.

[9] A. L. Nichols and W. D. Odell, *Competitive Protein Binding Assay Systems*, in *Nuclear Medicine*, W. H. Blahd (ed.), McGraw-Hill, New York, 1971.

[10] S. G. Korenman, L. E. Perrin and T. P. McCallum, *J. Clin. Endocrinol. Metab.*, 1969, **29**, 879.

[11] W. D. Odell, G. T. Ross and P. L. Rayford, *J. Clin. Invest.*, 1967, **46**, 248.

[12] J. J. Langone, H. Van Vunakis and L. Levine, *Acc. Chem. Res.*, 1975, **8**, 335.

[13] F. M. Segars and J. M. Kinkade, Jr., *J. Immunol. Methods*, 1977, **14**, 1.

[14] J. R. Antunes, S. L. Dale and J. C. Melby, *Steroids*, 1976, **28**, 621.

[15] J. M. Garel and J. P. Barlet, *J. Physiol. (Paris)*, 1976, **72**, 249.

[16] J. London and A. C. Moffat, *Analyst*, 1976, **101**, 225.

[17] N. S. Young, J. G. Curd, A. Eastlake, B. Furie and A. N. Schechter, *Proc. Natl. Acad. Sci. USA*, 1975, **72**, 4759.

[18] S. L. Jeffcoate, *Pathol.-Biol.*, 1975, **23**, 903.

[19] G. S. Linsley, *Radiography*, 1976, **42**, 89.

[20] A. Massaglia, U. Barbieri, C. Siri-Upathum and R. Vitali, *Int. J. Appl. Radiat. Isotopes*, 1973, **24**, 455.

[21] J. C. Baehr, Ph. Pradelles, C. Lebreux, P. Cassier and F. Dray, *FEBS Letters*, 1976, **69**, 123.

[22] W. R. Dixon and A. Wood, *German Patent*, 2604991, 19. 8. 1976.

[23] A. E. Bolton and W. M. Hunter, *Biochem. J.*, 1973, **133**, 529.

[24] G. Spira, N. Biswal and G. Dreesman, *Appl. Microbiol.*, 1974, **28**, 329.

[25] L. Knight, K. A. Krohn and M. J. Weleh, *Radiopharmaceuticals*, Society of Nuclear Medicine, New York, 1975, p. 149.

[26] B. K. Park, T. A. Rance and P. D. G. Dean, *FEBS Letters*, 1976, **71**, 18.

[27] J. J. Langone, H. B. Ajika and H. Van Vunakis, *Biochemistry*, 1973, **12**, 5025.

[27a] U. Rosa, *Pathol.-Biol.*, 1975, **23**, 853.

[28] E. Knoll and H. Wisser, *Z. Anal. Chem.*, 1976, **279**, 119.

[29] Z. Spirer, V. Zakuth, N. Bogair and M. Fridkin, *Eur. J. Immunol.*, 1977, **7**, 69.

[30] M. G. Forest, *Pathol.-Biol.*, 1975, **23**, 869.

[31] R. P. Ekins, *Tracer Methods in Organic and Biochemical Analysis* in *Radiochemical Methods in Analysis*, D. I. Coomber (ed.), Plenum Press, New York, 1975.

[32] P. Sizaret, *J. Immunol. Methods*, 1974, **5**, 283.

[33] E. Ulrich, U. Schneyer and J. Haedecke, *Endokrinologie*, 1974, **63**, 364.

[34] N. Levin and J. A. Fries, *French Patent Doc.*, 2283439/A, 25. 8. 1975.

[35] M. Andre, D. Boucher and L. Thieblot, *Compt. Rend. Soc. Biol. Paris*, 1976, **170**, 353.

[36] M. J. Giphart, E. Doyer and J. W. Bruning, *Immunochemistry*, 1977, **14**, 139.

[37] R. Lamerz and H. Ruider, *Z. Anal. Chem.*, 1976, **279**, 105.

[38] M. J. Barrett, *German Patent*, 2544366/A, 8. 4. 1976.

[39] J. R. Brown A. H. Cavanaugh and W. E. Farnsworth, *Steroids*, 1976, **28**, 487.

[40] M. A. Fletcher, T. M. Lo, B. A. Levey and W. R. Graves, *J. Immunol. Methods*, 1977, **14**, 51.

[41] T. Lehtinen and H. Adlercreutz, *J. Steroid Biochem.*, 1977, **8**, 99.

[42] B. R. Ziola, M.-T. Matikainen and A. Salmi, *J. Immunol. Methods*, 1977, **17**, 309.

[43] W. P. Vann and S. Yaverbaum, *US Patent*, 39755111, 17. 8. 1976.

[44] K. O. Smith, J. T. Harrington and W. D. Gehle, *J. Immunol. Methods*, 1977, **15**, 17.

[45] A. Tax and L. A. Manson, *J. Immunol. Methods*, 1976, **11**, 281.

[46] J. Esztergalyos and J. Miller, *Izotoptechnika*, 1975, **18**, 472.

[47] S. Kushinsky and M. Mirrasoul, *Steroids*, 1976, **28**, 805.

[48] L. David, D. V. Cohn and C. S. Anast, *Pathol.-Biol.*, 1975, **23**, 833.

[49] R. P. Ekins, *Curr. Top. Exp. Endocrinol.*, 1971, **1**, 1.

[50] W. D. Odell and W. H. Daughaday (eds.), *Principles of Competitive Protein Binding Assays*, Lippincott, Philadelphia, 1971.

[51] R. T. Dunn and L. B. Foster, *Clin. Chem.*, 1973, **19**, 1101.

[52] E. P. Frenkel, S. Keller and M. S. McCall, *J. Lab. Clin. Med.*, 1966, **68**, 510.

[53] J. L. Raven, P. L. Walker and P. Barkhan, *J. Clin. Pathol.*, 1966, **19**, 610.

[54] P. Wunderwald, G. Jurz and G. Michal, *Anal. Biochem.*, 1974, **59**, 468.

[55] E. Heftmann, *Modern Methods of Steroid Analysis*, Academic Press, New York, 1973.

[56] P. Maróy, J. Vargha and K. Horváth, *FEBS Letters*, 1977, **81**, 319.

[57] Report of an International Atomic Energy Panel, *Int. J. Appl. Radiat. Isotopes*, 1974, **25**, 145.

[58] G. Thamer and B. Kommerell, *J. Lab. Clin. Med.*, 1976, **87**, 734.

[59] S. Matzku, W. Rapp and H. Wesch, *Eur. J. Nucl. Med.*, 1976, **1**, 78.

[60] J. P. Brown, J. M. Klitzman and K. E. Hellström, *J. Immunol. Methods*, 1977, **15**, 57.

[61] A. Reuter, Y. Gevaert and P. Franchimont, *Ann. Biol. Clin. (Paris)*, 1976, **34**, 191.

[62] D. E. H. Llewelyn, S. G. Hillier and G. F. Read, *Steroids*, 1976, **28**, 339.

[63] R. A. Yeates and B. M. Ogilvie, *J. Immunol. Methods*, 1976, **12**, 57.

[64] F. Santoro, P. Wattre, J. P. Dessaint and A. Carpon, *J. Immunol. Methods*, 1977, **15**, 201.

[65] H. Patscheke, M. Breinl and E. Schäfer, *J. Immunol. Methods*, 1977, **16**, 31.

[66] J. P. Felber, *Helv. Med. Acta*, 1966, **33**, 367.

[67] J. P. Felber, A. J. Moody and A. Vannotti, *Helv. Med. Acta*, 1966, **33**, 378.

[68] G. Ayrey and K. L. Evans, *Liquid Scintillation Counting*, Vol. 2, M. A. Crook, P. Johnson and B. Scales (eds.), Heyden, London, 1972, p. 303.

[69] M. Kenny, *Am. J. Med. Technol.*, 1976, **42**, 318.

[70] E. Cerceo, *Lab. Pract.*, 1976, **25**, 685.

[71] J. Woo and D. C. Cannon, *Am. J. Clin. Pathol.*, 1976, **66**, 854.

[72] M. Wiedemann, A. Wirtz, H. J. Karl and L. Raith, *Z. Anal. Chem.*, 1974, **272**, 195.

[73] G. Bodem, *Prakt. Anaesth.*, 1976, **11**, 97.

[74] A. I. Barkalaya, *Radiobiologiya*, 1975, **15**, 127.

[75] Y. Nakai, *INIS Atomindex*, 1977, **8**, No. 301299.

[76] N. P. Kubasik, J. L. Hall and H. E. Sine, *Clin. Chem.*, 1976, **22**, 1745.

[77] R. Stupnicki and A. Madej, *Endokrinologie*, 1976, **7**, 68.

[78] I. Marschner, F. Erhardt and P. C. Scriba, in *Radioimmunoassay and Related Procedures in Medicine, Proc. Symposium, Istanbul, 1973*, Vol. 1, IAEA, Vienna, 1974, p. 111.

[79] A. Tembo, M. A. Schork and J. D. Wagner, *Steroids*, 1976, **28**, 387.

[80] P. England and O. Cain, *Clin. Chim. Acta*, 1976, **72**, 241.

[81] G. E. Abraham, *Pathol.-Biol.*, 1975, **23**, 889.

[82] H. Maier-Hüser, *Bull. Inform. Sci. Tech. (Paris)*, 1970, No. 47, 37.

[83] F. Franěk and J. Doskočil, *Eur. J. Immunol.*, 1975, **5**, 216.

[84] P. Pfannenstiel, *Diagnosis of Thyroid Diseases*, Byk-Malinckrodt Radiopharmaceutika-Diagnostika, Dietzenbach-Steinberg, Frankfurt/Main, 1975.

[85] P. Jackson, S. Lees, M. Pow, G. W. Willing and D. Watson, *J. Clin. Pathol.*, 1976, **29**, 902.

[86] L. R. Johnson, *German Patent*, 2612948, 21. 10. 1976.

[87] G. Brooker, W. L. Terasaki and M. G. Price, *Science*, 1976, **194**, 270.

[88] A. Castro, I. Prietto, C. Wunsch, G. Ertingshausen and H. Malkus, *Clin. Chem.*, 1976, **22**, 1655.

[89] V. Ullmann and M. Závada, *Čas. Lék. Čes.*, 1975, **114**, 1392.

[90] H. G. Eckert, *Laboratoriumsbl. Med. Diag. E. v. Behring*, 1975, **25**, No. 2, 56.

[91] D. Neumeier, W. Vogt and M. Knedel, *Z. Anal. Chem.*, 1976, **279**, 107.

[92] B. A. Weissmann, H. Gershon and C. B. Pert, *FEBS Letters*, 1976, **70**, 245.

[93] G. Ya. Bakalkin, K. N. Yarygin and V. A. Isachenkov, *Problemy Endokrinol.*, 1976, **22**, No. 6, 72.

[94] B. A. Sanford and K. O. Smith, *J. Immunol. Methods*, 1977, **14**, 313.

[95] K. zum Winkel, M. D. Blaufox and J.-L. Fuinck-Brentano (eds.), *Radionuclides in Nephrology, Proc. IIIrd Int. Symp.*, Berlin, 1974, Thieme, Stuttgart, 1975.

SELECTED BIBLIOGRAPHY

A. H. Beckett and J. B. Stenlake, *Practical Pharmaceutical Chemistry*, Part 2, *Physical Methods of Analysis*, Athlone Press, London, 1976.

W. P. Collins and J. F. Hennam, *Molecular Aspects of Medical Science: Radioimmunoassay and Reproductive Endocrinology*, Pergamon Press, Oxford, 1976.

B. K. Hartman and S. Udenfried, *Pharmacol. Rev.*, 1972, **24**, 311.

K. E. Kirkham and W. M. Hunter (eds.), *Radioimmunoassay Methods*, Churchill Livingstone, Edinburgh, 1971.

C. A. Pasternak (ed.), *Radioimmunoassay in Clinical Biochemistry*, Heyden, London, 1975.

D. S. Skelley, L. P. Brown and P. K. Besch, *Clin. Chem.*, 1973, **13**, 146.

L. Raith, *Introduction to Radioimmunoassay*, Byk-Mallinckrodt Radiopharmaceutika-Diagnostika, Dietzenbach-Steinberg, Frankfurt/Main, 1975.

A. F. Malentchenko, E. G. Matveenko, I. B. Lifshitz, V. V. Seregin and V. G. Spesivtseva, *Radioizotopnye metody issledovaniya v endokrinologii*, Nauka i tekhnika, Minsk, 1976.

7

Radiochemical methods for determination of biological activity of enzymes*

7.1 INTRODUCTION

Enzymes are proteins which can specifically catalyse biochemical reactions. The determination of the catalytic activity of enzymes in biological materials, such as plasma, blood, urine and tissue extracts or homogenates facilitates diagnosis of various diseases, enables their progress to be followed and the effectiveness of therapeutic treatment to be evaluated. As a rule, there is an enhancement of enzyme activity in an organism if a substantial quantity of its cells is damaged or destroyed. In that case, the permeability of the cell membranes is increased and, as a result, an increased diffusion of enzymes into the plasma takes place. On the other hand, in normal conditions the equilibrium between enzyme formation and decomposition is preserved. Thus we can collect valuable information about the progress of diseases and their cure, by following enzyme activity changes [1–3].

Radioanalytical determination of enzyme activity belongs to the group of non-saturation analytical methods. These methods have a number of common specific features which distinguish them from methods used for the determination of biochemical activity.

*By L. Kronrád and M. Kyrš.

A comparison between the radiochemical method and the method of concentration-dependent distribution, which also belongs to the group of non-saturation analysis methods, is given in Table 7.1.

Table 7.1 — The comparison of methods of non-saturation analysis: A — concentration-dependent distribution; B — radiochemical determination of enzyme activity.

	A	B
Value determined	Concentration of a substance	Catalytic activity of an enzyme
Labelled substance	Usually an inorganic reagent	Substrate
Reaction proceeds:	Usually to equilibrium	Only a small fraction of substrate is reacted
Concentration ratio; analyte/labelled substance	Usually $\gg 1$	Usually < 1
First used	1925	1955

7.2 DEFINITION OF ENZYME ACTIVITY UNITS

During the evolution of biochemistry various units of enzyme activity were used, which often made it impossible to compare corresponding samples. To avoid such misunderstandings, the Commission for Enzymes of the International Biochemical Society, in 1964 recommended use of the following definition [4]: "The International Unit is such a quantity of enzyme which catalyses the transformation of one micromole of a substrate in one minute under defined conditions" [5]. The 'defined conditions' include temperature (as a rule up to 30 °C), pH, substrate concentration and quantity and composition of buffers and substances used as inhibitors or activators of the reactions.

7.3 MEASUREMENT OF ENZYME ACTIVITY

Quite a number of enzymes allow employment of classical, especially colorimetric, analytical methods. Nevertheless, during the last twenty years radiochemical methods for the determination of enzyme biological activity have been developed because of their sensitivity and specifity.

Enzymes, being homogeneous catalysts, have the following properties which are directly or indirectly important for the determination of their biochemical activity:

— they exhibit chemical effects even in small quantities and at low concentrations,

— they do not undergo changes during the reactions catalysed,

— being present in low concentration in the reaction mixture (relative to that of the substrate), they do not affect the reaction equilibrium, but increase the reaction rates.

Radiochemical determination of enzyme activity is based on catalytic transformation of a substrate, labelled with radioactive nuclide (^{14}C, ^{3}H, ^{35}S, ^{32}P), into a labelled product. The substrate is a substance, the chemical transformation of which is specifically catalysed by enzymes. The rate of the enzymatic reaction is then determined by measuring the radioactivity of either the labelled product or the substrate residue, or both, after their quantitative separation from the reaction mixture.

It is advisable to base the determination of biochemical activity on measurement of the initial reaction rate and not just on the quantity of substrate transformed during a particular time interval. This latter procedure, which was formerly more widespread, can be successful only if the reaction rate is known to be constant. The initial reaction rate is, as a rule, directly proportional to the enzyme concentration. However, in a number of cases a non-linear relationship between these two values has been found [5]. It is therefore necessary to define conditions under which this dependence is linear for each new method of enzyme activity determination.

Knowledge of the exact value of the specific activity of the labelled substrate is one of the main conditions which have to be fulfilled in this type of determination. The radioactivity of the product or residual substrate and the value of the specific activity are necessary for calculating the weight of substrate transformed by the enzymatic reaction, and thus determining the enzyme activity. It is further essential that the conditions of each enzymatic reaction, i.e. pH, substrate concentration, quantity of activators (substances which facilitate the reaction) and temperature, are kept constant at the optimal values.

A typical radiochemical enzyme determination comprises the following steps.

(1) Preparation of a radioactive substrate with known specific activity.
(2) Preparation of the sample containing the enzyme to be determined.
(3) Incubation of enzyme and substrate reaction mixture under optimum conditions.
(4) Incubation of several 'blank' samples, either without enzyme or with enzyme inactivated by boiling.
(5) Quenching the reaction.
(6) Separation of the products from unreacted substrate.
(7) Separate radioctivity measurements of isolated parts of the sample.

(8) Calculation of quantity of product by using the equation

$$P = (C - C_0)/a$$

where C is the number of disintegrations per minute in the isolated product fraction, C_0 the number of disintegrations per minute in the corresponding fraction of blank, a the specific activity (disintegrations/μmole), P the quantity of the product.
(9) Determination of the enzyme activity from P by using a calibration curve.

A detailed evaluation of the individual factors influencing an enzymatic reaction has been given by Dixon and Webb [6].

7.4 ADVANTAGES OF THE RADIOCHEMICAL METHODS

The radiochemical methods have numerous advantages over classical methods:
— high sensitivity,
— the possibility of verifying the specificity of the reaction by isotope dilution analysis,
— minimum interference of other substances,
— possibility to determine low activities of enzymes even in homogenates,
— possibility to follow the competitive influence of inhibitors,
— possibility to minimize the influence of inhibitors by dilution of the reaction mixture.

The high sensitivity of radiochemical methods implies some other advantages. The biological activity of enzymes can be determined over a broader range of enzyme and substrate concentrations than with colorimetric methods. For this reason radiochemical methods are suitable for the measurements of Michaelis constants as well as for the evaluation of inhibitor influence at very low concentrations. They also make possible the use of very small (μg) samples of biopsy tissues, which exhibit low enzyme activity.

The main factor influencing the sensitivity of the radiochemical methods is the half-life of the radionuclide used, which affects the specific activity and the autoradiolysis of the labelled substrate. The theoretical and practical limits of sensitivity when employing radiometric instruments with a detection threshold of 3.7 Bq are summarized in Table 7.2 [7].

Table 7.2 — Sensitivity of analysis (compiled from data in [7]).

Radionuclide	Specific activity (Bq/gram-atom)	Minimum amount to be determined (mole of product)	
		theoretically	practically
^{14}C	2.4×10^9	3×10^{-12}	3×10^{-12}
^{3}H	1.1×10^{12}	3×10^{-15}	1×10^{-14}
^{35}S	5.6×10^{13}	6×10^{-17}	1×10^{-12}
^{32}P	3.4×10^{14}	1×10^{-17}	1×10^{-12}
^{36}Cl	4.1×10^7	1×10^{-10}	5×10^{-10}

Substrates with optimum specific activity, labelled at only one position of the molecule, are preferable for the radiochemical determination.

Too high specific radioactivity of the substrate leads to its rapid autoradiolysis and could result in erroneous experimental results.

Lower, but still sufficient, specific radioactivities of the substrates are advantageous owing to the reduced autoradiolysis rate, reduced influence of inhibitors, and enhanced accuracy of determining substrate specific activity.

The specificity of each enzymatic reaction, i.e. formation of a single radioactive product in the given fraction after separation, can easily be demonstrated either by isotope dilution or radiochromatography.

A further great advantage of radiochemical methods is the possibility of direct determination of enzyme activity in various homogenized tissues, which often contain natural substances with similar chemical structures to the analyte. If a classical colorimetric method is employed, discrimination from these substances causes a considerable difficulty, whereas radiochemical determination eliminates their interference.

7.5 DISADVANTAGES OF THE RADIOCHEMICAL METHODS

The main disadvantages of published radiochemical methods for the determination of enzyme biological activity are:
— the relatively long period necessary for separation of the unreacted substrate from the product,
— the necessity for accurate determination of the labelled substrate specific activity,
— the necessity to test the purity of the labelled substrate,
— the possibility of a decrease in labelled substrate specific activity caused by endogenous substrates,

— the difficulty of developing a continuous analytical method,
— the possibility of adverse influence of the isotope effect, especially
  with substrates labelled with ^3H.

The accuracy of activity determination is dependent on the quantitative separation of unreacted substrate from the product. The known separation processes in use for this purpose (described below) are relatively laborious. It is likely that further development of separation techniques will bring about a significant shortening of the time needed for the determination.

The specific activity of the labelled substrate is not always known, so each batch should be examined. These measurements, as well as purity testing of labelled substrates, are both complicated and laborious, especially on samples with high specific radioactivity.

Endogenous substrates, present particularly in tissue homogenates, can significantly lower the specific radioactivity of the added labelled substrate. This effect may be eliminated either by letting the homogenate pass through a Sephadex column where undesirable substrates will be absorbed (as in the determination of L-histidine decarboxylase [8]) or by using high concentrations of the labelled substrate.

The substrates labelled with tritium can exhibit an isotopic effect, especially if the carbon–tritium bond takes part in the enzymatic reaction. The effect occurs, for example, in the condensation of dihydroxyacetone phosphate (tritium-labelled) with glycerol-3-phosphate at pH 7 [9]. On the other hand, in the determination of the biochemical activity of thyroxine hydroxylase no marked difference was found when thyroxine-^{14}C or thyroxine-3,5-^3H were used as substrates [10].

7.6 SEPARATION METHODS USED
IN DETERMINATION OF ENZYME
BIOCHEMICAL ACTIVITY

The radiochemical determination of enzyme activity is largely dependent on quantitative separation of the reaction product from the unreacted substrate. In most determinations 10^{-12}–10^{-9} mole of substrate is used in a volume of several ml. The separation of such small amounts of substances is usually very difficult. In order to facilitate it, it is possible to add carriers (non-radioactive form of the substrate or product) after the quenching of the enzymatic reaction. The most frequently used separation procedures are given in Table 7.3.

Table 7.3 — Separation methods used in the radiochemical determination of biological activity of enzymes.

Method	Use, characteristics
Precipitation, sorption, ion exchange	Distribution between a solid and a liquid phase
Solvent extraction	Distribution between two immiscible liquid phases
Separation of radioactivity in volatile form	Distribution between a gaseous and a condensed phase
Chromatography, dialysis, gel filtration	Separation as the result of repeated elementary processes
Transformation into another chemical form	Complicated treatment of product before separation
Reverse isotope dilution	Increasing amounts separated
Labelling of product formed	Substrate non-radioactive

7.6.1 Precipitation

Several enzymes catalysing the synthesis of macromolecules, such as polymerases of nucleic acids [11, 12], or involved in the synthesis of polypeptides [13], polysaccharides [14] and aminoacyl RNA [15], have been determined by using labelled substrates and measuring the radioactivity of the fraction soluble in acids or organic solvents. Sometimes it is useful to measure the radioactivity of precipitates, which for this purpose can be dissolved by means of enzymes, hyamine or formic acid.

If the product or substrate is a macromolecular substance, very often perchloric acid is used as the precipitant [16]. It is evident that before measurement the precipitate must often be additionally purified or modified.

A simple method enabling a simultaneous treatment of several samples of limited size includes the following steps. The samples are placed on filter paper and the reaction is stopped by immersing the paper in a suitable cooled solvent. The macromolecular product is precipitated directly on the filter and the soluble substrate is removed by washing. After drying, the radioactivity remaining on the filter is measured [17].

Transparent (glass fibre) filters increase the efficiency of scintillation measurements by up to 10 % [18]. Moreover, the fibres hold the precipitate very strongly, so that even several days immersion in washing solution does not release it. Especially rapid and complete separation of precipitates occurs on Millipore filters (pore-size 0.45 µm) [19].

It is evident that if the substance decomposed by the enzyme is a macromolecular compound, precipitation of the unreacted portion of the substrate, followed by centrifugation or filtration, can be used for the determination. In the case of endonucleases, where relatively large oligonucleotides are produced, the danger of their partial simultaneous precipitation with the substrate renders this separation method less suitable [20].

7.6.2 Separation of volatiles

This method is used rather widely and is applicable whenever the product or the substrate of the enzymatic reaction is volatile or can easily be transformed into a volatile form. The most frequently used volatile compounds are elemental tritium or tritiated water [21], for instance in the case of amino-acid hydroxylases [22], thymidylate synthetase [23] and cobamide-dependent ribonucleotide reductase [24]. For the separation of tritiated water from the reaction mixture, vacuum distillation is usually used [25]. Another very useful volatile substance in this respect is CO_2. The determination of decarboxylases is based on the measurement of the amount of $^{14}CO_2$ formed from the labelled substrate [26]. The carbon dioxide is absorbed in alkaline solutions [27] or on filter paper impregnated with such a solution. Other possibilities include absorption by phenylethylamine [28] and especially by hyamine hydroxide [29, 30]. A very rapid method of sampling during the reaction is based on the absorption of the released $^{14}CO_2$ by 2-phenylethylamine sorbed on an inorganic sorbent placed in a series of columns into which, one by one, the released gas is introduced at 10-sec intervals. Up to 97–99 % of the total amount of radioactivity is absorbed. In several cases, where ^{14}C-labelled carbonate was used as the substrate, the unreacted portion was removed by adding acid and increasing the temperature, and the radioactivity of the product remaining in the reaction mixture was measured (malate dehydrogenase [31], pyruvate decarboxylase and pyruvate carboxylase [32]).

If the labelled radioactive product is an unstable oxy-acid, it can be stabilized by conversion into its 2,4-dinitrophenylhydrazone [33] or transforming its oxalacetate into malate by means of a different enzyme. After the reaction of ^{14}C-acetylcoenzyme A with oxalacetate and its separation, the citric acid formed is oxidized by bromine with the release of $^{14}CO_2$ [34].

Besides the basic procedures above, which rely on the direct use of a volatile product or substrate, methods have been based on a relatively easy transformation of either the product or substrate into a volatile form. Thus the simple transformation of acetyl-1-^{14}C-quinoline into free acetic acid was used for the determination of acetylcholine hydrolase [35]. A similar method was applied in the determination of enzymes transforming the acetyl group of acetylcoenzyme [36] and also for the determination of aspartate carbonyltransferase [37]. An enzymatic reaction in which volatile products are released has also been used for the determination of oxalic acid [38] and putrescine [39].

In addition to tritium, tritiated water and carbon dioxide, other volatile products can be used, e.g. ^{35}S-dimethylsulphide in the determination of trimethylsulphonium-tetrahydrofolate methyltransferase [40].

7.6.3 Extraction

The separation of the product from the substrate by solvent extraction is a very simple and widely used procedure. It can be performed directly in a centrifuge tube, because centrifugation significantly accelerates the phase separation and, consequently, the total analysis.

For extraction, chemically active organic solvents are frequently used, such as butanol [41], ethyl acetate [42], anisole or chlorinated alkanes [43]. Very often chloroform [44] is used, which also serves for preliminary product separation into fractions before subsequent chromatographic isolation [45].

An important shortening of the analysis is achieved by using as the extraction solvent one component of the scintillation mixture, e.g. benzene [46] or toluene [47], so that the extract can immediately be used for radioactivity measurement [48]. The time necessary for the analysis can also be shortened by immersion of the centrifuge tube in a cooling bath, so that the aqueous phase freezes.

A complete separation of substrate from product in one solvent extraction step is rare. Usually the extracted radioactive substance is accompanied by about 1 % of other components, which lowers the sensitivity of this procedure [49]. Therefore repeated extraction has to be applied.

7.6.4 Separation with ion exchangers

This method for separation of substrate from product is also rather frequently used. The simple procedure involves frontal analysis with a column of suitable size which retains only one of the components [50]. If several products are formed in the enzymatic reaction they can be separated by successive elution with different eluents [51]. If the radioactive substance to be measured is not sorbed on the ion exchanger it may be of advantage to add the necessary amount of ion exchanger directly to the reaction mixture in the centrifuge tube and to centrifuge after attainment of equilibrium [52]. The most widely used column-loading material is Dowex 50. The best developed procedure is the separation of the reaction mixture components by filtration through DEAE-cellulose filter discs [53]. Phosphorus-containing products are particularly well sorbed onto the discs. After removal of the remaining substrate from the disc by washing, the radioactivity of the product is measured directly on the disc. This procedure is used, for example, for the determination of kinases, in which ammonium formate is applied as a selective eluent for nucleosides [54]. A modification of the procedure was used for the determination of nucleotide kinases. In this case the reaction catalysed by the kinase is stopped after the sample is taken, then the sample is treated with phosphatase for the dephosphorization of the nucleoside monophosphate, the di- and tri-phosphates present remaining unchanged. The phosphorus-containing products are separated by filtration through the disc of DEAE-cellulose [55].

Ion-exchanger foils or papers have been used for the determination of some kinases [56] and transferases [57].

Besides Dowex 50, Amberlite IRA 400 [58] or Amberlite SA-2 ion-exchange paper is frequently used [59].

7.6.5 Chromatography
In the determination of enzyme activity without use of radioactivity, chromatographic methods are used only in exceptional cases, mainly because they are time-consuming and the individual substances must be quantitatively eluted from the chromatograms or electrophoretograms before the determination itself.

The introduction of radioactive substrates has substantially accelerated and simplified the analysis, because of the possibility of measuring the radioactivity of individual spots directly on the chromatograms and using it to determine the stoichiometry of the reaction. The high sensitivity of radioactivity measurements also makes it possible to find on the chromatograms very small amounts of substance formed in side-reactions or unexpected reactions. For separation of the product from the substrate, paper chromatography [60] or liquid chromatography [61] is mostly used after preliminary removal of the macrocomponent by some other separation method [62]. If the specific radioactivity is high enough, thin-layer chromatography [63] can be used, which is far less time-consuming. On the other hand, with preparations having low specific radioactivity, column chromatography [64] commends itself because significantly larger amounts of substances can be isolated. If the product or the substrate remains on the start of the chromatogram and the other component can easily be removed, so that the radioactivity of the remaining spot can be measured, the analysis is more rapid.

Separation by paper electrophoresis is used mainly for the determination of kinases [65] and hydrolases [66].

7.6.6 Sorption

Sorption on active charcoal is used mainly for the determination of nucleoside phosphatases. It is advantageous to use a ^{32}P-labelled substrate, so that the radioactivity retained on charcoal [67] or that remaining in the solution can easily be measured [68]. In order to prevent partial sorption of inorganic phosphates on active charcoal leading to erroneous determination, it is advisable to add a certain previously assessed amount of phosphate as carrier. Sorption on active charcoal is often combined with other methods, such as precipitation [10] or separation on columns filled with Norite or Celite 535, with successive elution of individual fractions [69]. Owing to the high auto-absorption of ^{3}H and ^{14}C, sorption is used in analyses with substrates labelled in this way only if the radioactivity measured is that remaining in the solution. Otherwise the radioactivity must be desorbed with eluents such as ethanolic ammonia [70] or formic acid [72] which is, as a rule, not quantitative, so the introduction of correction factors becomes mandatory.

7.6.7 Separation after conversion into different chemical form

This procedure is mostly rather complicated and time-consuming. The complications stem mainly from the fact that the conversion into a different chemical form is not complete even if carriers are added, so losses occur of the original product of the enzymatic reaction. The actual separation is achieved in most cases by repeated precipitation, to secure sufficient purity of the material.

This step is usually the source of further losses, which necessitates the introduction of several corrections into the calculation of the result. These methods are used for the determination of some kinases, by utilizing the possibility of precipitating the phosphorus-containing product in alcoholic solution with heavy metals [71]. For the precipitation of phosphates the reaction of molybdophosphate with triethanolamine is also frequently used [72]. Oxy-acids are usually transformed into their 2,4-dinitrophenylhydrazones [73], the precipitates being collected and measured after dissolution. In the case of adenylnuclease the product 3′,5′-cyclic adenosine monophosphate is separated from the substrate on an ion exchanger. The last portions of the substrate are removed by precipitation with barium hydroxide and zinc sulphate [74]. A similar procedure serves for the determination of amylsulphatase [75]. In the determination of choline acetyltransferase, precipitation with tetraphenylborate [76] or reineckate [77] is recommended.

In some cases the product is transformed into other compounds by oxidation in order to facilitate the separation. For example, in the determination of dopamine β-hydroxylase the product formed — octopamine — is converted into ^{14}C-p-hydroxybenzaldehyde [78]. The oxidation of ^{14}C-indole (substrate) with iodate into ^{14}C-indole-3-aldehyde was used in the determination of tryptophan synthetase [79].

Products of some enzymatic reactions can be transformed by other enzymatic reactions into derivatives which can be much more easily separated [80].

7.6.8 Labelling the product formed
In a procedure of this type the actual enzymatic reaction occurs with a non-radioactive substrate and only the product formed is then radioactively labelled to enable its measurement. It is evident that this procedure can only exceptionally be used, mainly to give an important increase in sensitivity.

An example of this type of procedure is the investigation of the reaction of the hydroxylation of N-acetylserotonin with the formation of unstable dihydroxy derivatives [81].

7.6.9 Dialysis and gel filtration
A common drawback of these methods is the long time necessary for the analysis, the latter method being somewhat the faster of the two [82]. These procedures are mainly used for the determination of enzymes catalysing the formation of macropolysaccharides [83].

7.6.10 Reverse isotope dilution
The principle of this method (see p. 55) consists in adding to the radioactive product a relatively large and known amount of the inactive form of the substance. After isolation the specific activity of the isolated fraction is determined. The drawback is the necessity of measuring the radioactivity and also determining the yield by an independent analytical procedure. The procedure was used in the determination of inositol formed from ^{14}C-glucose-6-phosphate [84] and of the biochemical activity of cellulose synthetase [85]. The determination of cholesterol-7-hydrolase is based on measuring the specific activity of ^{3}H-acetic anhydride used for acetylation of the product of the enzymatic reaction.

7.7 GENERAL CHARACTERIZATION
OF THE RADIOCHEMICAL METHODS,
AND RECOMMENDATIONS

The flow of papers dealing with this topic has been growing in recent years, which shows the keen interest in these procedures and their accuracy and sensitivity. Maximum utilization of this advantage is possible only if optimum conditions for the individual determination are fulfilled. In many cases an erroneous assumption is made that using a substrate with an extremely high specific activity leads to an enhanced sensitivity. In fact a high specific activity of the substrate can be advantageous only if the concentration of the enzyme in the sample is comparable to that of the substrate.

7.7.1 The choice of substrate concentration
A high concentration of substrate ensures zero order of the enzymatic reaction over a wide range of concentrations, which significantly simplifies the procedure. Moreover, other enzymes present in the sample then influence the determination only very feebly. On the other hand, the danger of small amount of impurities and/or substrate passing into the product during the separation process is increased with increasing concentration of substrate.

If another reaction, that is not catalysed by the enzyme, proceeds in parallel to the enzymatic reaction, then increasing the concentration of the substrate accelerates the other reaction without influencing the rate of the reaction catalysed by the enzyme. Therefore, it is useless to employ high concentrations of substrates, the price of which may not be negligible, if the activity of the enzyme is low.

On the other hand, the determination of enzyme activity with a low substrate concentration exhibits characteristic features. In this case the rate of the enzymatic reaction in terms of relative amount of substrate converted into product per time unit is maximal. This procedure is recommended for expensive substrates or those difficult to obtain. The rate, in terms of number of moles converted per unit time is very low, which leads to low sensitivity of the determination. The kinetics of the reaction being very complicated, the exact determination of the initial reaction rate is much more difficult than in the previous case. Furthermore, the errors of the determination, caused by the presence of endogenous substrates and inhibitors as well as other enzymes and impurities reacting with the substrate, are much higher than for higher substrate concentrations. Therefore, the rational choice of optimum concentration of substrate is of great importance.

In the determination of low activities of enzymes it is desirable to choose the conditions of the enzymatic reaction in such a way that the rate of product formation is maximal and the blanks (enzyme absent) exhibit minimal rate of product formation. The first requirement calls for high concentrations of substrate, the second for low concentrations. The optimum substrate concentration depends on the relative importance attached by the analyst to the two reaction velocities and should be proposed in each particular case only after a thorough consideration of the factors mentioned.

7.7.2 The choice of optimum volume of the reaction mixture
The optimum volume of the reaction mixture is generally a compromise between two tendencies. The tendency to minimize the volume is based on the fact that the reaction rate decreases with increasing volume for a fixed amount of substrate and enzyme. This tendency prevails if the amount of enzyme available for the analysis is small and its concentration low. The tendency to increase the volume is based on the desirability of treating the substrate with as much enzyme as practicable. This tendency is important if the sample containing the enzyme is available in large amounts and the activity of the enzyme is low.

7.7.3 The choice of optimum specific activity of substrate
The main factor influencing the choice of specific activity of the substrate is the efficiency of the measuring equipment available. For most cases, the concentration of the enzyme to be determined is around $10^{-3}M$ and the sample volume is in the range 0.1–1 ml. From these figures, on the assumption that common commercial measuring apparatus is available, the specific activity recommended is in the range 0.74–74 MBq/µmole for ^{14}C.

7.7.4 Other recommendations for high sensitivity
For routine determinations use a substrate with minimum specific activity compatible with the given procedure, in order to minimize autoradiolytic decomposition of the substrate.

Store labelled substrates under conditions ensuring minimum chemical and radiochemical decomposition.

Before storage lyophilize substrates labelled with tritium.

Purify the labelled product by the method of separation to be used in the actual determination.

Check the influence of the buffer solutions, pH, oxygen, metal traces and complexing substances on the rate of possible non-enzymatic reactions that could interfere with the result of the determination.

Ensure that no undesirable products introducing errors into the determination are formed during the boiling of the reaction mixture containing the enzyme for the preparation of 'blanks'.

7.8 RADIOCHEMICAL METHODS
OF TRACE ANALYSIS WITH ENZYMES

This type of determination has recently started a very promising development. The methods resemble those described above in that the basis of the determination is an enzymatic reaction leading, as a rule, to a radioactive product. By using calibration curves it is possible to find from the radioactivity of the product the amount of the substance to be determined, if this is a substrate, a cofactor, a coenzyme, an inhibitor or an activator of the enzymatic reaction. This type of procedure has been used for the determination of down to 50 µg/ml chloramphenicol in blood [86], different optical isomers of tranylcypromine [87], glucose [88], adenosine triphosphate [89], choline and acetylcholine [90].

Biochemists have developed a highly sensitive enzymatic determination of S-adenosyl-L-homocysteine and L-homocysteine [91]. A micromethod for an enzymatic determination of serotonin has been described [92, 93].

An extremely high sensitivity (1 pmole) can be achieved when using carnitine acetyltransferase and citrate synthetase for the determination of acetylcarnitine and acetylcoenzyme A [94].

REFERENCES

[1] E. Schmidt and F. W. Schmidt, *Les enzymes en pratique médicale,* Boehringer, Mannheim, 1969.
[2] E. Schmidt and F. W. Schmidt, *Fundamentals of Diagnostic Enzymology,* Boehringer, Mannhein, 1968.
[3] H. V. Bergmeyer, *Methods of Enzymic Analysis,* Verlag Chemie, Weinheim, 1965.
[4] International Union of Biochemistry, *Enzyme Nomenclature,* Elsevier, Amsterdam, 1965.
[5] M. Dixon and E. C. Webb, *Enzymes,* 1st Ed., Longmans, London, 1958.
[6] M. Dixon and E. C. Webb, *Enzymes,* 2nd Ed., Longmans, London, 1964.
[7] D. J. Reed, *Advances in Tracer Methodology,* Vol. 4, S. Rothchild (ed.), Plenum Press, New York, 1968, p. 145.
[8] R. J. Levine and D. E. Watts, *Biochem. Pharmacol.,* 1966, **15,** 841.
[9] I. A. Rose, C. L. O'Connell and A. J. Mehler, *J. Biol. Chem.,* 1965, **240,** 1958.
[10] S. H. Pomerantz, *J. Biol. Chem.,* 1966, **241,** 161.
[11] D. A. Smith, R. L. Ratliff, D. L. Williams and A. M. Martinez, *J. Biol. Chem.,* 1967, **242,** 590.
[12] J. S. Paul, R. C. Reynolds and P. O. Montgomery, *Nature,* 1967, **215,** 749.

[13] J. M. Wilhelm and J. W. Corcoran, *Biochemistry*, 1967, **6**, 2578.

[14] H. De Wulf and H. G. Hers, *Eur. J. Biochem.*, 1967, **2**, 50.

[15] M. P. Stulberg, *J. Biol. Chem.*, 1967, **242**, 1060.

[16] T. Okazaki and A. Kornberg, *J. Biol. Chem.*, 1964, **239**, 259.

[17] F. J. Bollum, in *Procedures in Nucleic Acid Research*, G. L. Cantoni and D. R. Davies (eds.), Harper and Row, New York, 1966, p. 296.

[18] D. M. Gill, *Int. J. Appl. Radiat. Isotopes*, 1967, **18**, 393.

[19] I. P. Williamson and S. J. Wakil, *J. Biol. Chem.*, 1966, **241**, 2326.

[20] E. P. Geiduschek and A. Daniels, *Anal. Biochem.*, 1965, **11**, 133.

[21] J. H. Fellman, E. S. Roth and R. F. Mollina, *Anal. Biochem.*, 1969, **30**, 339.

[22] G. Guroff, J. Daly, B. Witkop and S. Udenfriend, *Proc. 7th International Congress Biochemistry*, Tokyo, August 1967.

[23] M. I. S. Lomax and G. R. Greenberg, *J. Biol. Chem.*, 1967, **242**, 109.

[24] R. H. Abeles and W. S. Beck, *J. Biol. Chem.*, 1967, **242**, 3589.

[25] R. L. Miller, *Anal. Biochem.*, 1972, **45**, 202.

[26] A. A. Shatalova and G. I. Meerov, *Biokhimiya*, 1969, **28**, 384 (English translation, 1969, **28**, 315).

[27] J. Drsala and I. M. Hais, *Chem. Listy*, 1973, **67**, 853.

[28] D. Aures and W. G. Clark, *Anal. Biochem.*, 1964, **9**, 35.

[29] G. D. Hodgen and R. J. Falk, *Int. J. Appl. Radiat. Isotopes*, 1971, **22**, 492.

[30] R. D. Jones, *Anal. Biochem.*, 1972, **49**, 147.

[31] C. R. Slack and M. D. Hatch, *Biochem. J.*, 1966, **103**, 660.

[32] A. D. Gounaris and L. P. Hager, *J. Biol. Chem.*, 1961, **236**, 1013.

[33] B. Keech and G. J. Barritt, *J. Biol. Chem.*, 1967, **242**, 1983.

[34] P. K. Dixit and R. Cuestas, *Clin. Chem.*, 1973, **19**, 1106.

[35] F. P. W. Winteringham and R. W. Disney, *Nature*, 1962, **195**, 1303.

[36] J. Schuberth, *Acta Chem. Scand.*, 1963, **17**, Suppl. 1, 233.

[37] J. E. Young, M. D. Prager and I. C. Atkins, *Proc. Soc. Exp. Biol. Med.*, 1967, **125**, 860.

[38] B. L. Bengtsson, *Anal. Biochem.*, 1967, **19**, 144.

[39] S. I. Harik, G. W. Pasternak and S. H. Snyder, *Biochim. Biophys. Acta*, 1973, **304**, 753.

[40] C. Wagner, S. M. Lusty, Jr., Hsiang-Fu Kung and N. L. Rogers, *J. Biol. Chem.*, 1966, **241**, 1923.

[41] P. Laduron and F. Belpaire, *Anal. Biochem.*, 1968, **26**, 210.

[42] T. Kosuge, M. G. Heskett and E. E. Wilson, *J. Biol. Chem.*, 1966, **241**, 3738.

[43] M. W. McCaman, R. E. McCaman and G. J. Lees, *Anal. Biochem.*, 1972, **45**, 242.

[44] R. E. McCaman and K. Cook, *J. Biol. Chem.*, 1966, **241**, 3390.

[45] M. A. Rahim and C. J. Sih, *J. Biol. Chem.*, 1966, **241**, 3615.

[46] K. Suzuki and L. J. Reed, *J. Biol. Chem.*, 1963, **238**, 4021.

[47] H. Y. K. Chuang, *Biochim. Biophys. Acta*, 1973, **321**, 546.

[48] A. C. Cuello, R. Hiley and L. L. Iversen, *J. Neurochem.*, 1973, **21**, 1337.

[49] S. H. Snyder and J. Axelrod, *Biochem. Pharmacol.*, 1964, **13**, 805.

[50] E. A. Irving and W. H. Elliott, *J. Biol. Chem.*, 1968, **244**, 60.

[51] L. Bauguet-Mathieu, R. Goutier and M. Semal, *J. Labelled Compounds*, 1966, **2**, 77.

[52] D. J. Reed, K. Goto and C. H. Wang, *Anal. Biochem.*, 1966, **16**, 59.

[53] J. W. De Jong and C. Kalkman, *Biochim. Biophys. Acta*, 1973, **320**, 388.

[54] T. R. Breitman, *Biochim. Biophys. Acta*, 1963, **67**, 153.

[55] N. B. Furlong, *Anal. Biochem.*, 1963, **5**, 515.

[56] J. R. Sherman, *Anal. Biochem.*, 1963, **5**, 548.

[57] W. G. Ng, W. R. Bergen and G. N. Bonnel, *Clin. Chim. Acta,* 1967, **15,** 489.

[58] X. Scott, *Austr. J. Dairy Technol.,* 1965, **20,** 36.

[59] S. Gabay and H. George, *Anal. Biochem.,* 1967, **21,** 111.

[60] W. C. Rowe, A. K. Huggins and E. Baldwin, *Anal. Biochem.,* 1970, **35,** 167.

[61] B. R. Sitaram, G. L. Blackman, W. R. McLeod, S. P. Son and G. N. Vaughan, *Anal. Biochem.,* 1981, **117,** 250.

[62] J. Kirchner, J. G. Watson and S. Chaykin, *J. Biol. Chem.,* 1966, **241,** 953.

[63] S. Balasubramaniam and K. A. Mitropoulos, *Biochem. J.,* 1971, **125,** 138.

[64] I. Diamond and E. P. Kennedy, *Anal. Biochem.,* 1968, **24,** 90.

[65] M. Marshall and P. P. Cohen, *J. Biol. Chem.,* 1966, **241,** 4197.

[66] B. Zícha, L. Buřič and J. Beneš, *Strahlentherapie,* 1969, **137,** 183.

[67] S. G. Nathenson, J. L. Strominger and E. Ito, *J. Biol. Chem.,* 1964, **239,** 1773.

[68] C. L. Krumdieck and C. M. Baugh, *Anal. Biochem.,* 1970, **35,** 123.

[69] D. M. Ireland and D. C. B. Mills, *Biochem. J.,* 1966, **99,** 283.

[70] W. G. Fiscus and W. C. Schneider, *J. Biol. Chem.,* 1966, **241,** 3324.

[71] S.-I. Hayashi and E. C. C. Lin, *J. Biol. Chem.,* 1967, **242,** 1030.

[72] P. H. Cartier and L. Thuillier, *Anal. Biochem.,* 1971, **44,** 397.

[73] M. O. Oser and N. P. Wood, *J. Bacteriol.,* 1964, **87,** 104.

[74] B. Weiss and E. Costa, *Science,* 1967, **156,** 1750.

[75] R. L. Metzenberg and S. K. Ahlgren, *Anal. Biochem.,* 1970, **35,** 523.

[76] F. Fonnum, *Biochem. J.,* 1966, **100,** 479.

[77] A. Alpert, R. L. Kisliuk and L. Shuster, *Biochem. Pharmacol.,* 1966, **15,** 465.

[78] T. Nagatsu, P. Thomas, R. Rush and S. Udenfriend, *Anal. Biochem.,* 1973, **55,** 615.

[79] T. E. Creighton and C. Yanofsky, *J. Biol. Chem.,* 1966, **241,** 980.

[80] N. Carulli, S. Kaihara and H. N. Wagner, Jr., *Anal. Biochem.,* 1968, **24,** 515.

[81] J. W. Daly, J. K. Inscoe and J. Axelrod, *J. Med. Chem.,* 1965, **8,** 153.

[82] R. A. Eisenman, A. S. Balasubramaniam and W. Marx, *Arch. Biochem. Biophys.,* 1967, **119,** 387.

[83] D. A. Kalbhen, K. Karzel and R. Domenjoz, *Med. Pharmacol. Exp.,* 1967, **16,** 185.

[84] F. Eisenberg, Jr., *J. Biol. Chem.,* 1967, **242,** 1375.

[85] G. A. Barber, A. D. Elbein and W. Z. Hassid, *J. Biol. Chem.,* 1964, **239,** 4056.

[86] A. L. Smith and D. H. Smith, *Clin. Chem.,* 1978, **24,** 1452.

[87] J. A. Fuentes, M. A. Oleshansky and N. H. Neff, *Biochem. Pharm.,* 1975, **24,** 1971.

[88] H. B. Pollard, S. S. Stopak, C. J. Pazoles and C. E. Creuz, *Anal. Biochem.,* 1981, **110,** 424.

[89] C. Gonzales and J. Garcia-Sancho, *Anal. Biochem.,* 1981, **114,** 285.

[90] W. D. Reid, D. R. Haubrich and G. Krishna, *Anal. Biochem.,* 1971, **42,** 390.

[91] N. M. Kredich, H. E. Kendall and F. J. Spence, Jr., *Anal. Biochem.,* 1981, **116,** 503.

[92] I. Hammel, Y. Naot and E. Ben-David and H. Ginsburg, *Anal. Biochem.,* 1978, **90,** 840.

[93] R. D. Hurst, K. Lund and R. W. Guynn, *Anal. Biochem.,* 1981, **117,** 339.

[94] V. Pande and M. N. Caramancion, *Anal. Biochem.,* 1981, **112,** 30.

Index